AF791144

Rethinking Pathways to a Sustainable Built Environment

This book aims to provide insights into rethinking pathways in the transition to sustainable built environments in the wake of the pandemic and COP26. It examines our abilities and capabilities to leverage resources to the best use for achieving the universal sustainable development goals. The goal is to identify fresh thinking to make the goals of sustainable built environment more achievable, particularly to align with the national and international targets set by COP26. The book will help address the need for mainstreaming sustainability into the core of decision-making of buildings and infrastructure projects throughout the life cycle from planning, to design, construction and operation. This book consolidates a comprehensive body of knowledge of sustainable development that will equip industry professionals, educators, scholars and students with knowledge and skills to deliver sustainability practice within the built environment. Through theoretical underpinning and presentation of best practices, the book offers solutions to advance the development of sustainability practices in the context of the built environment. The book covers the following content:

- Sharing of best practice and case studies
- Review of Sustainability Contemporary Practices in the Built Environment
- Innovative net-zero/carbon-neutral solutions and strategies
- Sustainable building assessment and certification systems
- Key Sustainability Deliverables in the Built Environment
- Social Transition towards Sustainability

The book will be of value to interested scholars and practitioners who are involved in sustainable design and engineering practice within the built environment.

Dr. Cheng Siew Goh is a professional surveyor and academic. Her subject knowledge lies in sustainability, green buildings, life cycle management, risk management, construction project management and construction innovations. She has undertaken several individual and joint research projects to produce high-quality research outputs and her works have been published in peer-reviewed journals, conference proceedings and books. Her expertise is highly recognised where she has been invited to serve as a reviewer for several peer-reviewed journals and international conferences.

Dr. Heap-Yih (John) Chong obtained his PhD in Civil Engineering (Construction Engineering and Management) from Universiti Teknologi Malaysia in 2011. He brings with him a sound understanding of the construction sector. He has a strong track record of coherent research outputs in the top refereed journal articles and has successfully secured a number of competitive research grants from funding bodies in Australia, Malaysia, China, the Middle East and Japan. He has a growing profile and is actively involved in professional bodies, industry engagements and international collaboration. His research focuses on the interdisciplinary area of construction engineering and management.

About CIB and the CIB series

CIB, the International Council for Research and Innovation in Building and Construction, was established in 1953 to stimulate and facilitate international cooperation and information exchange between governmental research institutes in the building and construction sector, with an emphasis on those institutes engaged in technical fields of research.

CIB has since developed into a worldwide network of over 5000 experts from about 500 member organisations active in the research community, in industry or in education, who cooperate and exchange information in over 50 CIB Commissions and Task Groups covering all fields in building and construction related research and innovation.

https://cibworld.org/

This series consists of a careful selection of state-of-the-art reports and conference proceedings from CIB activities.

Public Private Partnerships
A Akintoye et al.

Court-Connected Construction Mediation Practice
A. Agapiou and D. Ilter

Advances in Construction ICT and e-Business
S. Perera et al.

Clients and Users in Construction
K. Haugbolle and D. Boyd

Integrating Information in the Built Environments
A. X. Sanchez et al.

Rethinking Pathways to a Sustainable Built Environment
C.S. Goh and H. Chong

Rethinking Pathways to a Sustainable Built Environment

Edited by Cheng Siew Goh and Heap-Yih Chong

Routledge
Taylor & Francis Group
LONDON AND NEW YORK

First published 2025
by Routledge
4 Park Square, Milton Park, Abingdon, Oxon OX14 4RN

and by Routledge
605 Third Avenue, New York, NY 10158

Routledge is an imprint of the Taylor & Francis Group, an informa business

© 2025 selection and editorial matter, Cheng Siew Goh and Heap-Yih Chong; individual chapters, the contributors

The right of Cheng Siew Goh and Heap-Yih Chong to be identified as the authors of the editorial material, and of the authors for their individual chapters, has been asserted in accordance with sections 77 and 78 of the Copyright, Designs and Patents Act 1988.

All rights reserved. No part of this book may be reprinted or reproduced or utilised in any form or by any electronic, mechanical, or other means, now known or hereafter invented, including photocopying and recording, or in any information storage or retrieval system, without permission in writing from the publishers.

Trademark notice: Product or corporate names may be trademarks or registered trademarks, and are used only for identification and explanation without intent to infringe.

British Library Cataloguing-in-Publication Data
A catalogue record for this book is available from the British Library

ISBN: 978-1-032-33021-1 (hbk)
ISBN: 978-1-032-33043-3 (pbk)
ISBN: 978-1-003-31789-0 (ebk)

DOI: 10.1201/9781003317890

Typeset in Times New Roman
by KnowledgeWorks Global Ltd.

Contents

About the Editors

Cheng Siew Goh

Dr. Cheng Siew Goh is a professional surveyor and academic. Her subject knowledge lies in sustainability, green buildings, life cycle management, risk management, construction project management and construction innovations. She has undertaken several individual and joint research projects to produce high-quality research outputs and her works have been published in peer-reviewed journals, conference proceedings and books. Her expertise is highly recognised where she has been invited to serve as a reviewer for several peer-reviewed journals and international conferences.

Heap-Yih (John) Chong

Dr. Heap-Yih (John) Chong obtained his PhD in Civil Engineering (Construction Engineering and Management) from Universiti Teknologi Malaysia in 2011. He brings with him a sound understanding of the construction sector. He has a strong track record of coherent research outputs in the top refereed journal articles and has successfully secured a number of competitive research grants from funding bodies in Australia, Malaysia, China, Middle East and Japan. He has a growing profile and is actively involved in professional bodies, industry engagements and international collaboration. His research focuses on the interdisciplinary area of construction engineering and management.

Contributors

Saeed Banihashemi

Dr. Saeed Banihashemi is the Associate Professor, Founding Director of Digital Building and Built Environment Hub (BIM Hub) and Postgraduate Program Director in the School of Design and Built Environment, Faculty of Arts and Design; University of Canberra (UC). Saeed is the associate editor of the Journal of Frontiers in Built Environment and an editorial board member of the journals of Smart and Sustainable Built Environment, PLOS One and Buildings. He is also an executive member of the Australian Institute of Building (MAIB), a member of the Australian Institute of Project Management (MAIPM) and the Fellow of the UK Higher Education Academy (FHEA). He has published three books and authored and co-authored more than 80 research articles, book chapters and non-traditional research outputs in the areas of sustainable built environment, BIM and advanced technologies in design and construction. Saeed's work has been cited more than 2650 times (H-index 27) in frontline journals, dissertations and conferences, in knowledge areas relating to his research expertise.

Emmanuel Boateng

Dr. Emmanuel Boateng is a Lecturer and the Academic Program Director for the undergraduate Occupational Health and Safety programme in the School of Health and Society at the University of Wollongong, Australia. His research interests are in the areas of occupational health and safety, artificial intelligence and construction management.

Albert Chan

Professor Albert Chan is currently Dean of Students, Associate Director of the Research Institute for Sustainable Urban Development and Chair Professor of Construction Engineering and Management at The Hong Kong Polytechnic University. He is also a Chartered Construction Manager, Engineer, Project Manager and Surveyor by profession. He earned his MSc degree in Construction Management and Economics from the University of Aston in Birmingham and his PhD degree in Project Management from the University of South Australia. His research interests include project management and project success, construction procurement and relational contracting, public-private partnerships and construction health and safety.

Amos Darko

Dr. Amos Darko is currently an Assistant Professor in the Department of Construction Management at the University of Washington, Seattle, WA, USA. His research interests cover sustainable built environment, sustainable construction, green building, modular and offsite construction and digital applications including building information modelling and artificial intelligence.

Harry Deane

Harry Deane is a professional Quantity Surveyor working in the commercial real estate market within London. Deane has obtained his BSc (Hons) in Quantity Surveying and Commercial Management from Nottingham Trent University (NTU). Since graduating from university Harry has been working for global consultant "Turner & Townsend" for three years and has been an RICS member for over a year. Mr. Deane has worked on some of the largest projects in London for clients such as Citigroup, Blackstone and Societe General. Mr. Deane has a particular interest in sustainability within the construction industry and has recently worked with architects advising them on the cost options for retrofitting office spaces in London. As well as sustainability, Deane has interests in topics such as; embodied carbon, retrofitting, procurement and life cycle costing (LCC).

Kumar Biswajit Debnath

Dr. Kumar Biswajit Debnath is a Research Associate in Passive Building Design at the Hub for Biotechnology in the Built Environment (HBBE), School of Architecture, Planning and Landscape (APL) at Newcastle University, UK. As an architect, he investigates bio-integrated (and bio-climatic) passive building design for climate change adaptation and mitigation. He has 12 years of extensive experience in biomimetic design, bio-based material development, building physics and energy modelling. He is also interested in decarbonising energy systems, especially within the built environment in developing countries.

Caleb Debrah

Caleb Debrah is a PhD. student and Research Associate at the Department of Building and Real Estate of the Hong Kong Polytechnic University. Mr. Debrah's research focuses on "green finance in green building". He is a member of the Ghana Institute of Construction and a student member of the Hong Kong Green Building Council. Caleb is also a certified IFC EDGE Expert and Project Management Professional (PMP®).

Xin Dong

Xin Dong, PhD candidate at the School of Architecture and Urban Planning, Chongqing University, actively engages in research as a member of the Centre for Climate-Resilient and Low-Carbon Cities. Additionally, he holds membership in the Architectural Society of China as a student affiliate. His scholarly pursuits encompass a range of topics including net-zero-carbon buildings, comprehensive life cycle assessments of green roofs and strategies for urban heat mitigation and adaptation. Presently, his primary focus lies in meticulously evaluating the cooling efficacy of green roofs on urban microclimates.

Yahong Dong

Dr. Yahong Dong is an associate professor in the Department of Environmental Science and Engineering, Macau University of Science and Technology. She obtained her PhD degree from the Department of Civil Engineering at the University of Hong Kong. Her research areas include life cycle assessment, carbon accounting and sustainability assessment. Dr. Dong serves as an editorial board member for the International Journal of Sustainable Engineering and a reviewer for over ten international journals. Accredited with BEAM Pro by HKGBC, Dr. Dong is a managing director of the China Green Building and Energy Saving Association in Macao. As a principal investigator, she has supervised and managed research projects funded by the Science and Technology Development Fund (FDCT) of Macao and the Joint Funding Scheme of NSFC-FDCT. Her projects include research on carbon peak and neutrality for the Lingshan Island Nature Reserve, developing a zero-carbon campus and assessing the life cycle sustainability of buildings.

Adji Achmad Rinaldo Fernandes

Adji Achmad Rinaldo Fernandes is a lecturer at Brawijaya University, Malang. He obtained his bachelor's in Statistics from the Department of Mathematics, Faculty of Mathematics and Natural Sciences, Brawijaya University (2004). He earned a Master of Science in Statistics from the School of Information and Technology, Hasselt University, Diepenbeek, Belgium (2007). He obtained his Doctoral degree in Statistics from the Department of Statistics, Faculty of Mathematics and Natural Sciences, Sepuluh Nopember Institute of Technology, Surabaya (2016).

Karla Fox-Reynolds

Karla is Hassel's Sustainable Design Leader in Perth and is part of the design practice's global sustainability team. She participates in national industry working groups as well as the GBCA Sustainable Precincts Expert Reference Panel and co-chairs the MECLA working group 6 (Residential). Prior to joining Hassell, she was a Principal Innovation Project at Climate-KIC Australia, where she led the Retrofit Australia portfolio of work. Karla has broad knowledge and industry experience, having also worked for Curtin University, Lendlease and Cundall.

Hui Hwang Goh

Hui Hwang Goh (Senior Member, IEEE) received the BEng (Hons.) and MEng degrees in electrical engineering and the PhD degree in electrical engineering from Universiti Teknologi Malaysia, Johor Bahru, Malaysia, in 1998, 2002 and 2007, respectively. He is currently a Professor of electrical engineering with the School of Electrical Engineering, Guangxi University, Nanning, China. His research interests include embedded power generation modelling and simulation, power quality studies, wavelet analysis, multicriteria decision-making, renewable energies and dynamic equivalents. He is also a Fellow of the Institution of Engineering and Technology (IET), UK, the ASEAN Academy of Engineering and Technology

(AAET) and The Institution of Engineers, Malaysia (IEM), a Chartered Engineer under the Engineering Council United Kingdom (ECUK) and a Professional Engineer under the Board of Engineers, Malaysia (BEM). He is also the foreign fellow of the Chinese Society for Electrical Engineering (CSEE).

Kai Chen Goh

Ts. Dr. Kai Chen GOH is currently an Associate Professor in the Department of Construction Management at Universiti Tun Hussein Onn Malaysia (UTHM). With a PhD in built environment and engineering, he was named a Young Lecturer for the Malaysian Ministry of Higher Education's National Blue Ocean Strategy-Joint Initiative Programme (NBOS-PILS) in 2015. Dr. Goh is an active member of several professional organisations. He is a Fellow of the ASEAN Academy of Engineering and Technology (FAAET), a Professional Technologist (Building and Construction) registered with the Malaysian Board of Technologists, a Chartered Construction Manager, a Member of The Chartered Institute of Building in the United Kingdom (MCIOB) and an accredited facilitator of the Green Building Index (GBI) Malaysia. His research interests include construction financial management, sustainability, management of sustainable infrastructure, multicriteria decision-making, decision support systems, disaster-resilient infrastructure and renewable energy.

Aso Haji Rasouli

Aso Hajirasouli is an accomplished academic with a rich background in digital and resilience architecture, design and academia. Throughout her career, Aso has held various teaching and research positions, showcasing her commitment to academia. Currently, she serves as a Senior Lecturer in Architecture at Western Sydney University (WSU) and is a core member of the Urban Transformation Research Centre (UTRC) and contributes to the university's External Advisory Committee. In the realm of research, Aso has established herself as a leading figure, with a focus on digital design, resilient design and smart design. Her interdisciplinary approach and successful collaborations have led to numerous peer-reviewed publications and research grants, including projects funded by the Department of Climate Change, CSIRO and others. Aso's expertise extends beyond academia, as evidenced by her roles as a Visiting Professor at the University of Florence and her involvement in prestigious conferences like CONVR. She has also contributed significantly to the development of architectural education focused on regenerative and resilient design, having led curriculum development projects and designed innovative pedagogical frameworks. Beyond her academic and research endeavours, Aso is actively engaged in professional organisations and community initiatives. She is a member of the Australian Institute of Architects and has received several awards and recognitions for her outstanding contributions to the field.

Chris Harty

Professor Chris Harty is the Dean of the School of the Built Environment and Architecture and a Professor of Technology and Organization at London South Bank University. Prior to joining the university, he held the position of Head of the School of Built Environment at the University of Reading. Since 2019, he has chaired BSRIA (Building Services Research and Information Association), previously serving as a non-executive Director. Chris's research interests encompass understanding cultural and strategic change in organisations, implementing new technologies and working practices, future scenario development and the interplay between physical space and organisation. Over the years, he has led or served as the principal investigator on research grants totalling over £3.5 million. Additionally, he has held visiting professorships at Copenhagen Business School, Chalmers University of Technology and the University of Stuttgart.

Bao-Jie He

Bao-Jie He is a (Full) Professor of Climate Change and Sustainable Built Environment at the School of Architecture and Urban Planning, Chongqing University, China. He is the leader of the Centre for Climate-Resilient and Low-Carbon Cities at Chongqing University and a Research Fellow of NERPS at Hiroshima University (Japan). Baojie focuses on studies on heat-resilient urban planning and design. Baojie has published more than 150 peer-reviewed journal papers (Scopus H-index: 47) and has been leading/involved in about 20 research projects in China and Australia. Baojie is a Highly Cited Researcher (Clarivate) in 2022 and 2023. Baojie acts as Topic Editor-in-Chief, Leading Guest Editor, Associate Editor, Editorial Board Member, Conference Chair, Sessional Chair and Scientific Committee in various reputable international journals and conferences. Baojie received the Green Talents Award (Germany) in 2021 and was ranked as one of the single-year and career 100,000 global scientists (2%) in 2020, 2021, 2022 and 2023.

Benny Hutahayan

Benny Hutahayan is a lecturer in the Doctoral Program of Administrative Sciences at Brawijaya University. He has a Bachelor of Law Degree from the Christian University of Indonesia and a Master of Legal Studies degree from the Christian University of Indonesia, and he completed his Doctoral Program in Administrative Sciences from Brawijaya University in 2014. Before becoming an academician, he had 20 years of experience working for a multinational company, British Petroleum.

David Jenkins

Professor David Jenkins leads the Urban Energy Research Group at Heriot-Watt University and has a background in building physics and energy modelling. Recent work has focussed on the application of statistical techniques for enriching the understanding of energy demand data, and modelling, of buildings. He was the lead of the CEDRI project, using a range of multi-disciplinary techniques to capture the relationship between occupants, buildings and energy networks for energy demand reduction strategies in Indian communities and also the UK lead on the Horizon-funded crossCert project, investigating next-generation Energy Performance Certificates. He is/was co-lead on the H+C Zero Network, CESI and ReFLEX projects, exploring methods of aligning local renewable supply with energy demand and the role of energy modelling of buildings within this process. He has been on advisory panels for the Scottish and UK Governments (energy building policy) and leads the Energy Efficiency in Buildings theme of the Scottish Energy Technology Partnership.

Cen-Ying Lee

Dr. Cen-Ying Lee is an Assistant Professor in the Faculty of Society and Design at Bond University, Gold Coast, Australia. Before joining Bond University, she held a position as a lecturer in the Built Environment department at the University of Reading Malaysia, where she also served as Programme Lead for BSc Quantity Surveying. Her research interests primarily revolve around digitalisation and sustainability in the built environment, as well as construction contract and cost management. Her scholarly articles have been published in reputable journals indexed in Web of Science, ISI and Scopus, including Archives of Computational Methods in Engineering, Journal of Cleaner Production, Journal of Management in Engineering, Journal of Construction Engineering and Management, and Technological and Economic Development of Economy. Her publications have also contributed to various research projects in Australia, Malaysia and China.

MK Leung

MK joined RLP in 2008 and has been the firm's Director of Sustainable Design since 2012. Under his leadership, his team has delivered numerous high-profile projects, among them China's first civilian-use zero-carbon building, CIC Zero Carbon Park; a pioneering sustainability master plan for Hong Kong Science Park Phase 3 aiming for carbon neutrality from the individual building level and at a neighbourhood scale; a community-centric vertical campus for THEi which won a Sustainable Design and Performance Award at the 2020 WorldGBC Asia Pacific Leadership in Green Building Awards; and many consultancy studies conducted for the Hong Kong government on low-energy residential building design, infrastructure design and construction under extreme temperatures, sponge city design and construction for public buildings and a revamp of the Building (Planning) Regulations for the Hong Kong SAR.

Iina Lohi

Iina has worked on landscape architecture projects in Queensland, New South Wales, Western Australia, Northern Territory and the United Arab Emirates. Her experience covers urban and regional master planning and public works; environmental design for parks, playgrounds and infrastructure, as well as residential, industrial and commercial developments. She works amongst other built environment specialists providing input ranging from the liminal to the as-constructed form. Her work considers disaster management, wildfires, water, culture, heritage, aesthetics, ecology, liveability, thermal comfort, practicality, sustainability and just good design. She is currently pursuing PhD research with the University of Canberra in the development of a bushfire-resilient home retrofit toolkit for older Australians, with an emphasis on helping to identify and preserve the personal stories and values of living in the landscape sustainably. This has involved identifying the legacies in how human and building adaptation preparedness for wildfire disasters is occurring both locally and internationally, across multi-hazard scenarios and different jurisdictions.

Md Asrul Nasid Masrom

Dr. Md. Asrul Nasid Masrom holds the position of Associate Professor at the Faculty of Technology and Business Management, Universiti Tun Hussein Onn Malaysia (UTHM). Currently, he serves as the Deputy Dean (Research, Development and Publication) and is a Principal Researcher at the Centre for Sustainable Infrastructure and Environmental Management (CSIEM). Dr. Masrom has been recognised with academic awards such as the QUT East West Recognition Award in 2009 and the Outstanding Higher Degree Research (HDR) Student for February 2011 Awards organised by the Built Environment and Engineering Faculty, Queensland University of Technology (QUT). His research interests revolve around satisfaction levels, both theoretical and practical, quality measurement, project performance, procurement strategies, community engagement and decarbonisation in green technology in the construction industry. Furthermore, he actively engages in community service activities, sharing his knowledge at both local and international levels.

Peter McCallum

Dr. Peter McCallum is a mechanical engineer, specialising in numerical methods and data-driven modelling. Having built on ten years of industrial experience working between the construction and low-carbon technology development sectors, he has been a researcher for the last seven years working between Heriot-Watt University and the University of Edinburgh. The foundation of Peter's expertise lies in HVAC design and on-site implementation, having been involved with around 40 large-scale construction projects across Europe, the Middle East, the US and Australia, including schemes upwards of £200M in construction value. Many of these involved heat networks with the incorporation of low- and zero-carbon technologies. Following his PhD, Peter has specialised in software and cloud-based model development, creating GIS-based computational models for assessing heat demand from residential buildings, as well as a wide range of other bespoke analysis techniques to interpret energy demand.

Roberto Minunno

Dr. Roberto Minunno is a senior lecturer, researcher and Course Coordinator of the Master of Environment and Climate Emergency at Curtin University in Perth, Australia. His passion, research streams and teaching focus on the circular economy of buildings, life cycle assessment, carbon accounting and modular buildings. Roberto has received prestigious awards for his research and teaching, including the Best Early Career Researcher of the Year for Business and Societies, in recognition of his cutting-edge project, the Legacy Living Lab. This building, which he designed and collaborated on building, showcases applications of circular economy concepts to buildings.

Sulzakimin Bin Mohamed

Ts. Gs. Dr. Sulzakimin Bin Mohamed serves as the Head of Department for Bachelor of Technology Management (Construction) and is a Principal Researcher at the

Centre for Sustainable Infrastructure and Environmental Management (CSIEM). He joined Universiti Tun Hussein Onn Malaysia, Malaysia as a temporary tutor from 2007 to 2013. In 2014, he obtained his PhD from Universiti Tun Hussein Onn Malaysia and worked as a full-time lecturer there onwards. Up until 2024, Dr. Sulzakimin has more than 60 journal papers, proceeding/conference papers, more than 10 books/chapters in the book and several articles in Malaysian mass media to his name in areas of construction management, Local Governments and sustainable development. Apart from that, his research interest ranges from awareness of technical and vocational education, GIS's awareness, sustainable development and smart city planning and development.

Eric Ohene

Eric Ohene holds bachelor's and master's degrees in quantity surveying and construction Management, respectively from Ghana. Eric is currently a PhD student and Research Associate at The Hong Kong Polytechnic University with a research interest in building energy efficiency and net-zero buildings. He is a certified Project Manager (PMP®) and faculty advisor of the American Concrete Institute.

De-Graft Owusu-Manu

Prof. De-Graft Owusu-Manu is a Professor and Postgraduate Coordinator at the Department of Construction Technology and Management of the Kwame Nkrumah University of Science and Technology, Ghana. De-Graft is the president of Green Communities International and editor of the Journal of Built Environment. His research interests include construction financing, construction economics, project finance strategy, investment appraisals and development economics.

Dicken Poon

Dicken studied sustainable and environmental design at the Chinese University of Hong Kong. He has a background in environmental simulations. Dicken works with architectural studios in RLP to enhance the sustainability performance of projects through quantitative studies including building performance simulations. He is also familiar with green building certifications and facilitates their certification with the highest possible rating for some of the interiors and new building projects. Dicken has three years of experience in acting as a coordinator and internal auditor for the implementation of the Integrated Management System, especially for the ISO14001 standard.

John Rogers

Dr. John Rogers primarily undertakes SME building projects. This enables him to experiment with evolving digitalisation tools in GIS, BIM and computer vision and begin to optimise the AEC process and product at each stage of the whole building lifecycle. In 2009 John was a founding member of Malaysia's UN Global Compact, and sustainability remains a central principle in his works. He was a founding member of BuildingSMART Malaysia and an author of Malaysia's Construction Industry Transformation Plan (CITP) 2020 and the National BIM Guide. He has developed strategic initiatives for off-site, modular construction. He is an occasional academic and was trained as an Industrial Professor, by Heriot-Watt University, to act as a bridge between industry and academia. Prior to this, he managed the design and construction of airports, shopping centres and large construction projects and the digitalisation of telecoms and banking organisations.

Carolina Sanchez Salan

Carolina holds a BAppSc in Architectural Science from Curtin University (Australia), a Licenciatura in Economics from University Pontificia Comillas (Madrid), and she is a Chartered Accountant trained at PricewaterhouseCoopers (London). Carolina works on projects supporting the UN's SDG agenda for liveable cities, access to quality housing, creation of stimulating shared public spaces for all and climate adaptation and resilience strategies for Hong Kong. She has collaborated with the Architectural Services Department to develop a Sponge City Concept Handbook and on a study of Extreme Temperatures in HKSAR and their effect on buildings and people with a view to develop coping strategies for the warming future.

Martin Skitmore

Professor Martin Skitmore obtained his PhD from Salford University, UK, where he held Lecturer, Senior Lecturer, Reader and Personal Chair positions. He was subsequently a Professor, Head of School and Discipline Leader at the Queensland University of Technology (QUT). He is currently a University Professorial Fellow at Bond University and Adjunct Professor at QUT. He has published around 480 papers in peer-reviewed journals. His research interests include construction/project management, urbanisation and sustainability. See https://staff.qut.edu.au/staff/rm.skitmore and https://research.bond.edu.au/en/persons/martin-skitmore/.

Matt van Someren

Matt is a climate change and sustainability consultant for a Sydney-based strategic advisory firm. He combines a deep knowledge and passion for sustainability and climate change with extensive communications, stakeholder engagement and project management experience gained in both the private and public sectors. Matt has advised businesses across a range of industries to help them navigate the net-zero transition, understand their climate-related risks and opportunities and develop robust management approaches to dealing with the impacts of climate change. He has a Bachelor of Business (Western Sydney University) and a Masters in Sustainability and Climate Policy (Curtin University).

Thomas Tang

Dr. Thomas Tang has been working as a sustainability practitioner for over 25 years. He is currently based in Hong Kong where he has worked extensively on projects for the government and the private sector. He has in-depth corporate sustainability knowledge from working as the sustainability director for AECOM, a large US multinational engineering company, as well as advisory consulting experience in a wide range of industry sectors. His interests are in livable cities, resilient cities, smart cities, sustainable communities and the role of technology as an enabler towards a more sustainable world.

Bruno Lot Tanko

Dr. Bruno Lot Tanko is the Head of Built Environment at the University of Reading, Malaysia Campus. Holding a PhD in Quantity Surveying, an MSc in Construction Management and a First-Class Honors bachelor's degree in quantity surveying, Dr. Bruno has been actively engaged in teaching and training in the fields of quantity surveying, cost engineering and construction project management both in Nigeria and internationally since 2009. Recognised for his outstanding contributions to education, he is a Fellow of the Higher Education Academy, UK (FHEA). As a professional, he is a member of the Royal Institution of Chartered Surveyors (MRICS), a member of the Nigerian Institute of Quantity Surveyors (MNIQS) and a Registered Quantity Surveyor (RQS). His scholarly work has gained recognition, with articles listed in Web of Science, ISI and Scopus journals. His research focus spans Smart Buildings, Sustainable Construction, Building Information Modelling and Value Management.

Mershack Opoku Tetteh

Dr. Mershack Opoku Tetteh is a Lecturer in Construction Management at the School of Infrastructure and Sustainable Engineering, Aston University, Birmingham, United Kingdom. His research interests are construction management, international construction, joint ventures, green or sustainable construction, modular and offsite construction, building information modelling, artificial intelligence and green retrofitting of existing buildings.

Serik Tokbolat

Dr. Serik Tokbolat is an Assistant Professor in Sustainability and Sustainable Construction Engineering at the Department of Civil Engineering, Faculty of Engineering, University of Nottingham. He previously worked as a Senior Lecturer at the School of Architecture, Design and Built Environment, Nottingham Trent University. Before joining NTU, Dr. Tokbolat worked as a postdoctoral researcher at the School of Engineering and Digital Sciences, at Nazarbayev University. Dr. Tokbolat obtained his PhD in Energy and Process Engineering from the Norwegian University of Science and Technology. He holds an MSc degree in Management in Civil Engineering from Nottingham University as well as a MEng degree in Civil Engineering from Eurasian National University named after L.N. Gumilyev. Dr. Tokbolat's research interests include topics such as sustainable construction; integration of renewable energy technologies into the built environment; life cycle assessment (LCA); building energy simulations; building information modelling (BIM); recycling of carbon fibre composites and circularity.

Lois C Towart

Dr. Lois C Towart is an Adjunct Fellow in the School of Built Environment at the University of Technology Sydney. She has over 30 years' valuation and research experience specialising in Retirement Housing and Residential Aged Care.

Seow Ta Wee

Seow Ta Wee is a professor in environmental and solid waste management at Universiti Tun Hussein Onn Malaysia, Malaysia. His transdisciplinary and interdisciplinary research covers aspects of environment management, solid waste management, geography, construction sociology, TVET and social return on investment (SROI). Many of his research projects have significantly impacted the country's socio-economic development and at the level of the ASEAN region. Through these research works, he also managed to win several national and international awards. He has more than 380 publications including reputed international journals, proceedings, books, book chapters and policy papers.

Karl Welsch

Karl is currently a technical officer in the National Australian Built Environment Rating System (NABERS) programme, using his technical background in engineering to deliver sustainability solutions to the built environment. After a career as a biomedical engineer, and then an extended paternity leave running a family business, Karl returned to university to complete a master's degree in sustainability at Curtin University. Here, he developed a keen interest in the built environment and sustainable urban design - and their significance to the challenge of reducing carbon emissions.

Wiratno

Wiratno is an expert staff member of the Minister for Restoration and Conservation Partnership at the Ministry of Environment and Forestry of the Republic of Indonesia. Before serving as an Expert Staff to the Minister, Wiratno also held the position of Director General for the Conservation of Natural Resources and Ecosystems at the Ministry of Environment and Forestry. He has extensive experience in the field of conservation. He has researched several National Parks, including Gunung Leuser National Park, Bukit Duabelas National Park and Lore Lindu National Park.

Mingqiang Yin

Mingqiang Yin is a PhD candidate at the School of Architecture and Urban Planning at Chongqing University, China. Meanwhile, Mingqiang is a student member of the China Institute of Architects. Mingqiang's research focuses on urban heat mitigation and adaptation, heat-resilient planning and urban ventilation corridors. Mingqiang has published ten peer-reviewed papers in reputable journals and delivered oral presentations at several high-quality conferences. Mingqiang once worked as a design assistant in two urban planning and design institutes in China. As a conference assistant, Mingqiang participated in the operation and organisation of the first, second and third sessions of the Urban Climate and Urban Design Academic Symposium. Mingqiang won the Outstanding Postgraduate Award from Chongqing University, the First Prize from Chongqing Municipal Planning and Design Award and the Second Prize from the Guangdong Provincial Survey and Design Award.

Preface

Sustainability imperatives are placed as a key agenda in the transformation of the built environment. Numerous sustainability goals and carbon emission targets have been set for construction projects and organisations. International alliances such as the Global Alliance for Building and Construction are working on a commitment to "Buildings as Critical Climate Solution (BCCS)" to advance building sector measures at the national and international levels for creating the appropriate enabling environment towards zero-emission, efficient and resilient buildings. The United Nations Climate Change Conference 2023 in Dubai (COP28) has highlighted the importance of creating a more sustainable built environment in response to the climate emergency. The COP28 was momentous as it set to conclude the first-ever global stocktake – an inventory to see where the world stands on climate actions and where we are not.

Carbon emissions and climate change are hot topics in both research and professional activities, following the call for greater actions in COP28. Net zero and sustainability have become central to the building and construction industry, and it is important to map sustainability competency into the construction professional disciplines. Nonetheless, embracing sustainability in the built environment appears to be off pace and not aligned with the international and national targets. Net zero or carbon neutrality appears to be too ambitious or aggressive a move for construction practitioners. Moreover, the outbreak of the COVID-19 pandemic and the recent world tensions in Gaza, Sudan and Ukraine have put sustainability deliverables further out of reach, following the unprecedented energy crisis. Some paradoxes and gaps in delivering sustainability in the built environment have been amplified, taking into consideration the impacts of the pandemic and supply chain disruptions on the construction industry. It is therefore necessary to revisit our sustainability measures and review the contemporary sustainability research and practices in the built environment to seek out the most innovative and cutting-edge in order to push the agenda forward.

The book presents a broad sustainability application in buildings and infrastructure projects with regard to technologies and innovative solutions to tackle climate change issues and support the delivery of sustainable development goals. The book aims to provide valuable insights into rethinking our pathways in a transition to sustainable built environments. It examines our abilities and capabilities to leverage resources to the best use for achieving the universal sustainable development goals and exploring best practices. While searching for new ideas and resolutions, it is also vital to identify fresh thinking to make the goals of a sustainable built environment more achievable, particularly to align with the national and international targets set in COP26. The book helps address the need for mainstreaming sustainability into the core of decision-making of buildings and infrastructure projects throughout the life cycle from planning, to design, construction and operation. The book is of value to interested scholars and practitioners who are involved in sustainable design and engineering practice within the built environment.

The book consolidates a comprehensive body of knowledge on sustainable development that equips industry professionals, educators, scholars and students with knowledge and skills to deliver sustainability practice within the built environment. Through theoretical underpinning and best practice sharing, the book offers solutions to advance the development of sustainability practices in the context of the built environment. This book offers insights into challenging the status quo of sustainable built environments by stressing the cutting-edge, pushing the boundaries and implementing innovative solutions of sustainability in practice.

The book starts with an introduction to the recent changes in sustainable initiatives and a review of carbon policies and contemporary sustainable building practices. The book is structured into five parts and a special collection.

Chapter 1 provides a snapshot of the status of climate policies and the progress of carbon emission reduction on a global scale. It examines different issues of sustainable construction and highlights the need for rethinking pathways in upholding a sustainable built environment.

Part I is a rich source of innovative sustainable solutions and strategies for net zero. Chapter 2 explores the use of transdisciplinary integrated design for carbon-neutral high-rise office buildings in a high-density built environment. A wide range of innovative net-zero strategies was reviewed in this chapter for their environmental, social and economic impact potential versus cost, technical, regulatory and cultural challenges. This chapter identified three strategies: (1) "Quick-win" strategies, (2) "Exemplary" strategies and (3) "Visionary" strategies by examining their impacts and challenges to implementation. The chapter presents an outcome-based transdisciplinary net-zero-ready building design framework to provide notable and innovative guiding principles for urban sustainability.

Chapter 3 explores pathways for energy efficiency retrofitting towards net-zero building operations. It identifies the net-zero potential of Australia's ageing residential buildings and provides several strategies for retrofitting ageing homes towards net zero. The strategies are grouped under three main themes: technical, social sustainability, political and regulatory. A harmonised approach encompassing technical, social sustainability and political and regulatory elements is underscored for a successful transition of ageing homes towards net zero.

Chapter 4 reviews a range of renewable energy sources available in tropical countries and explores the potential for implementing building-integrated photovoltaic systems (BIPV) in the Malaysian housing construction sector. Apart from describing the prospects and barriers of adapting BIPV, this chapter also discusses recent development endeavours of BIPV and different government policies in relation to BIPV.

Part II covers processes and strategies involved in planning, design, operation and decommissioning within the sustainable built environment. Proper planning and design is critical to ensure sustainable development is well adapted to buildings and neighbourhoods to avoid increased vulnerabilities of people to climate changes. To support the adoption of passive design for built environment decarbonisation, Chapter 5 provides a comprehensive analysis of passive design technologies, in aspects of site space and form, building types, envelope properties, ventilation performance, lighting, renewable energies and others. This chapter identifies various carbon source/sink scenarios and assesses the decarbonisation potential of passive design for carbon neutrality within the built environment.

Existing buildings represent a substantial opportunity to decrease societal adverse environmental impacts. Chapters 6 and 7 highlight the critical role of retrofitting existing building stocks to achieve sustainable building operations. In Chapter 6, a retrofitting initiative based on four hierarchical upgrades is proposed to transform existing Australian homes to be healthier and more sustainable housing. The proposed retrofitting scheme emphasises the need for a holistic intervention beyond quick solutions such as installing solar panels to ensure sustainable operations. Chapter 7 examines public sentiment towards energy efficiency retrofitting of existing buildings (EEREB) to effectively promote a widespread adoption through policy interventions. It used sentiment to analyse social media data to assess the overall needs, concerns and perceptions of the general public regarding EEREB. This chapter offers a solid basis for modelling energy efficiency decisions in existing buildings, thereby improving existing retrofit policies by tailoring them to the public's preferences.

Part III reviews sustainable building-related assessment approaches and systems. Sustainability assessment standards have become increasingly important in accelerating the sustainable transformation within the built environment. Energy performance is a key focus in most of the sustainability rating systems. Chapter 8 discusses the performance of Energy Performance Certificates (EPC) and exemplifies building energy analysis using a very specific case study occurring during the period of the COVID lockdown. By using dynamic building simulations, this chapter brings in value empirical energy consumption data to understand the dynamics and intricacies of energy consumption and generation.

Chapter 9 explores the state-of-the-art life cycle assessment (LCA) in the building and construction sector by employing bibliometric analysis and critical review. It identifies challenges of LCA application in the building and construction sector with respect to methodology, data and integration with other techniques. This chapter also uncovers the evolving transition from single-dimensional environmental assessment (LCA) to multi-dimensional sustainability assessment – Life Cycle Sustainability Assessment (LCSA).

Part IV is a collection describing various social transitions towards creating a more sustainable and resilient built environment. The power, motivations and influences of various stakeholders in sustainable transition are indispensable. Chapters 10 and 11 seek ways to invigorate stakeholder engagement and revolutionise decisions for sustainability by creating a bottom-up approach with proper alignment of social considerations. Chapter 10 attempts to establish the fundamental social needs for a more sustainable and resilient built environment following unprecedented global crises such as climate change and the COVID-19 pandemic. This chapter identifies six social transitions to empower people through engagement, interaction, transactions and recreation. The proposed six transitions, if materialised, can help to build resilience in society and promote sustainable living.

Chapter 11 identifies the main attributes of transformative adaptation to develop the adaptive capacity of stakeholders in employing sustainable design and construction practices. This chapter reviews the social challenges and the relevant transformative adaptive theories that influence the adaptive capacity of built environment stakeholders. A framework is then proposed to guide practitioners and researchers to formulate strategies for implementing sustainability practices effectively, hence creating a better synergy of sustainable construction and enabling stakeholders to innovate more environmentally sustainable and resilient practices.

Ageing communities are disproportionately affected by climate change, and it is of paramount importance to create more resilient and sustainable housing for older people. Chapter 12 therefore explores the implications of people ageing in place and how the resilience of older people is affected by climate change and wildfires. This chapter investigates barriers to older people improving the bushfire resilience of their homes, using examples from the areas affected by the 2019 Black Summer Bushfires in Australia.

Part V presents sustainability roadmaps, sustainable policies and financial regulatory frameworks in the building and construction sector. Green finance represents a strategic policy blueprint for enabling the transition towards sustainability in the built environment. Chapter 13 provides a comprehensive understanding of green finance for green buildings and its implementation potentials. It assesses the opportunities and risks associated with green finance for sustainable buildings. Green finance offers stable funding opportunities such as green bonds, green securitisations and green credit facilities to increase accessibility to finance and overcome economic challenges for sustainable transformation.

Chapter 14 evaluates the approaches used by the UK government and assesses how successful they are in promoting the implementation of sustainable construction practices. Using a mixed methods approach, this chapter unveils that the industry struggles to comply with the Government's requirements for net zero and sustainable construction. This chapter critically reviews the current approach taken by the UK government towards sustainability and offers recommendations to provide a more fitting adaption of sustainable practices for the construction industry.

Chapter 15 utilises a knowledge framework – Integrated Knowledge-Based View to provide a snapshot of knowledge capabilities with regard to sustainability in the Malaysian construction industry. The chapter also describes how knowledge in general, and sustainability knowledge in particular, is critical components of the pathway to a more sustainable built environment.

Chapter 16 is a special addition to the book that highlights the use of ecotourism to reinvigorate a village to support a harmonious relationship between humans, settlements and nature. This chapter demonstrates how ecotourism can offer ecological, social and economic benefits to local communities while preserving the nature and culture.

In conclusion, this book offers a new and broad sustainability practice in buildings and infrastructure projects to tackle climate change issues and support the delivery of sustainable development goals. The book is beneficial for anyone who seeks to develop their knowledge and skills in sustainability in the built environment, as well as those who intend to move to sustainable construction. The sharing of case studies and best practices reflected in the book would draw readers closer towards sustainable built environment practice and promote the diffusion and adoption of sustainability at the local as well as the international levels.

1 Rethinking Our Pathways towards Sustainable Built Environment

Are We Failing?

Cheng Siew Goh and Heap-Yih Chong

1.1 Introduction

Climate emergency is no longer a future problem – it is happening now. Rising sea levels, heatwaves, wildfires, droughts, extreme weathers, floods are recent examples of climate change impacts. According to European Parliament (2023), the average global temperature is now 0.95–1.20°C higher than it was at the end of the 19th century. The last decade (2011–2020) was reported as the warmest decade on record, and 2022 is the second warmest year on record and recorded the hottest summer (European Parliament, 2023). The Emissions Gap Report 2033 published by UNEP also revealed that September 2023 was the hottest month ever and the global average temperature was at 1.8°C above the pre-industrial levels, exceeding the previous record by an unprecedented 0.5°C (United Nations Environment Programme [UNEP], 2023). These extreme events are a meek beginning of global climate crises.

It is evident that the Earth is now in crisis and the temperature could increase by between 2.2 and 3.5°C by 2100, if no further decisive action is taken (UKGBC, 2024). The Intergovernmental Panel on Climate Change (IPCC) has highlighted the climate emergency and urged the global community to mobilise and scale available solutions for decarbonisation. The urgency to combat climate change has never been greater to deliver a sustainable and resilient future for all.

The introduction of the Brundtland Report in 1987 set a wake-up call for the sustainable development movement. At the United Nations Summit on Sustainable Development Summit at the UN Headquarter in New York on 25–27 September 2015, the majority of worldwide countries formally approved to adopt a new agenda – Transforming Our World: The 2030 Agenda for Sustainable Development by 2030 that included a declaration, 17 sustainable development goals (SDGs) and 169 targets. Two months later, a political agreement known as the Paris Agreement was signed and then ratified by 187 parties. This landmark Paris Agreement which entered into force on 4 November 2016 agreed to hold the global average temperature increase to well below 2°C above the pre-industrial level and pursue efforts to limit the temperature increase to 1.5°C. The United Nations' 17 SDGs have since then compelled jurisdictions in transition to sustainable development.

DOI: 10.1201/9781003317890-1

1.2 Global Stocktake – Worldwide Current Progress, Trends and Gaps of Delivering Carbon Emission Targets

According to Ritchie et al. (2023), the global warming has not been equally distributed across the world, where the poles experienced the strongest warming and the Northern Hemisphere has warmed more than the Southern Hemisphere. Human-induced carbon emissions and greenhouse gas emissions are the primary sources of global warming and climate change. Carbon management is a credible pathway to limit global warming alongside ongoing efforts to creating inclusive and sustainable economic growth. This calls for a global coalition of countries, states, businesses and entities to take rigorous and immediate actions to reduce carbon footprints in transition to a sustainable world.

There are different metrics used to determine the carbon footprints, including annual carbon emissions by country, per capita carbon emissions, consumption-based carbon emissions and carbon emissions adjusted for trade. The complexities of carbon emissions have led to contentious debates about using a single metric to reflect a nation's contribution to the continuous increase of the global total emissions, since several factors such as population size, national economic development, historical contributions and cumulative emissions should be taken into consideration.

Annual carbon emissions are the carbon emissions aggregated and reported at the national level. In 2022, China recorded the highest total annual carbon emissions with 11.396 billion tonnes, followed by the US (5.057 billion tonnes), India (2.829 billion tonnes), Russia (1.652 billion tonnes) and Japan (1.053 billion tonnes) (Ritchie et al., 2023). In terms of annual carbon emissions per capita, Qatar topped the 2022 rank with 37.60 tonnes, followed by United Arab Emirates (25.83 tonnes), Bahrain (25.67 tonnes), Kuwait (25.58 tonnes), Brunei (23.95 tonnes) and Trinidad and Tobago (22.42 tonnes) (Ritchie et al., 2023).

Per capita carbon emissions refer to the carbon emissions of an average person in a country or region. As displayed in Figure 1.1, over 26 countries have over 10 tonnes per capita carbon emissions and most of them are high-income countries. High-income countries have recorded an average of 10.13 tonnes annual carbon emissions per capita (Ritchie et al., 2023). The Group of 20 or G20, which comprises a majority of the world's largest economies, accounts for about 76% of global greenhouse gas emissions (average of 7.9 tCO_{2e}), while the least developed counties averaged 2.2 tCO_{2e} (United Nations, 2023).

A new record of global greenhouse gas emission of 57.4 $GtCO_{2e}$ was reported in 2022, an increase by 1.2% from 2021 (UNEP, 2023). Since the beginning of industrial revolution, industrial countries which represent approximately 20% of the world's population account for nearly three quarters of cumulative carbon dioxide emissions. In terms of cumulative emissions, the US has emitted the highest amount of carbon emissions to date. The US accounts for about 17% of global warming from 1850 to 2021, although it represents 4% of the current world population (UNEP, 2023). The visualisation in Figure 1.2 shows inequality of in consumption-based emissions where G20 countries account for approximately 80% of historical cumulative carbon emissions, with the largest contributions from

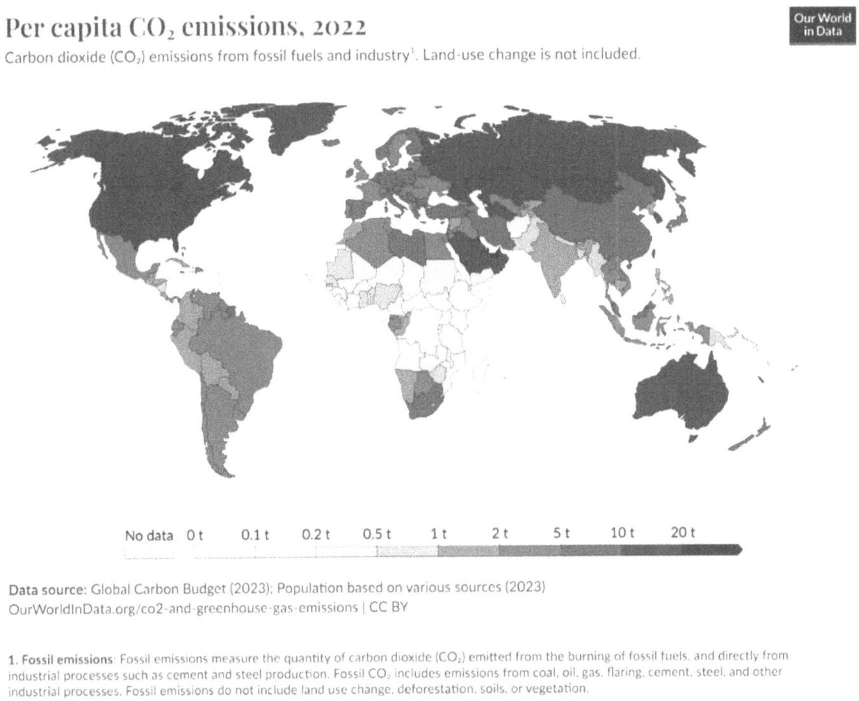

Figure 1.1 Worldwide per capita carbon emissions

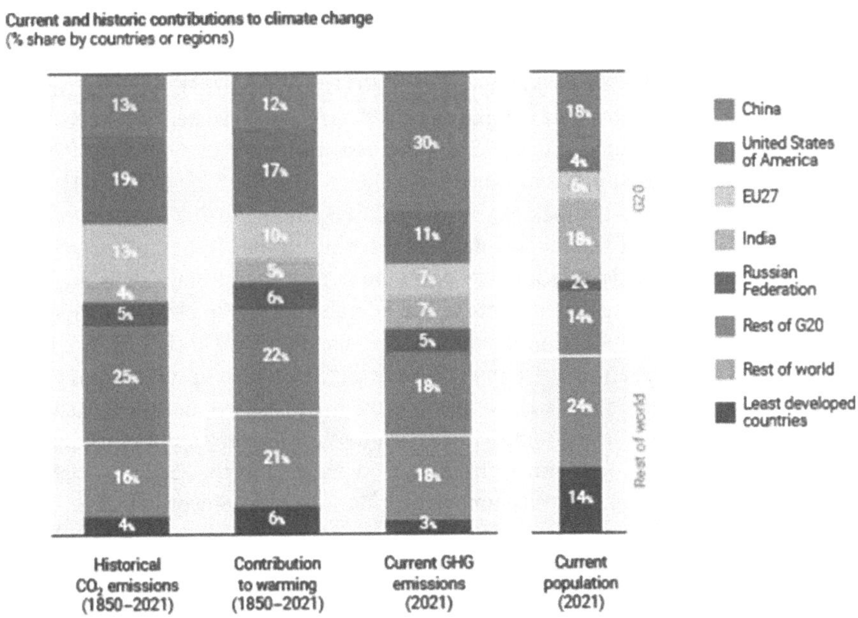

Figure 1.2 Current and historic contributions to climate change by countries/regions

China, the US and the European Union (EU) (UNEP, 2023). In the meantime, the bottom 50% of the world's population is responsible for only 12% of total emissions (UNEP, 2023).

The UK set the Climate Change Act 2008 to reach net zero by 2050 and developed five-yearly carbon budgets to cap the total greenhouse gas emissions the UK can emit in a five-year period. In the 2023 Progress Report issued by the Climate Change Committee (CCC) to Parliament, greenhouse gas emissions had been reduced by 46% relative to 1900 levels – a slight increase compared to the previous year due to progress in the transport sector (Burnett et al., 2023). The UK reduced 44% of its carbon emissions between 1990 and 2019 whilst growing the economy by 78% (BEIS & ESNZ, 2022). The provisional government estimates show UK greenhouse gas emissions dropped 5.4% to 383 Mt in 2023 – the first time fell below 400Mt CO_{2e} since 1879 (Evans & Viisainen, 2024).

By comparison, net EU greenhouse gas emissions between 1990 and 2021 fell by 30% and preliminary estimates show EU can reach a potential further decrease of 1.9% for 2022 as compared to the 2021 level (European Environment Agency, 2023). Energy crisis in 2020 has resulted in a partial switch to more coal generation, but at the same time spiking gas prices also decreased energy outputs and reduced carbon emissions in the building industry. As advocated by European Environment Agency (2023), the current policies and measures taken by the EU Member States would project a reduction of 41%, but this still leaves a gap of 7% to meet the target level of −310 Mt CO_{2e} by 2030. Considering the pace of carbon emission reductions observed in the past ten years, EU needs to accelerate the average annual rate of absolute greenhouse gas emissions by more than three times to meet the carbon target in 2030.

In the US, greenhouse gas emissions in 2021 reached 6340 million metric tonnes which was reduced by 2% as compared to 1990 (US EPA, 2023). It is noteworthy to note that US greenhouse gas emissions have not been consistent between 1990 and 2021 and it reached a peak of 15.8% increase above 1990 levels in 2007 (US EPA, 2023). The emission has also been on roller coaster post the COVID-19 pandemic, plunging during the lockdowns of 2020, rebounding in 2021 and plateauing in 2022 (increased by 1% after accounting for sequestrations from the land sector). Even though some analyses show US emission is projected to fall by as much as 3% in 2023, analysis suggested to make considerable efforts to close the huge gap which calls for a 50% reduction in emissions by the end of 2030 (US EPA, 2023).

Similar to the US, an upward and downward trend is also observed in the historical Australia's greenhouse gas emissions over the past two decades. Australia is one of the world's largest emitters per capita and its per capita emissions in 2022 were more than four times the global average (Morton, 2023). Australia recorded the highest carbon emissions per capita from coal power in G20 in 2022 and doubled China's per capita electricity (Morton, 2023). Australia reduced greenhouse gas emissions by 29% from 640.64 Mt $CO_{2\text{-}e}$ in 1990 to 432.90. Mt $CO_{2\text{-}e}$ (Department of Climate Change, Energy, the Environment and Water Australia, 2024).

China has a sharp increase of carbon emissions in the past four decades. The total carbon emission increased from 2426 million tonnes in 1990 (about 11% of the world's total emissions) to 11,680 Mt in 2020 (about 32% of the total world emissions), attaining 3.8 times increase within three decades. China's total greenhouse gas emissions without Land Use, Land Use Change and Forestry (LULUCF) stood at 3649.6 million tonnes in 1994 and it grew to 7467.09 million tonnes, 10,544 million tonnes and 12,301 million tonnes in 2005, 2010 and 2014, respectively (United Nations, 2023). Coal power emissions per capita in China raised by 30% over seven years since the growth of electricity consumption outpaced the growth of renewable energy generation (Morton, 2023). Renewable energy capacity in China has increased to 670 gigawatts since 2015, which is nearly a third of the world's solar and wind energy (Morton, 2023). With a rising share of non-fossil fuels in energy production, carbon emissions in China fell by 3.8% from the 2020 level and 50.8% from the 2005 level (United Nations, 2023).

India's net carbon emission in 2019 was 2.6 billion tonnes, marking an increase by 4.56% from the 2016 level and 115% from the 1994 level (Sehr Raheja, 2024). The average emission reduction rate in India increased to 3% annually in the period of 2016–2019, as compared to about 1.5% in the period of 2014–2016 (Singh, 2023). Although the reduction was the fastest (primarily attributable to the governmental push towards renewables), fossil fuel power plants continue to dominate the energy sector in India. India recorded a 10% rise of emissions arising from fossil fuel consumption for power generation in 2019. Energy represented the largest sector of India's greenhouse gas emission inventory over the two decades, where it stood at 75.8% of the total greenhouse gas emission in 2019 (Sehr Raheja, 2024).

1.2.1 Reneging on Climate Pledges and Recent Backtracking of Carbon Targets and Sustainability Policies

The uptake of sustainability is, to a significant extent, influenced by international and domestic policies and legislation frameworks. Sustainability and climate policies provide us with a clear roadmap in facilitating the deliverables of sustainability, along with the processes for monitoring and ensuring accountability. The legislative decisions and policies reduce the growth of carbon emissions and warming collectively relative to a world without climate policies, though these policies may still leave us far off track of reaching SDGs (European Parliament, 2023). Strengthening policies and measures has a definitely important role to play in meeting SDGs and reaching climate neutrality. There is a need for developing more enabling sustainable transition policies and regulatory frameworks to meet the mid-term and long-term SDGs.

Yet, recent backtracking on net zero policies has been observed in several major economies and this has sent a further alert to the global community that we are off track to meet the SDGs and have lost our momentum to the sustainable transformation. Various governmental and NGO reports have shown that we are not only falling much behind the progress towards net zero emissions but drifting away from the SDGs.

In the US, the Trump administration dismantled major climate policies and officially reversed, revoked or otherwise rolled back nearly 100 environmental rules, leaving an unprecedented legacy to the climate combat (Popovich et al., 2021). Although Biden's administration reestablished leadership and rejoined international efforts against climate change by signing the Inflation Reduction Act into law in August 2022 – the most comprehensive legislation in the US, four years of US absence during the Trump administration left a huge gap in the international credibility towards climate action. Being the largest historical greenhouse gas emitter by far, the US, who sat on the sidelines in the previous years, needs to advance more credible climate plans at the state and national levels to regain momentum on climate fight.

Australia had been seen as a laggard on climate change over the past three decades, as Australia was the first to dismantle a national carbon pricing scheme (Pickering, 2022; Slezak, 2021). During the Abbott and Turnbull government, Australia's carbon target was to reduce emissions 26–28% from 2005 levels – a target that was viewed as too conservative and insufficient even by the government's own Climate Change Authority (Slezak, 2021). Despite enshrining a pledge by boosting the 2030 emission target to a 43% reduction and meeting the net zero target by 2050, the Albanese government approved several coal mine and fossil fuel projects, continuing to expand fossil fuel production in the country and ballooning coal and gas export globally.

The UK is the first major economy to create a legally binding target and pass a law to achieve net zero by 2050, as a response to the CCC (an independent statutory body established under the Climate Change Act 2008) and overwhelming public support. Nonetheless, the UK has recently softened green policies by introducing "a more pragmatic, proportionate and realistic approach" to meeting net zero without burdening on working people. On 20 September 2023, the prime minister introduced several new approaches, including watering down the plan to phase out gas boilers by 2035, pushing back the ban on new petrol and diesel cars by five years to 2035, setting an exemption to the proposed 2035 phrase out of all fossil fuel boilers and scrapping policies to force landlords to upgrade energy efficiency of their properties (Burnett et al., 2023). The CCC noted that the UK government's plans remained insufficient to meet the emissions targets, considering all recent developments. The CCC then decreased its confidence for the UK government in meeting the mid-term targets, including the 2030 nationally determination contribution and the fifth and sixth carbon budgets.

Being one of the front runners in climate combat, the EU established the European Climate Law to set the binding target of reducing greenhouse gas emissions by at least 55% in 2030 compared to 1990 and reaching climate neutrality by 2050. The European Green Deal is the roadmap to guide the EU to reach climate neutrality by 2050. The concrete legislation for meeting the Green Deal targets is laid out in the Fit for 55 legislative package in July 2021 (European Parliament, 2023). However, several EU countries have faced growing pushback against climate policies and their sustainable agenda started to fray. Italy's right government reneged on 2021 climate pledge by signing a allowing fossil fuel support to continue long past the 2022 deadline. The Italian government also demanded the EU dilute a directive targeted at increasing energy efficiency of buildings, reevaluate plans to phase out combustion engine cars and question a drive to slash industrial emissions. On the other side, Germany passed a bill in September 2023 to phase out gas and oil heating systems but the bill was watered down from the original draft in April. The new bill is estimated to leave a gap to meet the emission goal by 2030.

Back to the G20 meetings in Bali in the year of 2022, some of the world's largest economies failed to agree on a joint communique, where the G20 bloc failed to reach a consensus on climate finance and language to reduce methane emissions. The global leaders had a long-running dispute over whether to emphasise limiting global warming well below 2°C or to 1.5°C above pre-industrial levels. At a preliminary meeting in India in July 2023, the G20 again failed to reach an agreement on phasing down fossil fuels and tripling renewable energy capacity by 2030. The outcomes of the 18th G20 Summit held in September 2023 (after 200 hours of negotiations and 300 bilateral meetings) presented a final declaration to the watering down of carbon neutrality, in which G20 major economies failed twice to agree on phrasing out the use of fossil fuels and on setting concrete targets on cutting carbon emissions (Singh, 2023). Although all countries agreed to "revisit and strengthen" the 2030 climate targets by the end of 2022 at COP 26, few countries have done so (Lo, 2022).

Carbon neutrality appears to be subterfuge when most developed nations which have the highest influences to mobilise carbon reduction solutions appear to stall the progress towards sustainable development. UNEP (2023) stated that high-income and high-emitted countries (which are responsible for the majority of historical emissions) failed to stringently reduce emissions, while low- and middle-income countries (which assume responsibility for the majority of current emissions) failed to limit emission growth.

Climate crises coupled with the COVID-19 pandemic and the Russia-Ukraine war have exacerbated the already faltering progress towards sustainable development. In the wake of Russia-Ukraine war, the international fossil fuel market has been fundamentally altered and the world has been scrambling to find new sources of fuel to reduce overreliance on Russian supply. This translated into huge boosts for fossil fuels, in which five developed countries (i.e. the US, UK, Canada, Australia and Norway) are responsible for 51% of fossil fuel expansion planned between 2023 and 2050 (Paddison, 2023). Public financial funds for fossil fuels doubled between 2019 and 2022, to over $1.4 trillion (OECD, 2022). A shift to unabated fossil fuels from renewable energy sources would leave us a wider gap in meeting carbon neutrality by 2050. It appears that rising energy prices could have led to a contradiction between the developed countries' stance at the international summits and their domestic policies.

Owing to energy crisis, committing to sustainable moves is now under huge pressure to ensure energy security and affordability, whilst maintaining the economic growth. There has been a delay or failure to stringently reduce carbon emissions despite pledges made by the governments and jurisdictions to meet the SDGs. Seeing policy rollbacks in many countries and jurisdictions, it appears that the divide between rhetoric and reality on sustainability is widening and increasing. It appears that the road to carbon neutrality is not socially and economically sustainable. This raised a question whether our goals of carbon emissions too ambitious or we are doing not enough to make this happen.

1.2.2 Risks of Greenwashing

Ambitious net zero pledges are only valuable if they are followed up with stringent implementation with tangible measures. It underscores that concrete actions must be taken in line with the climate pledges made by countries. A lack of implementation of stringent emissions reductions starting in 2020 has implied that the world has already depleted limited remaining carbon budget (UNEP, 2023).

Failure to meet the sustainability pledges and enact (and execute) plans swiftly with better governance and transparency might leave governments and leading economies open to greenwashing allegations. A critical analysis of government policies and relevant data about sustainable move and carbon reduction initiatives is needed at the international, regional, national, state and local levels. Although well-intended environmental legislation is promulgated by the government of leading developed nations, there may not be effective implementation at the local level. The implementation gap needs to be studied critically to identify what hampered the progress.

Not only dubious corporate green practice shall be rejected, but all governments and state actors shall also communicate accurate carbon reporting and track their progress towards the commitments made for carbon reductions and sustainability. All entities including public and private organisations need to complete carbon auditing and reporting to ensure integrity and transparency. Greater disclosure of sustainability accountability and climate risks would help move the needle in areas the policy lags behind. With this, the global coalitions and investors can mobilise new sources of finance to support both technological and socioeconomic transition towards sustainable development.

1.3 Roles of the Built Environment in Sustainable Transformation

Climate challenges have exacerbated a range of societal issues such as energy poverty, social disparities, public health and productivity loss. A drastic domino effect on the built environment such as urban heat island effects is triggered following the loss of equilibrium in the ecosystem due to climate change. The built environment offers required infrastructure and services to support the people's needs. It offers an enabling environment to deliver fundamental socioeconomic needs, improve quality of life and allow the people to thrive. Considering its encompassing ability to enhance people, places and the natural environment, the built environment has a role to combat climate change, hence improving sustainability and resilience.

The building and construction sector consumes a plethora of resources. The built environment represents a huge sectoral contribution to climate change, being responsible for 37% of global carbon emissions, 34% of energy demands and 50%

of material consumption (WGBC, 2023). Building energy demands in 2021 has grown to 135 EJ, exceeding the previous peak in 2019 by 3% and rising by about 4%. Compared to the 2020 level. Carbon emissions from buildings also recorded an all-time high of 10 GtCO_2, an increase by 5% from 2020 and by 2% from 2019 (UNEP, 2022). Both energy demands and carbon emissions from buildings registered the biggest increase in 2022 over the past decades (UNEP, 2022).

In the UK, the built environment is the second largest source of territorial greenhouse gas emissions after surface transport (UKGBC, 2023). Nearly 40% of UK carbon emissions are attributable to the built environment and around 25% of which are direct emissions resulted from building heating as well as energy consumption to support building operation and the operation of transport infrastructure (UK Government Commercial Function, 2022). It is expected that 80% of buildings to be occupied in 2050 in the UK already exist, thereby decarbonising existing building stocks is a major priority in the sustainability agenda.

Buildings (both residential and commercial buildings) account for 31% of total US emissions and residential buildings emitted more than half of this amount (U.S. Environmental Protection Agency, 2022). In Germany, buildings are also recognised as one of the main factors leading to a miss in meeting climate target in 2030. The building sector emitted 112 million tonnes of greenhouse gases in 2022, contributing to 15% of Germany's total emissions (Alkousaa & Kraemer, 2023). In New Zealand, around 9.4% of domestic emissions were building related, including emissions from building operation and emissions from the manufacture, transport, use and disposal of building materials throughout the life span.

The building and construction industry also has a ripple effect of carbon footprints along its supply chains. Construction-related activities account for 50 million tonnes of carbon emissions and more than half of which are construction materials production particularly steel and cement, which generated nearly 15% of global carbon emissions (UK Government Commercial Function, 2022). According to UNEP and GlobalABC (2024), the manufacturing of cement, steel and aluminium increased the building industry's carbon footprints by another 2.5 Gt CO_2, while brick and glass production added a further 1.2 Gt CO_2. Embodied emissions resulted from construction process, maintenance and demolition of buildings contribute to approximately 40–50 million tonnes annual carbon emissions (Environmental Audit Committee, 2022).

Nearly 100 billion tonnes of waste is produced by construction-related activities, including construction, renovation and demolition and about 35% of which are sent to landfill (WBCSD, 2022). In the UK, about 60% of waste produced comes from the construction industry (UK Government Commercial Function, 2022). Developing a circular economy in the building and construction industry can lock in a huge potential of hazardous waste into landfill and potentially reduce greenhouse gas emissions that accelerates climate disasters.

The built environment has an indispensable role to play in meeting the worldwide SDGs and net zero targets by 2050. Being one of the largest sources of greenhouse gas emissions, the untapped potential in the built environment is greatly acknowledged. Net zero and sustainability have become mainstream in the building and construction sector in the past few years. Global leaders, businesses, industry associations and various stakeholder groups have been committed and undertake efforts to enable best practice for reducing carbon footprints within the built environment. COP 28 saw the launch of the Building Breakthrough that has gathered 20 countries to deliver near zero and resilient buildings by 2030. Organisations such as IEA, WorldGBC, WBCSD, World Resources Institute, C40 Cities, Climate Group, and Building to COP have joined forces to navigate, influence and instigate change for sustainability in the building and construction sector.

1.3.1 State-of-Play of the Built Environment in Sustainable Revolution

Worldwide efforts are underway to realise carbon neutral and SDGs within the built environment. Since 1987, extensive regional- and country-led efforts have been made in support of sustainable policies, plans, strategies and frameworks for the building and construction industry. A wide range of sustainable policies, industry standards and interventions have been put in place to create a net zero, sustainable and resilient built environment. According to World Green Building Council (WorldGBC) (2023), about 80% of countries refer to buildings as part of their nationally determined contributions (NDCs). A total of 158 countries set up comprehensive climate action plans covering buildings and construction in their NDCs as of 2021, which recorded a sharp rise from 88 in 2015 and a considerable increase from 135 in 2020 (WBCSD, 2022; UNEP, 2022). There are 81 and 77 national building energy codes enacted for residential and non-residential structures in 2023, and 8% out of which are mandatory (UNEP, 2024).

There is encouraging progress with a substantial increase in investment in green technologies and energy efficient solutions in the built assets over the past decades. Investment in building energy efficiency reached USD 237 billion in 2021, recording a rise by 16% than the 2020 level (WBCSD, 2022). In 2022, global investment to decarbonise built environment rose again by 14% to USD 285 billion and this was largely driven by Europe's response to energy security and public investment (UNEP, 2024). Credible advancement of green technologies in the built environment has also demonstrated evident progress on boosting decarbonisation of the built environment, seeing a 3.5% reduction in building energy intensity since 2021 (UNEP, 2024).

On the other hand, national and regional governments have also enacted sustainable policies, plans, strategies and regulatory frameworks for addressing the climate impacts at the local, national, regional and worldwide levels. In European Union, the Energy Performance of Buildings Directive that aims to achieve a fully decarbonised building stock by 2050 in Europe was approved by the European Parliament in December 2023, after two years of negotiations. The Inflation Reduction Act passed by the US Congress in 2022 also includes six sections aiming to reduce embodied carbon of construction and building materials. India which attributes 40% of carbon emissions to buildings also enacted legislation to initiate carbon trading and sustainable building codes in 2022 (UNEP, 2024).

However, there is a tug of war between growth of built assets and improved energy efficiency interventions. The global floor area of buildings has risen by more than 31% since 2010 and reached to 250 billion square metres, increasing the total energy demand and offsetting the improved energy intensity (IEA, 2022). The global floor area of buildings is estimated to experience a further 15% increase, which is about 40 billion square metres by 2030 for meeting the needs of the population and economic growth (IEA, 2022; UNEP, 2024).

Although climate policies and green technologies continue to expand, gaps remain. The 2024 Global Status Report for Buildings and Construction revealed a concerning gap between the current state and the desired decarbonisation path in the built environment, amounting to 40 decarbonisation points (UNEP, 2024). There is a lack of enforceability of sustainability policies and legislation to effectively facilitate sustainable transformation in the built environment (Goh et al., 2020). The expanding policy landscape is not yet translating into actions where only seven NDCs align with net zero operational emissions (UNEP, 2024). There is also a lack of building coverage in the national and regional commitments for building sustainability, where only 11 out of the 194 countries and the EU have expanded on building-related actions for their NDCs' updates (UNEP, 2024).

In a recent progress report – Net Zero Whole Life Carbon Roadmap published by the UKGBC (2023), the UK-built environment falls short in meeting the 19% carbon reduction target between 2018 and 2022, leaving a gap of 6% which is equivalent to 11 $MtCO_{2e}$. The significant off track from the trajectory would pose challenges to the UK in meeting the national carbon targets. A substantial and critical gap in the UK's national policies needs enormous and bold efforts to meet its statutory climate targets.

According to World Green Building Council (2023), 4.2 million square metres of green building space had been certified through the worldwide members of Green Building Councils. The proliferation of sustainable building spaces however warrants neither superior environmental performance nor achievement of the SDGs within the built environment. Goh (2022) found that green construction or sustainable buildings do not always live up to the people's expectations and some even significantly deviate from their original sustainable targets when they are in use. The deployment of green technologies in sustainable buildings is found to have little or no impact on the building's average energy use intensity, and some green buildings even perform poorer than their conventional counterparts (Goh & Chong, 2023). Notwithstanding a lack of constant positive progress on achieving the environment goals, a growing number of operation-oriented studies suggested no strong evidence to link sustainable buildings to consistent success in delivering social sustainability (Goh & Chong, 2023). Some fall short in addressing privacy, occupant well-being, thermal comfort and indoor air quality (Goh, 2022).

In the built environment, sustainability is intertwined with people (society), nature (environment) and self-sustaining value (economy). We need a systemic approach for an immediate concerted effort from all parties to support the transition towards sustainable development. Policy makers, business leaders and decision makers are urged to take firm and innovative actions to decarbonise the built environment.

1.3.2 Revisit and Rethink Our Pathways to Creating Sustainable Built Environments

The built environment's sheer importance in terms of the contributions to sustainability cannot be overstated. To accelerate roadmap development and achieve sustainability and resilience in the built environment, there is a call to revisit and reevaluate the existing policies and applications of sustainability initiatives to set the stage for a more successful transition. An enabling environment and positive interventions by the built environment are indispensable to mitigate the climate change impacts. A strengthened methodology is required to assess the size and uncertainty of remaining carbon budgets to narrow the gap between the actual performance of the built environment and the decarbonisation pathway.

Successful transition to sustainability involves two strands of actions: mitigation of the extent and adaptation to the impacts. By far, extensive efforts have been put to mitigate the climate impacts within the built environment, by integrating low carbon materials, increasing energy efficiency, implementing circular economy and optimising resource efficiency. It seems that the ability of the built environment to adapt to the impacts of climate change have been overlooked. UKGBC (2021) found that the UK-built assets are inadequately prepared for mounting climatic risks, resulting in an "adaption gap" to protect communities, businesses and investments in the long term. The government and industry leaders need to embed proactive measures to foster climate resilient structure throughout the entire value chains of the built environment to mitigate maladaptation risks and decrease potential energy poverty (UNEP, 2024). A sustainable built environment shall be self-sustaining and designed for future-proofing, flexibility, adaptability, assembly and disassembly, reuse and recoverability, and resilience.

Building climate adaptation and resilience in the built environment shall be given attention to proactively address and respond to growingly intense climate change impacts, hence meeting the SDGs. At the current stage, it appears that mitigation measures are given much greater emphasis than adaptation measures to tackle climate risks (Scott et al., 2022). Adaptation in the built environment appears to be in infancy stage and more concerted efforts need to be mobilised at the local and national levels. It is important to integrate local climate adaptation measures to develop a conducive environment to improve climate resilience. Adaptation measures require a robust and geographically tailored evidence. Local landscape, governance capacity and vulnerability assessment are required to be taken into consideration for making the built environment resilient to the current and future climate challenges.

The building and construction sector is fragmented and highly heterogeneous. Not a single actor can drive change for sustainability alone. Greater social, political and organisational leadership is needed to support decarbonisation and sustainability transition of the built environment. Deep collaboration across the entire value chain of the construction and building sector is essential to tackle the climate challenges. Governments in collaboration with businesses, non-profit institutions, financial institutions and wider communities shall deliver sustainability actions at speed and scale to drive radical transformation towards sustainability in the built environment. Public agencies and private sector shall mobilise finance and scale up investments in sustainable built assets to create more effective and efficient blended finance vehicles for long-term sustainable solutions in the built environment.

Concerted capacity building at the national and local levels to strengthen sustainability knowledge and capacity is critical for informed decisions in the built environment. Sustainability and climate literacy should be embedded in the whole value chain to address climate emergency within the built environment. Embedding sustainability and climate literacy as standards for built environment professionals can empower professionals to achieve the sustainability aspirations by understanding the key competences for delivery and contextualising the implementation to their roles and responsibilities (Construction Industry Council, 2023). Sustainability literacy would increase the capacity and capability to drive and enable change for sustainable practice in the built environment. In addition to government, non-state actors such as authorities and local communities hold the key for placemaking of sustainable built environments. Sustainability literacy should be expanded to non-state actors and wider communities to equip people with knowledge and motivation for taking effective and transformative actions towards sustainable built environment.

Humans are at the very foundation of sustainable cities and communities, and the built environment holds the key in shaping and influencing social dynamics (Kawamura & Brady, 2023). An increasing number of studies and professional reports (Goh, 2022; Goh & Chong, 2023; IEA, 2022; Kawamura & Brady, 2023) suggested that behavioural changes and community participations are critical to drive and enable sustainable transformation of the building and construction sector. Wider communities and end-users have a crucial role to play in determining the ultimate sustainable performance and carbon footprint of a built environment. Contextual factors such as politics and legislation, culture, local habits, lifestyles and the capability to control and react to the system would shape the ways users engage and interact with the built environment, hence affecting the effectiveness of sustainable solutions (Goh, 2022). Human-centred approaches that place humans at the core of designing built environments would help create more synergies between humans and the environment, thereby increasing agility of built assets for climate adaptation and resilience.

Applying life cycle thinking is integral to enable greater consistency in delivering high quality and sustainable outcomes across the life cycle of built environments (UK Government Commercial Function, 2022). Life cycle-based approaches go beyond the traditional boundary at the construction finish line and extend consideration to the entire life cycle of built assets, i.e. from cradle to grave. The whole life cycle approach assesses the implications of sustainability in a more holistic and systematic way. Life cycle thinking allows the integration of carbon costs in decision making across the full value chain of construction. Costs and benefits of carbon reductions and environmental goals are assessed alongside the associated social and economic implications. The interconnections between economic, social and environmental sustainability can therefore be examined thoroughly across the whole value chain (Goh, 2017).

On the account of various mid-term climate goals set out in Paris Agreement as well as Agenda 2030, the year of 2030 could be viewed as an inflection point for the climate fight. Various international consortiums and governments have also referenced 2030 as the year of a focus for their carbon pledges and climate agenda. The milestone in 2030 reflects the near-term actions towards the carbon goal by 2050 and would conclude the worldwide efforts on climate combat for nearly four decades since the establishment of the United Nations Framework Convention on Climate Change (UNFCCC) in 1992. This critical year would give us a clear signalling on the alignment of our sustainability initiatives against the long-term goals, positioning ourselves better towards a robust pathway for creating a net zero, climate resilient, healthy, and socially and economically inclusive built environment for all.

1.4 Conclusion

This chapter provides a snapshot of the status of climate policies and the progress of carbon emission reduction on a global scale. It examines different issues of sustainable construction and highlights the need for rethinking of pathways in upholding sustainable built environment.

The built environment is at the forefront of sustainable transformation on the account of its contribution to the global carbon emissions. In the wake of a tumultuous political and economic period, progress of decarbonising the built environment has been glacially slow, particularly some high-profile legislative decisions have moved in the opposite direction. Developed and emerging economies are slipping on their commitments to decarbonisation. There is need to revisit existing climate policies and applications of current sustainability initiatives to set the stage for a more successful transition towards sustainability and resilience in the built environment.

Creating a sustainable and resilient built environment for meeting carbon targets is not pie-in-the-sky. To drive a positive systemic transformation in the built environment, all stakeholders including governments, businesses, financial institutions, academia, voluntary sector organisation and wider communities need to recognise the urgency of creating a resilient and sustainable built environment and actively engage in the sustainable transition. Capacity building is of value to strengthen sustainable knowledge and skills across the value chain and empower non-state actors and motivating wider communities to engage more actively in sustainability.

A coherent and enforceable strategy is essential to create the momentum for change towards sustainable built environments. Apart from climate mitigation measures, adaptation and resilience in the built environment become increasingly important. It calls for a shift to adopt a whole life cycle-based strategy in assessing carbon footprints of the built environment. Human-centric approaches shall also be integrated in design and planning of built environment for increased agility to climate resilience.

References

Alkousaa, R., & Kraemer, C. (2023). 'German Parliament Passes Law to Phase Out Gas and Oil Hearing'. Retrieved from https://www.reuters.com/world/europe/german-parliament-passes-watered-down-heating-law-2023-09-08/

BEIS & ESNZ (2022). 'Policy Paper 1. Why Net Zero'. Retrieved from https://www.gov.uk/government/publications/net-zero-strategy/1-why-net-zero

Burnett, N., Edwards, T., & Watson, N. (2023). 'The UK's Plans and Progress to Reach Net Zero by 2050'. Retrieved from https://researchbriefings.files.parliament.uk/documents/CBP-9888/CBP-9888.pdf

Construction Industry Council (2023). 'Embedding Sustainability and Climate Literacy in Education and Professional Qualification: A Construction Industry Council (CIC) Toolkit'. Retrieved from https://www.cic.org.uk/uploads/files/Embedding-Sustainability-and-Climate-Literacy-in-Education-and-Professional-Qualification-A-CIC-Toolkit_FINAL.pdf?v=1690368966

Department of Climate Change, Energy, the Environment and Water Australian (2024). 'National Greenhouse Gas Inventory Quarterly Update: December 2023'. Retrieved from https://www.dcceew.gov.au/climate-change/publications/national-greenhouse-gas-inventory-quarterly-update-december-2023

Environmental Audit Committee (2022). 'Building to Net Zero: Costing Carbon in Construction'. Retrieved from https://publications.parliament.uk/pa/cm5803/cmselect/cmenvaud/103/report.html

European Environment Agency (2023). 'Total Net Greenhouse Gas Emissions Trends and Projects in Europe'. Retrieved from https://www.eea.europa.eu/en/analysis/indicators/total-greenhouse-gas-emission-trends#:~:text=Net%20greenhouse%20gas%20(GHG)%20emissions,%2Dyear%20reduction%20of%201.9%25.

European Parliament (2023). 'EU Action on Climate Change'. Retrieved from https://www.europarl.europa.eu/topics/nl/article/20180703STO07129/eu-actie-tegen-klimaatverandering

Evans, S., & Viisainen, V. (2024). 'Analysis: UK Emissions in 2023 Fell to Lowest Level Since 1879'. Retrieved from https://www.carbonbrief.org/analysis-uk-emissions-in-2023-fell-to-lowest-level-since-1879/

Goh, C. S. (2017). Towards an Integrated Approach for Assessing Triple Bottom Line in the Built Environment. In *SB-LAB 2017: International Conference on Advances on Sustainable Cities and Buildings Development*. Green Lines Institute.

Goh, C. S. (2022). Unlocking Human Factors for More Resilient and Sustainable Built Environments: Human Centric Solutions. In *IOP Conference Series: Earth and Environmental Science* (Vol. 1101, No. 7, p. 072011). IOP Publishing.

Goh, C. S., & Chong, H. Y. (2023). Opportunities in the sustainable built environment: Perspectives on human-centric approaches. *Energies 16*(3), 1301.

Goh, C. S., Jack, L., & Bajracharya, A. (2020). Qualitative study of sustainability policies and guidelines in the built environment. *Journal of Legal Affairs and Dispute Resolution in Engineering and Construction 12*(2), 04520016.

IEA, (2022). 'Residential Behaviour Changes Lead to a Reduction in Heating and Cooling Energy Use By 2030.' IEA, Paris https://www.iea.org/reports/residential-behaviour-changes-lead-to-a-reduction-in-heating-and-cooling-energy-use-by-2030, Licence: CC BY 4.0

Kawamura, S., & Brady, C. (2023). 'Social Impact Across the Built Environment'. Retrieved from https://worldgbc.org/wp-content/uploads/2023/12/C23.9525-WGBC-Creating-Better-Places-for-People_FINAL.pdf

Lo, J. (2022). 'Collapse of G20 Talks in Bali Spark Fears of 'backtracking' on Climate Pledges'. Retrieved from https://www.wri.org/insights/cop26-climate-pledges-tracking-progress

Morton, A. (2023). 'Australia Has Highest Per Capita CO2 Emissions from Coal in G20, Analysis Finds'. Retrieved from https://www.theguardian.com/environment/2023/sep/05/australia-has-highest-per-capita-co2-emissions-from-coal-in-g20-analysis-finds

OECD (2022). 'Fiscal Cost of Support Measures for Fossil Fuels Almost Doubled in 2022'. Retrieved from https://www.oecd.org/coronavirus/en/data-insights/fiscal-cost-of-support-measures-for-fossil-fuels-almost-doubled-in-2022

Paddison, L. (2023). 'The World Has Ground to Make Up at Climate Summit in Dubai after a Year of 'backsliding''. Retrieved from https://edition.cnn.com/2023/11/30/world/cop28-climate-backsliding/index.html

Pickering, J. (2022). 'Laggard to Leader? Labor Could Repair Australia's Tattered Reputation on Climate Change, if it Gets These Things Right'. Retrieved from https://theconversation.com/laggard-to-leader-labor-could-repair-australias-tattered-reputation-on-climate-change-if-it-gets-these-things-right-182860

Popovich, N., Albeck-Ripka, L., & Pierre-louis, K. (2021). 'The Trump Administration Rolled Back More Than 100 Environmental Rules. Here's the Full List'. Retrieved from https://www.nytimes.com/interactive/2020/climate/trump-environment-rollbacks-list.html#:~:text=total%20rollbacks-,The%20Trump%20Administration%20Rolled%20Back%20More%20Than%20100%20Environmental,Here's%20the%20Full%20List.&text=Over%20four%20years%2C%20the%20Trump,water%2C%20wildlife%20and%20toxic%20chemicals.

Ritchie, H., Rosado, P., & Roser, M.. (2023). 'CO_2 and Greenhouse Gas Emissions'. Published online at OurWorldInData.org. Retrieved from https://ourworldindata.org/co2-and-greenhouse-gas-emissions [Online Resource]

Scott, M., Burns, L., Lennon, M., & Kinnane, O. (2022). 'Built Environment Climate Resilience and Adaptation'. Retrieved from https://www.epa.ie/publications/research/climate-change/Research_Report_418.pdf

Sehr Raheja, F. Z. (2024). '5% Rise in India's GHD Emissions Since 2016, Driven by Energy and Industrial Sectors'. Retrieved from https://www.downtoearth.org.in/blog/pollution/5-rise-in-india-s-ghg-emissions-since-2016-driven-by-energy-industrial-sectors-94076

Singh, S. C. (2023). 'India Succeeds in Reducing Emissions Rate by 33% Over 14 Years – Sources'. Retrieved from https://www.reuters.com/world/india/india-succeeds-reducing-emissions-rate-by-33-over-14-years-sources-2023-08-09/#:~:text=India's%20average%20rate%20of%20reduction,to%20dominate%20the%20energy%20mix

Slezak, M. (2021). 'Australia Looks Like a Climate Laggard, as Other Countries Ramp-Up Efforts'. Retrieved from https://www.abc.net.au/news/2021-11-03/will-australia-meet-and-beat-2030-emissions-target/100568416

U.S. Environmental Protection Agency. (2022). 'Inventory of U.S. Greenhouse Gas Emissions and Sinks'. Retrieved from https://www.epa.gov/ghgemissions/inventory-us-greenhouse-gas-emissions-4and-sinks

UK Government Commercial Function (2022). 'Promoting Net Zero Carbon and Sustainability in Construction'. Retrieved from https://assets.publishing.service.gov.uk/media/631222898fa8f54234c6a508/20220901-Carbon-Net-Zero-Guidance-Note.pdf

UKGBC (2021). 'Financing the Built Asset Adaptation Gap: Short Report'. Retrieved from https://ukgbc.org/resources/financing-the-built-asset-adaptation-gap-short-report/

UKGBC (2023). 'Built Environment Must Decarbonise Nearly Twice as Fast by 2025, Says UKGBC Analysis'. Retrieved from https://ukgbc.org/news/progress-report-launch/

UKGBC (2024). 'Climate Change Mitigation'. Retrieved from https://ukgbc.org/our-work/climate-change-mitigation/

United Nations (2023). 'China's Policy Strategies for Green Low Carbon Development: Perspective from South-South Cooperation'. Retrieved from https://unctad.org/system/files/official-document/gds2023d6_en.pdf

United Nations Environment Programme (UNEP) (2022). '2022 Global Status Report for Buildings and Construction: Towards a Zero-emission, Efficient and Resilient Buildings

and Construction Sector'. Nairobi. https://www.unep.org/resources/publication/2022-global-status-report-buildings-and-construction
United Nations Environment Programme (UNEP) (2023). 'Emissions Gap Report 2023'. Retrieved from https://www.unep.org/interactives/emissions-gap-report/2023/#section_0
United Nations Environment Programme (UNEP) and Global Alliance for Buildings and Construction (GlobalABC) (2024). 'Global Status Report for Buildings and Construction: Beyond Foundations: Mainstreaming Sustainable Solutions to Cut Emissions from the Buildings Sector'. Nairobi. https://doi.org/10.59117/20.500.11822/45095
US EPA (2023). 'Retrieved from Climate Change Indicators: U.S. Greenhouse Gas Emissions'. US EPA
WBCSD (2022). 'COP27: Built Environment Accelerates Climate Action'. Retrieved from https://www.wbcsd.org/Pathways/Built-Environment/News/COP27-Built-environment-accelerates-climate-action
World Green Building Council (WorldGBC) (2023). 'Sustainable Building Certifications'. Retrieved from https://worldgbc.org/sustainable-building-certifications/#:~:text=By%202021%2C%204.2%20billion%20square,and%20firmly%20supports%20their%20use

Part I

Innovative Resolutions/ Strategies for Net Zero

2 Transdisciplinary Integrated Design for Carbon-Neutral High-Rise Commercial Buildings

Man Kit Leung, Dicken Poon, and Carolina Sanchez Salan

2.1 Background

2.1.1 Climate Change Impact on Tropical/Subtropical High-Density Cities

The 2007 Intergovernmental Panel on Climate Change (IPCC) Fourth Assessment Report (AR4) showed that Earth's climate system was warming (IPCC, 2007). Global averaged surface warming (for 2090–2099 relative to 1980–1999) ranges from 1.1°C to 6.4°C, with sea level rise predicted at 18–59 cm (IPCC, 2007). In tropical/subtropical coastal zones this means that temperatures will increase, rainfall will fluctuate, and sea levels will rise; extreme storms and tropical cyclones will be more frequent and severe leading to storm surges and flooding. Climate change is likely to affect human health in cities (IPCC, 2007). The IPCC's AR4 concluded that climate change had already contributed to increased disease and premature deaths, altered infectious disease vectors distribution, and increased heatwave-related deaths (IPCC, 2007; Dodman, 2009).

Density is a major factor impacting how cities will be influenced and affected by climate change. Increased density as a strategy for urban development without evaluating other factors, such as location of employment opportunities, transportation systems, and residents' quality of life, may not result in lasting sustainability or resilience. High urban densities can bring localised climatic effects such as the urban heat island (UHI) effect (Coutts et al., 2008). Other vulnerabilities made worse by density are coastal erosion, high levels of outdoor and indoor air pollution, and impaired sanitation in low-lying areas (Campbell-Lendrum and Corvalan, 2007). These vulnerabilities give opportunities for addressing human health and cutting greenhouse gas emissions (GHGe) derived from transport, urban form, planning, building regulations, and household energy supply (Dodman, 2009). By making efforts to reduce polluting activities that exacerbate climate change, e.g., traffic congestion, we can also improve health aspects such as better air quality, improved commuting experience, and increased opportunities for physical exercise. Well-planned and managed, vertically dense cities can help to limit GHGe and limit biodiversity loss (Dodman, 2009; Yu and Shaw, 2018).

Globally, buildings generate nearly 40% of annual global CO_2 emissions (United Nations Environment Programme [UNEP], 2020). In 2019, China's CO_2 emissions

DOI: 10.1201/9781003317890-3

exceeded those of all developed countries combined and accounted for about 27% of the global total (Larsen et al., 2021). China's construction and building sector is also the largest in the world, making up about 50% of the total national CO_2 emissions (Yang and Li, 2021), 28% of which are from embodied carbon from materials and construction, and 22% of which come from operational carbon. These figures underline the urgency and importance of moving China's building sector towards net zero.

2.1.2 *Whole Life Carbon Framework*

The EN 15643 family of standards is used in the sustainability assessment of buildings (British Standard [BS], 2011). Within this family, the EN 15978: 2011 standard gives the framework (Figure 2.1) for appraising the environmental impacts of the built environment (BS, 2011). It provides guidance for calculating the whole life impact on the environment from built projects based on a life cycle assessment (LCA). Whole life carbon assessments should be done sequentially during the design, procurement, construction, and post-completion stages of a project and should start at the conceptual design stage (BS, 2011).

This chapter focuses on whole life carbon: the embodied carbon during the product and construction stages in the diagram above (A1–A5), plus the operational carbon during the in-use stage (B6). Building design plays a major role in these stages. Referring to EN 15978: 2011 and the diagram above, we focus on:

- The product stage [A1, A2, and A3] – The carbon emissions attributable to the cradle-to-gate processes, such as raw material supply, transport, and manufacturing (BS, 2011).

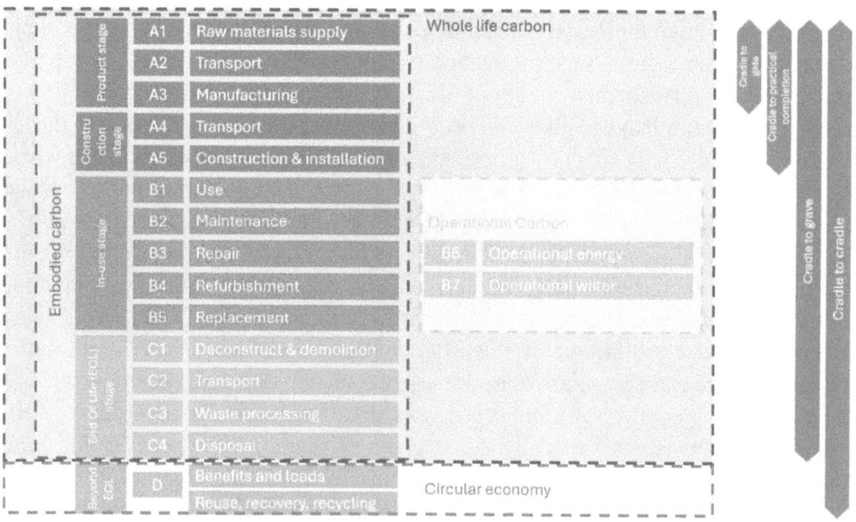

Figure 2.1 Modular information for carbon emission assessment as per EN 15978 (BS, 2011)

- The construction process stage [A4 and A5] – The emissions associated with the transportation of the materials and components from the factory gate to the project site and their assembly into a building (BS, 2011).
- The operational energy in-use stage [B6] – This stage includes all energy used in:
  - Building-integrated systems, i.e. energy "regulated by building regulations" (BS, 2011). It includes heating, air conditioning, ventilation, lighting, and auxiliary systems.
  - Further building-integrated systems (lifts, safety, security, and communication installations) should be included but reported separately where possible.
  - Carbon emissions from non-building-related systems, i.e. "unregulated energy use" (BS, 2011), such as information and communications technology (ICT) equipment, cooking appliances, and specialist equipment, can represent a significant part of the total operational emissions. Emissions from these systems should be included where possible to give a complete picture of the life cycle impacts but reported separately under B6 (BS, 2011).

2.1.3 The HKGBC ANZ Ideas Competition and Treehouse

Advancing Net Zero (ANZ) is a global project initiated by the World Green Building Council that aims to promote the creation of 100% net-zero carbon buildings by 2050 and at least 40% less embodied carbon for new buildings, infrastructure, and renovations by 2030. The Hong Kong Green Building Council (HKGBC) supports the ANZ project, and in 2021, organised the first-ever international ANZ Ideas Competition to generate ideas and solutions for carbon-neutral high-rise office buildings in a subtropical, high-density urban context (Hong Kong Green Building Council [HKGBC], 2021).

Treehouse won the "Future Building" category of the competition. The design, led by Ronald Lu & Partners (RLP), adopted a transdisciplinary approach. The team comprised architects; mechanical, electrical, and plumbing (MEP) engineers; structural engineers; climate engineers; sustainability engineers; smart building analytics engineers and service providers; and quantity surveyors.

Treehouse is a 220-m tall, Grade A office building, with a site area of 4,238 m^2, and a total gross floor area of 94,144 m^2. The design responds to a hot and humid climate in a mixed-use high-density district of Hong Kong. The energy use intensity (EUI) for the operational energy and the upfront carbon was estimated as 51 kWh/m^2/year and 342 kg CO_{2e} per m^2 of construction floor area, respectively, resulting in a substantially reduced carbon footprint of 9.3 million tonnes of CO_{2e}, which is 74% smaller than the business-as-usual baseline.

2.2 Transdisciplinary Design for Future-Ready Workplace

To efficiently weave deep decarbonisation and resilience into a commercial project, all disciplines involved need to align their goals and strive to attain a full understanding of the challenge at hand. The design for Treehouse offers a new model for the workplace, offering integrated solutions to address socio-environmental sub-systems including the vertical microclimate, new ways of working, and diverse human behaviours.

2.2.1 Transdisciplinary Design

Building professionals need to work together to tackle the scale and complexity of the challenges involved in decarbonising our built environment. This requires the involvement of multiple disciplines working synergistically from the early design stages to deliver solutions that transcend the individual disciplines. Such an approach is often called "transdisciplinary design". The objective of the approach is to build a deeper, common understanding of a complex system from the perspectives of different disciplines with the objective of developing holistic solutions (Chou and Wong, 2015). This is fundamentally different from the traditional hierarchical approach that emphasises a clear division of work scopes among various disciplines, a single-discipline-focused and fast-paced design process as shown in Figure 2.2.

Transdisciplinary design goes beyond multidisciplinary design, which is primarily concerned with drawing knowledge from different disciplines while staying within the boundaries of one discipline. It also transcends cross-disciplinary and interdisciplinary design, which also seeks to analyse, synthesise, and harmonise fields of knowledge. Transdisciplinarity not only views one discipline from the perspective of another and synthesises links between disciplines in a coordinated and coherent whole. It integrates, within a humanities context, natural, social, and building sciences, transcending their boundaries in the process.

Our transdisciplinary design process involved all the disciplines working together from the pre-design stage and providing inputs to the other disciplines to achieve a common goal, chiefly, greater sustainability in the form of a reduced carbon footprint for the whole life cycle.

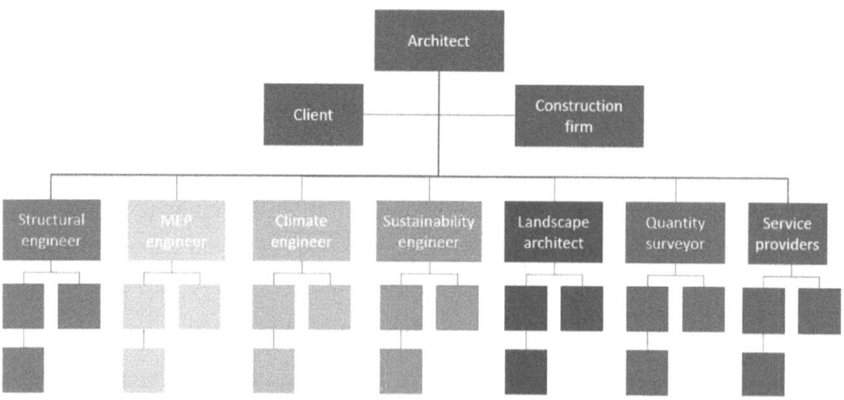

Figure 2.2 Traditional hierarchical design process

The driving force behind transdisciplinary design is an integrated process that harnesses the desire to find synergies among systems and components, the mutual advantages that can help achieve high levels of building performance, human comfort, and environmental benefits. Therefore, team members must collaborate to enhance the efficiency and effectiveness of every system, not just their own speciality.

In the figure above the large, dotted circles represent the major areas of work or key tasks, e.g., foundation and structural system, façade, internal layout, MEP systems, at-grade and elevated landscaping, etc. The coloured dots represent the different disciplines in the team, e.g., structural engineer in red or MEP engineer in ochre. As Figure 2.3 shows, the expertise of all team members is employed in the various tasks and they overlap with interrelated activities.

An understanding of building system interrelationships allows the team to uncover opportunities for innovative design and greater building performance. Synergies between systems help the team to save time and money and optimise resource use by avoiding delays and costs resulting from design changes during the construction document phase.

Before the schematic design, the team meets for a design charrette to discuss options to optimise the integration of green strategies across all aspects of building design, construction, and operations, drawing on the expertise of all participants. It

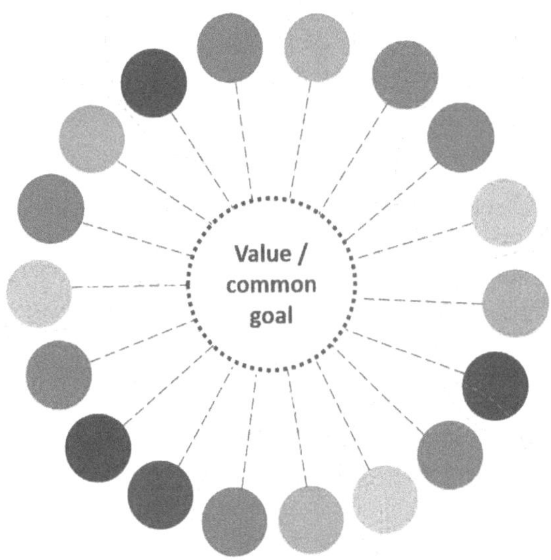

Figure 2.3 Integrated design process for transdisciplinary design

is important to conduct background research on likely energy and water consumption patterns. This is a discovery step that uses simple-box energy modelling analysis that explores how to reduce energy loads and water consumption in the building and accomplish related sustainability goals by questioning default assumptions.

Next, the team engages the project owner in a team workshop to determine the project goals, including budgets, schedules, functional programmatic requirements, scope, quality, performance, and occupants' expectations. Understanding the owner's goals promotes creative problem-solving and encourages interaction. In order to better understand the owner's goals, the team workshop objectives should be to:

1 Share initial background and analysis findings.
2 Discover the owner's and stakeholders' values and aspirations.
3 Establish functional and programmatic goals.
4 Establish initial principles, benchmarks, metrics, and performance targets.
5 Generate potential integrative strategies for achieving performance targets.
6 Determine the questions that must be answered to support project decisions.
7 Identify initial responsibilities and deliverables.
8 Initiate documentation of the owner's project requirements (OPR).

The schematic design stage will start after the completion of the pre-design tasks. Unlike in a conventional situation, a transdisciplinary design process incorporates all the collective understandings of system interactions that were found during discovery and pre-design. Having identified opportunities to achieve synergies across disciplines and building systems during the pre-design, the schematic design will continue to develop the design and team members will continue to meet regularly to confirm the design.

2.2.2 Future Workplace: A Place for Hybrid Work Mode

While the physical elements are there, the way we work and what can be considered a workspace needs to be reimagined. Organisations had to address the changes brought by the pandemic and the rapid pace of technological investment to enable remote and flexible work (Mugayar-Baldocchi et al., 2021). Computer-based office work needs moderate physical proximity to others and moderate interaction with others. This is the largest job sector in advanced economies, accounting for roughly one-third of jobs. Nearly all potential remote work is within this area (Lund et al., 2021).

Surveys found that about 20%–25% of the workforce in advanced economies could work from home three to four days per week without productivity loss (Steelcase, 2021; https://www.steelcase.com). This means four to five times more remote work than before the pandemic. Consequently, there could be a large change in the geography of work (Lund et al., 2021). The primary reason why employees eschew 100% home working is a lack of contact, a sense of team, and a blurring of the distinction between work-life (ISG, 2020). A survey of 278 executives by McKinsey in August 2020 found that, on average, they planned to reduce office space by 30%.

Demand for downtown food and drink, retail space, and for public transportation may decline as a result (Lund et al., 2021).

While work today is already increasingly mobile, the workplace of tomorrow will enter the metaverse. Businesses will occur in the virtual world, offering employees fully immersed digital experiences via extended reality (XR). As individuals step into non-traditional working modes, future workers equipped with virtual and augmented reality (VR/AR) tools remain in touch with the immediate team members and the wider business. The work areas with the highest levels of human interaction are likely to see the greatest acceleration in the adoption of automation and artificial intelligence (AI) (Lund et al., 2021).

2.2.3 Strategies: Positive Impact and Challenges

Many developers and owners of office buildings have embraced the business case for decarbonising and improving the overall wellness of building occupants. This is so because reductions in energy use, carbon emissions, water consumption, and waste production all lead to lower utility costs, lower maintenance costs, healthier buildings, and more satisfied tenants. In short, they improve the overall value of the asset (Swett et al., 2020).

However, a deep decarbonisation of office buildings will need a thorough investigation of the opportunities and challenges presented by many factors, including urban planning, microclimate, technological innovation and integration, regulatory barriers and economical considerations. For instance, to be effective, passive, and low-energy architectural design must be integrated with urban planning and must be based on an in-depth understanding of the urban climate (Bulkeley et al, 2010; Leadership in Energy and Environment Design [LEED], 2020; Sijmons, 2014). New and emerging technological advancements in architecture can achieve very high environmental quality, exemplary efficiency in resource use, and effective building management. Meanwhile, digital disruptions are accelerating the evolution of new ways of working (Trenerry et al., 2021). Stricter energy codes and standards will produce higher energy and emissions benchmarks, upgraded building systems, and more robust on-site renewable energy (RE) targets (Civic Exchange, 2020). The cost involved in deep decarbonisation will be compared with how performance can be optimised through integrated building design, energy price, the readiness of the supply chain, and numerous other factors (Civic Exchange, 2020).

We have evaluated and considered a range of strategic options based on their potential impact to achieve exemplary carbon emissions reduction and create sustainability values versus the foreseen challenges in implementing them. Strategies were reviewed for their positive impact in terms of whole life carbon performance, health and well-being, functionality and flexibility, ecological value, on the one hand, and challenges in terms of supply chain readiness, regulatory barriers, capital and life cycle cost, and resource use, on the other hand. "Quick-win" strategies create a moderate impact facing a moderate number of challenges. "Exemplary" strategies achieve greater impact but tackle surmountable challenges. "Visionary" strategies demonstrate admirable impact but require a further concerted effort by various stakeholders.

2.3 Quick-Win Strategies

Quick-win strategies are innovations with immediate benefit, albeit of moderate impact, implementable relatively quickly. Most quick-wins are easier to implement, less expensive, and have a narrow and focused area of action. They include a mixture of biophilic elements, structural, façade, MEP, and smart building innovations. These are:

1 Permeable pedestrian zone
2 Urban green canopy
3 Appropriate window-wall ratio and high-performance glazing
4 West-facing side core
5 Optimisation of pumping energy
6 Structural elements optimisation
7 Apparel attribute recognition

2.3.1 Permeable Pedestrian Zone

Breezeways are a crucial element of ventilation in dense urban areas (HKGBC, 2018). Adequate ventilation alleviates the UHI effect and is essential for thermal comfort and pollution dispersion at the pedestrian level. The Buildings Department (BD) of Hong Kong promulgated its Sustainable Building Design (SBD) Guidelines in 2011 (updated 2016). These require:

1 Overall 20%–33.3% permeability (Figure 2.4) for buildings in large sites or buildings with a long façade (greater than 60 m long), depending on their planes and height (BD, 2016).
2 Buildings abutting a street narrower than 15 m wide to be set back no less than 7.5 m from the centreline of the street (BD, 2016).

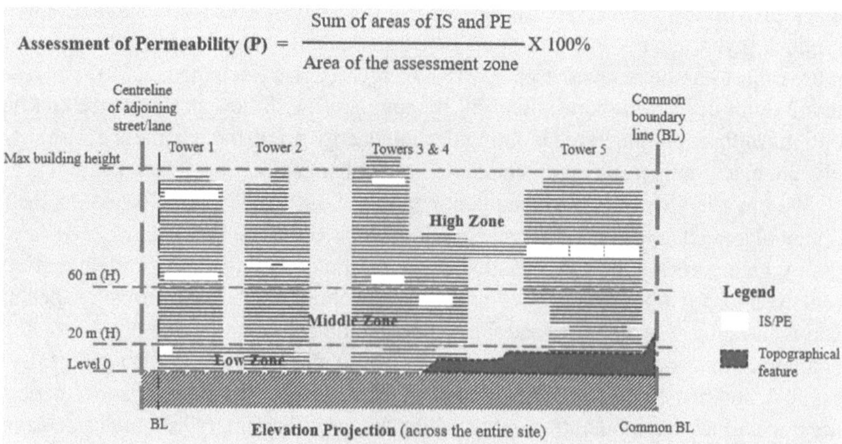

Figure 2.4 Illustration of building permeability (P) in the SBD guidelines (IS is intervening space, PE is permeable element, and BL is boundary line) (based on BD, 2016)

Treehouse's permeable elements, such as void decks, a large setback, and foot-bridges, improved its north-facing low-zone permeability from 29.6% in the base case to 35.3%. Moreover, it achieved a reduced footprint of around 30%, compared to the current building footprint. The more permeable low zone also increases wind availability in the north of the site from 15% in the base case to 45%.

2.3.2 Urban Green Canopy

Shade from tree canopies offers the most effective remedy to thermal discomfort caused by direct solar radiation at the pedestrian level. Prior research (Kong et al., 2017) demonstrated that trees with large crowns, short trunks, and dense canopies are effective at reducing average daytime temperatures at the pedestrian level on sunny summer days by values up to 5.1°C in open spaces. The SBD Guidelines require that, at a minimum, sites up to 20,000 m^2 provide green areas up to 20% of site coverage or 30% for larger sites.

To meet this requirement an urban woodland and an artificial wetland are proposed for the perimeter of Treehouse (Figure 2.5). Extensive greenery at grade also guides and cools the prevailing wind lessening the UHI effect. An "urban green canopy" integrates ethylene tetrafluoroethylene (ETFE) shading panels and greenery coverage over 40% of the site's open area. This canopy gives outdoor thermal comfort and all-weather protection around the building. The canopy over the urban woodland and elevated connection bridges foster a biophilic feeling when entering the site.

Acceptable thermal comfort at grade during typical summer days was tested with a microclimatic simulation. Physiological Equivalent Temperature (PET[1]) receptors were simulated in three locations (Pn, Pc, and Ps in Figure 2.6a) – corresponding

Figure 2.5 The proposed urban woodland for the Treehouse

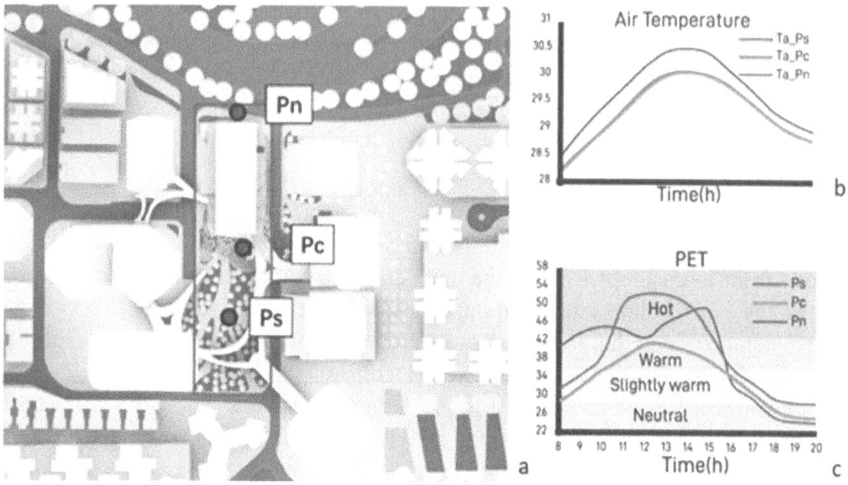

Figure 2.6(a–c) Microclimatic simulation results for the effectiveness of the urban canopy

to the northern pedestrian access (Pn), under the urban green canopy (Pc) and at an uncovered space within the urban woodland (Ps), respectively. The air temperatures were similar for the receptors under the urban canopy (Pc and Ps), while the northern access receptor (Pn) showed slightly hotter temperatures throughout the simulation period as shown in Figure 2.6b. The PET simulation (Figure 2.6c) showed that the urban green canopy (Pc) maintained outdoor thermal comfort within an acceptable range throughout the day, while the two uncovered receptors indicated temperatures outside of an acceptable range (from neutral to warm), especially from noon until later in the afternoon.

2.3.3 Appropriate Window-To-Wall Ratio and High-Performance Glazing

Commercial buildings usually have a higher window-to-wall ratio than residential ones, with glazed areas being weak points for heat transfer. But controlling solar heat gain and glare may sometimes lead to reduced visual transparency and transmittance in terms of views and daylight. The integrated façade system of Treehouse (Figure 2.7) finds a balance between optimising daylight and views and controlling solar heat gain and glare. Deep overhangs, over 1.5 m wide, and tilted glazing work in tandem with horizontal daylight reflectors to achieve an EUI reduction of 7 kWh/m^2 and excellent daylight performance.

The inclined glazing and the 1.5 m deep overhangs effectively block all direct sunlight and avoid the need for triple glazing, with its heavier dead load, high embodied carbon, and lower visual light transmittance.[2] This is coupled with occupancy sensors and light-emitting diode (LED) lighting designed to synchronise with circadian rhythms for general lighting for the office areas, supplemented by LED task lights.

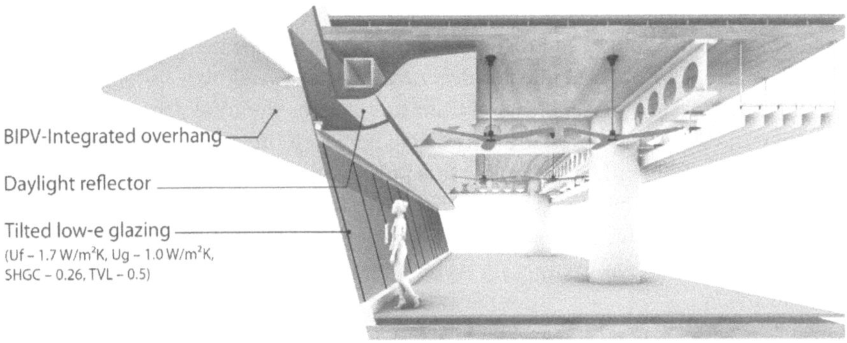

Figure 2.7 Sectional diagram of the integrated façade system

2.3.4 West-Facing Side Core

The average western annual solar radiation value is 650 kWh/m^2. It is challenging to reduce solar radiation entering the building because of a low-angled sun during sunset. Overhangs are not as effective as in a south-facing façade. We considered structural options as listed in Section 2.3.6.3. We opted for a side-core option as a solar heat buffer for the tenant-occupied spaces, especially in the upper and mid zones. We compared the structural options to the side-core option for achieving a balance between the performance requirements of both structure and building services. The major benefit of the side-core option is that it also enables cost-effective and energy-efficient free-cooling with reduced duct work, and hence less energy consumption over the operation of the building. Furthermore, the embodied carbon is optimised by side-core design (further details in Section 2.3.6.3). This reduction is still achieved despite the core location causing a slightly lower structural efficiency, i.e. eccentricity. A west-facing side-core option, however, offers a solar heat buffer for tenant-occupied spaces, especially in the upper and mid zones. Also, embodied carbon-intensive structural elements such as an outrigger or belt truss were avoided with a strengthened side-core-only system.

2.3.5 Optimisation of Pumping Energy

The Treehouse design team explored two ways to optimise water pumping energy and thus reduce energy use. One strategy is the use of distributed chilled water and potable water pumps to reduce the pressure loss from control devices and maximise the operating efficiency of the pumps. The second strategy is the reduction of proportional balancing by using balancing valves and throttling devices.

2.3.5.1 Distributed Pumps

Water is heavy and pumping it to a high point uses substantial energy, while the lower floors do not require additional pressure from gravity. The lower levels use pressure-reducing valves to address this excessive pressure. Having excessive pressure in the pumping system implies that the amount of pressure to be reduced could have been avoided and thus the energy used to generate that pressure could have been saved. The Treehouse is designed with a cascading booster pump with diaphragm pressure vessels. The bottom unit carries the total flow rate required for the building but at a pressure only to the height of the next booster pump. The next booster pump will have to carry less flow rate as water would have been consumed by the floors below it. This repeats with less and less water being pumped by subsequent booster pumps until the top floor of the building is reached. For fire safety reasons, the firefighting water supply would still be distributed using a water tower.

2.3.5.2 Reduced Pumping Energy

Traditional chilled water and condensing water systems are designed so that the supply pressure is sufficient to satisfy the furthest most demanding field equipment, called the index path, with sufficient spare pressure to allow for proper control authority. Therefore, every other field equipment other than the index path has excess supply pressure and thus requires balancing valves to throttle the flowrate to meet the design values. This is called proportional balancing. It implies that the additional energy used to generate the higher flow rate could have been avoided. This is so because, to reach the furthest most point, the pumping system must generate excessive pressure, which is not needed for equipment not on the index path. To prevent this loss of energy, Treehouse eliminated control valves. Flowrate control into field equipment is guaranteed by localised pumps and pumping loops. Such a design is possible due to the advancement in encapsulated variable speed-driven or electronic-cumulated pumps which are now common and affordable. Pump affinity laws dictate that multiple smaller pumps operating to achieve the same head and flowrates as a single larger pump would be much more energy-efficient during part load conditions. This design also decouples valve control authority from supply pressure as there are no control valves and therefore the system can modulate and maintain stability to a much lower flow and pressure regime.

2.3.6 Structural Elements Optimisation

The structural design had to integrate architectural, spatial, and building services design. The aim was structural adaptability to maximise flexibility to accommodate innovative approaches from other disciplines and future upgrades during the life of the building. To enhance occupant well-being and sustainability, the structural floor framing is of shallow-depth housing building services while maximising headroom and daylight.

2.3.6.1 Design Load

The team calculated optimal design gravity load allowances plus the foreseen future change of use as advised by other disciplines. The aim was to determine more accurately the design loads and avoid overprovision of primary structural components. Load distribution over the height of the building was considered at a very early stage, while the architectural massing was being developed. At this stage the primary issues were: avoiding heavy loads at high levels; adopting low-carbon, lightweight, and high-strength construction materials; and potential short/midterm change of local practice and regulation towards more innovative materials. Those concerns were investigated at early stage so to discover most of opportunities for improving structural efficiency and reducing structural materials and the front phase embodied carbon. In addition to design loads, a range of structural design parameters, such as material, stability system, and floor framing configuration were examined to fully understand the potential impact to embodied carbon of all options proposed by the design team.

2.3.6.2 Timber Canopy

We propose a lightweight, self-supported, and continuous hexagonal timber canopy starting from the ground floor to the rooftop as shown in Figure 2.8. This canopy provides shading to the upper levels and hosts the photovoltaic (PV) panels. We use timber primarily because it is very versatile, naturally renewable, aesthetically pleasing, and strong but with low embodied carbon. The canopy's gradient, shape, and size respond to the pedestrian level's microclimate: a dense area that benefits from increased ventilation and sunlight. We used a parametric design approach to optimise the structural member and pattern shape and size by considering sustainability features (shading, hosting greenery, and PV panels), as well as the actual design stress distribution, and thus minimise the material usage. All the timber elements used in this project are sourced from sustainable sources.

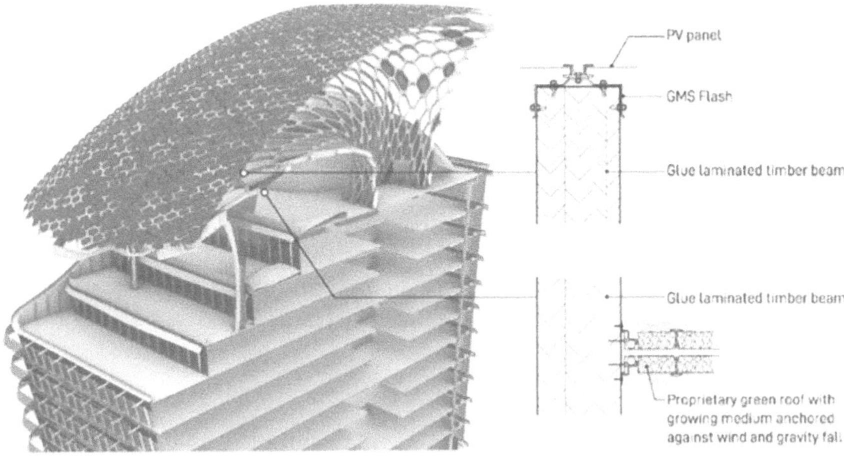

Figure 2.8 An engineered timber canopy as secondary structural components

2.3.6.3 Lateral Stability System

For a tall building, the choice of the structural lateral stability system significantly impacts on both construction cost and structural performance, and consequently on the embodied carbon. Especially if carbon-intensive structural elements like steel outriggers are in place. Various primary structural stability systems were evaluated against side-core options (Figure 2.9). These include:

1 central core only,
2 core and perimeter columns (i.e. tube in tube),
3 core and steel outrigger including belt truss,
4 exoskeleton, and
5 others.

The main lateral load resisting system for the Treehouse consists of reinforced concrete (RC) walls forming a rectangular core over the entire height of the tower. RC link beams were placed over large penetrations in the core walls to ensure the entire core is stiffened and behaves as one system. Side-core options can enable cost-effective and energy-efficient free-cooling with reduced duct work, and hence have lower energy consumption. A west-facing side-core option provided a solar heat buffer for tenant spaces, especially in the upper and mid zones.

The embodied carbon-intensive outrigger, belt truss, and exoskeleton structures options were avoided with a strengthened side core. Thus, to achieve both energy and carbon reduction, the team decided to adopt a west-facing side-core approach. The lateral stability of the building is provided by the RC core with a maximum lateral deflection under wind loads less than H/500 (Figure 2.9).

The inter-storey drifts were also reviewed, and the numbers are within the design limit. Therefore, a RC core is sufficient to provide lateral stability for the building. We specified low-carbon concrete and reinforcement with a high-recycled content (≥60%) for construction of the building structures including the core. The selected system is the most efficient solution and has the least embodied carbon emission.

2.3.7 Apparel Attribute Recognition

To promote adaptive comfort and engage occupants' adoption of "Super Cool Biz" dress code, especially during summer months, clothing levels of occupants are measured by Apparel Attribute Recognition (AAR) video technology at entrances. Necessary adjustments of the granular environmental control to suit the clothing levels of individuals may be executed as appropriate. Individuals recognised with "Super Cool Biz" outfit will be rewarded with incentives.

2.4 Exemplary Strategies

In this section we present strategies that deliver a higher level of impact than the quick-wins while addressing similar barriers to implementation. These strategies should be considered early in the design process given their enhanced benefits for

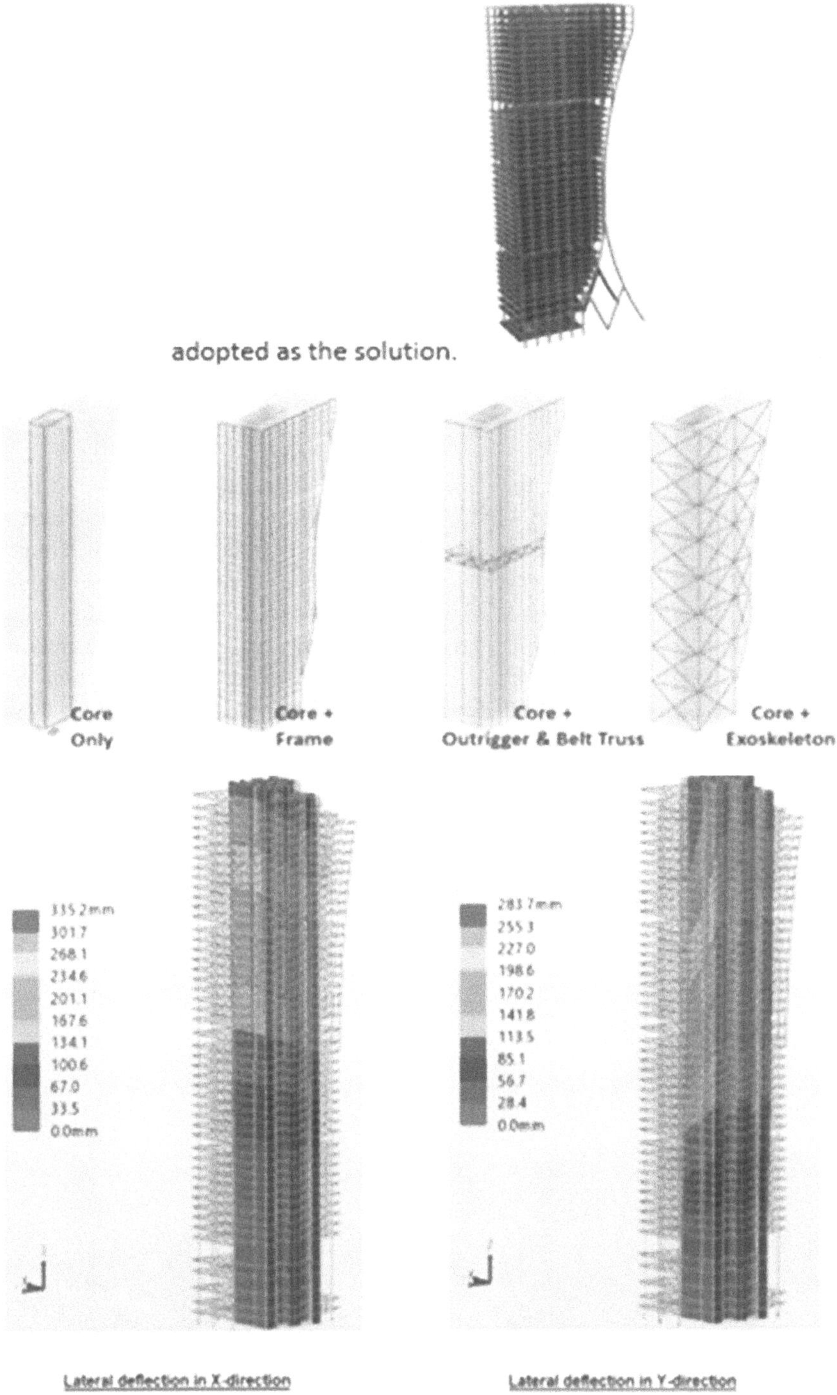

Figure 2.9 Structural lateral stability systems evaluated

similar efforts to achieve net zero. They include massing and façade aspects, MEP innovations, and smart materials:

1 Self-shading built form
2 Integrated façade for high spatial daylight autonomy (sDA) and visual comfort
3 Façade of high airtightness
4 Radiant cooling
5 Higher chilled water temperature for enhanced chiller efficiency
6 Maximised PV potential and efficiency
7 Phase change materials (PCM)
8 Regenerative twin-lift system
9 Ultra-low carbon structural elements.

2.4.1 Self-shading Built Form

The design team agreed that the built form of Treehouse should respond to variations in solar irradiance and wind availability in the different vertical zones. A humid subtropical climate requires solar shading on roofs and façades. The climate engineers advised that the upper and middle zones of the building were significantly exposed to solar heat gain, while the low zone was more sheltered from the sun by surrounding buildings but received less wind. The team worked through numerous proposals and after design iterations involving façade and structural disciplines, the team agreed to incline the southern upper zone of the tower by an angle of five degrees. This level of tilt achieves a 19.2% solar irradiance reduction compared to a straight façade, resulting in an overall EUI reduction of 4 kWh/m^2 (Figure 2.10). The tilt synergised effective self-shading with an enlarged rooftop

Figure 2.10 Southern façade solar irradiance simulations for base case (L) and proposed (R)

area for the installation of PV panels, and a reduced building footprint in the low zone improving the breeze availability for pedestrians.

2.4.2 Integrated Façade for High Spatial Daylight Autonomy and Visual Comfort

The Treehouse achieves excellent daylight performance thanks to an integrated façade with a wide overhang (refer to 3.3 appropriate window-to-wall ratio and high-performance glazing). The design achieves sDA of 69%, 72%, and 83% in the low, mid and upper zones (Figure 2.11).[3] Overall annual sun exposure (ASE) analyses showed excellent glare control near the south and north façades but revealed glare close to the perimeter zones of the west and east façades. ASE was lower in low/mid-levels compared to the upper floors due to shade from surrounding buildings. The potential glare areas lay mainly 0.5 m–1.8 m from the east and west façades, respectively. Simulations for floors without daylight reflectors showed a reduced sDA of 2% and less even distribution compared to the floors with reflectors. ASE increased by 2% mainly on the south façade, while the performance on the east and west façades remained the same. This demonstrates the effectiveness of the reflectors.

2.4.3 Façade of High Airtightness

A high riser's envelope is a key factor in sustainable design for hot and humid climates (Halawa et al., 2018). Airtightness measures the total air lost through leaks in a building's envelope, and it can account for up to 50% of cooling loss to the outside (Love et al., 2017). The more airtight a building is, the less energy will be

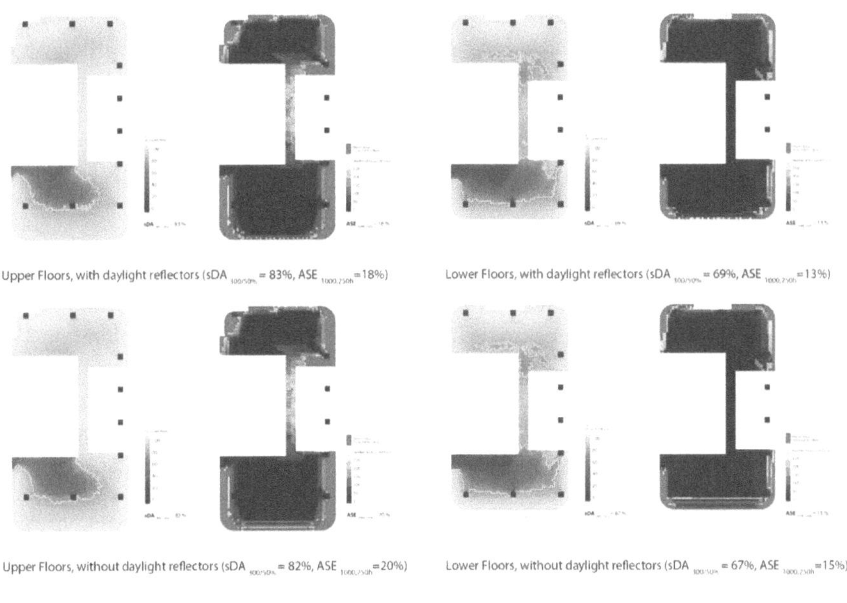

Figure 2.11 Simulation of daylight availability at different levels with and without reflectors

wasted through the leakage of conditioned air. However, an airtight façade may run contrary to other environmental features, such as natural ventilation. To mitigate these issues, it is preferable to provide operable window as part of an airtight façade system, giving occupants some level of control. This lowers the environmental impact of the building and saves energy. However, it is possible for buildings to be too airtight without adequate ventilation. This can lead to condensation and black mould growth. The Air Permeability Index (API), measured in $m^3/h\times m^2$ (air permeability rate per hour, per square meter of envelope area), measures how much air leaks through a building. An API between 3 and 5 $m^3/h\times m^2$ would usually be acceptable by building control (Love et al., 2017).

2.4.4 Radiant Cooling

While in the perimeter zone ceiling fans and air-conditioning allow a hybrid strategy, the core zones are cooled by decentralised air handling units (AHUs) and radiant cooling baffles. Radiant cooling works on the principle of exposed radiant temperatures. It will provide near-instantaneous cooling without moving bulk air hence requiring less distribution energy than mechanical movement in conventional air-conditioning systems. We estimate an EUI reduction of 4 kWh/m^2 by using radiant cooling in the core areas.

Integrating with the granular and user-centric environmental control (refer 2.5.4), the radiant panels adjust to individual thermal comfort preferences. The limitations of a radiant panel are a lower cooling output, possible obstruction of views and potential surface condensation risks. These limitations are mitigated with a high-performance envelope that limits heat gain (see 2.3.3), open and flexible spatial planning that keeps the working surface exposed (refer to 2.5.2 on design for adaptability). Condensation risk is eliminated by ensuring the chilled water temperature is above the dew point (no lower than 13°C) and the space humidity is controlled separately by a desiccant primary air-handling unit (PAU) system (refer to 2.4.5 on higher chilled water temperature).

2.4.5 Higher Chilled Water Temperature for Enhanced Chiller Efficiency

Decoupling of ventilation and space cooling will allow healthier ventilation. The decoupling of sensible and latent heat control systems would mean that the chiller plant does not need to produce low enough chilled water for dehumidification. The chiller plant can operate at a much higher chilled water temperature and therefore reduce refrigerant lift for sufficient cooling capacity. Energy consumption simulations indicate an EUI reduction of 5 kWh/m^2 because of the higher chiller water temperature setting with a high coefficient of performance (COP) chiller.

With a chilled water temperature supply of 16°C, returned at 23°C, the chiller COP can go as high as 17 in winter at part load, and 8 in summer at full load. The performances are much higher than those of a typical chilled water design that supplies 7°C, returning 12°C. For the Treehouse, the average chiller-only COP would be 12.6, and including the pumps and cooling towers, it would be 7.24 during a typical meteorological year. A high supply temperature is compatible with radiant-cooled ceilings and would not cause condensation issues. This system may struggle to maintain sufficient refrigerant lift to safely operate during the winter months. This is remediated using phase change thermal storage (refer to 2.4.7).

2.4.6 *Maximised PV Potential and Efficiency*

To maximise PV potential we incorporate innovations in the façade, roof, built form, and solar panels. A high-performance integrated façade with more than 2 m depth overhang cladded with lightweight building-integrated photovoltaics (BIPVs) in the upper and mid zones of the tower has an optimised tilted angle towards the south for capturing most of RE on the building façade.

Roofs represent a large part of a building's total surface area absorbing solar radiation. Uppermost sky terraces will be covered by BIPV with a green lattice structure. Its top skin comprises mono-crystalised PV and is anchored on glulam timber lattice. The bottom skin has custom-designed two-sided greenery infill panels and is separated from the PV top skin by a ventilated space (Figure 2.8). The bottom green skin will shield occupants from the radiant heat of the PV and provide cooling to the PV to increase its efficiency. The combined shading effect of the double-skin PV-clad tree canopy will also significantly reduce the heat gain through the roof.

The self-shading building form enlarges the footprint in the upper level providing a larger area for placing PV panels at the top, a location with the highest solar potential. Floor plates at the topmost three floors are also set back to create an inclined angle towards the south for the PV panels. Therefore, the tilted angle and size are maximised for the highest efficiency to create clean power. Besides the roof, every location on the building façade, including shading fins with >0.5 Sun in the mid and high zones of the tower, is designed with BIPV to further enhance the RE generation opportunities as shown in Figure 2.12 (refer to 2.5.9).

When the PV panels are facing the sun directly, panel temperature gets up to 65°C, thus the temperature difference is about 40K compared to standard test condition (STC) (at 25°C), which leads to a power reduction of about 16%. By applying a radiative cooling coating on PV surfaces on the roof, the operating temperature of PV panels can be reduced by 5K–18K. Assuming an average 10K temperature difference for this project after applying radiative cooling coating on the PV surface (i.e. only 30K temperature rise during operation instead of 40K), the power reduction is 12% or a 4% power generation loss recovered. To minimise energy generation reduction by dust and dirt on panels, the PV panels on the roof are tilted for self-cleaning by rainwater naturally. The coating is also self-cleaning by its super-hydrophilic properties.

2.4.7 *Phase Change Materials (PCM)*

In typical cooling load profiles, the cooling demand increases and decreases with solar exposure during different times of the day. Thus, in conventional cooling without thermal storage, the largest loads are generated during the period of the hottest period and the energy performance of the system is throttled. To mitigate

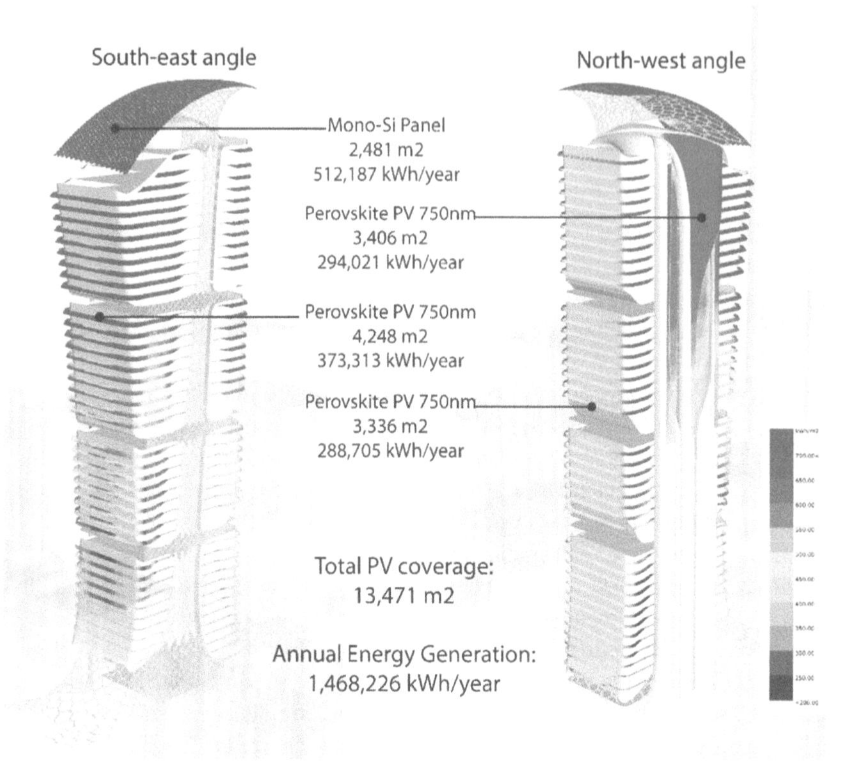

Figure 2.12 Simulation on PV-clad location and its performance

this, we propose a PCM thermal storage tank (TST) to defer cooling loads to night-time when the loads can be generated more efficiently with cooler outdoor conditions. The selected PCM will phase change at 14°C to allow chilled water draw-off at 16°C and recharging to be completed by 12°C chilled water. This PCM has a latent heat capacity of up to 200 kJ/kg (compared to ice at 334 kJ/kg and water, with a sensible heat capacity of 4.18 kJ/kgK).

A PCM thermal storage of 25 MWh, or approximately 500 tonnes, is installed in the basement, encapsulated in small tubes inside a water tank measuring 13 m wide by 11 m long by the full height of the room (Figure 2.13). The water tank will discharge cooler or warmer water to recharge or discharge the stored heat. At this size, the PCM thermal storage is expected to reduce the plant electrical energy consumption by approximately 160 MWh/yr. An EUI reduction of 2 kWh/m^2 is estimated by adopting PCM.

2.4.8 Regenerative Twin-Lift System

Regenerative twin lifts – two-lift cars operating independently in the same lift shaft – produce energy during operation and save floor space. The Treehouse only needs

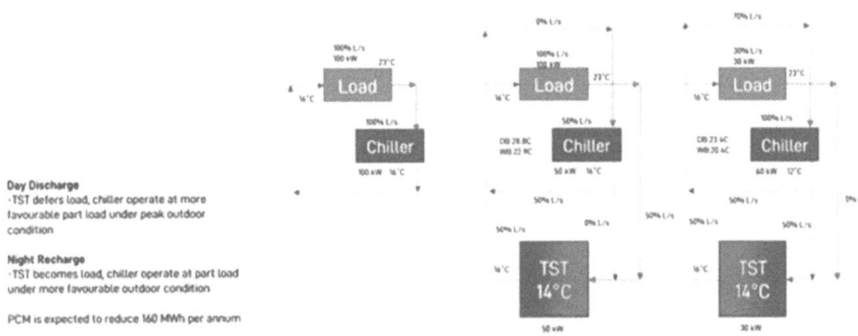

Figure 2.13 Schematic diagram of phase change material thermal storage tank

16 lift shafts to deliver the required lift capacity compared with 25 lift shafts in the base case (i.e. One Taikoo Place). The proposed twin-lift system is equipped with an energy recovery function that can feed about 30% of the energy generated when braking back into the building's power grid. An EUI reduction of 2 kWh/m^2 is estimated.

The twin-lift system is cost- and energy-efficient. Twin-lift systems can transport up to 40% more passengers with fewer shafts. Compared to the double-decker lift system, the twin system uses independent lift cars, while the double-decker cars move together. Twin lifts can park one cab, so when passenger volumes are low no energy is consumed moving empty cars. Labour and material expenses are drastically reduced because the elevators share a single shaft, landing doors, and guide rails.

2.4.9 Ultra-Low Carbon Structural Elements

Treehouse's structural grid is optimised to fit the designed floor usage. Each floorplate's structural span allows a large open space for office needs, while avoiding being over-stretched and without employing heavy and deep structures. The sizes of all structural members are optimised to achieve minimum material usage. This high structural and material efficiency helps reduce the overall associated embodied carbon in the structure. This high efficiency is also achieved by placing heavy MEP equipment in the basement and/or ground level, so to reduce the loading at the top. Thus, the upper floor columns can be thinner, and consequently have less embodied carbon in the structure.

In addition to the conventional iteration of structural optimisation, the design of Treehouse adopted the following approaches to reduce structural embodied carbon esteeming from structural materials:

1 Lightweight and high-strength concrete for structural elements, e.g., secondary beams, slabs, in addition to using them for only non-structural components as current local practice.
2 Cement replacement – by partially replacing traditional Ordinary Portland Cement (OPC) with ground granulated blast furnace slag (GGBS) or pulverised

fuel ash (PFA). The concrete mix with a 55%–75% GGBS cement replacement can achieve equivalent weight and strength as normal concrete while conferring significant embodied carbon reduction.

3 A hybrid structural solution with appropriate choices of RC, steel, composite, and timber as part of a holistic study. For example, avoid the use of outriggers which tend to induce a significant amount of embodied carbon; adopt steel composite beams, and cross-laminated timber (CLT) floor slabs for typical floors.
4 CarbonCure technology – CO_2 is injected during cement production, which transforms into nano-limestone and becomes permanently embedded in the concrete.
5 Replace RC slabs with steel composite CLT floor, due to its structural efficiency, lighter weight, inherited low-carbon feature and high repetition in the design, it has demonstrated significant potential to achieve net-zero carbon.
6 We specify steelwork and reinforcements with a high-recycled content (≥60%). Compared to virgin iron steelwork, the embodied carbon of recycled steelwork is approximately 75% lower. This helps to reduce the upfront embodied carbon significantly and build a circular economy.

In general, high-rise buildings often have substantial embodied carbon as presented in Figure 2.14. The design of structural, spatial, and building services systems is also highly interrelated and needs to be conceived in an integrated manner to achieve both high-performance and a low-carbon footprint. The largest reduction in embodied carbon comes from the superstructure core where OPC has been replaced by 55%–75% of GGBS cement or PFA in combination with CarbonCure and the use of high-recycled content steel reinforcement bars. Similarly, the largest carbon emission gain is in the pile cap for the substructure.

2.5 Visionary Strategies

Visionary strategies will deliver the greatest impact but also must overcome the greatest challenges to implementation. These strategies can be game changers in the race to achieve net zero in the construction sector. They include elements of passive design, technology, and human-centred innovations:

1 Reversing UHI
2 Design for adaptive comfort
3 Sky Common for skyrise active biophilic living
4 Granular user-centric environmental control
5 Holistic passive cooling system
6 AI and big data analytics for building energy management
7 Automated carpark
8 Lightweight CLT for floor slabs
9 Lightweight high-efficiency perovskite PV

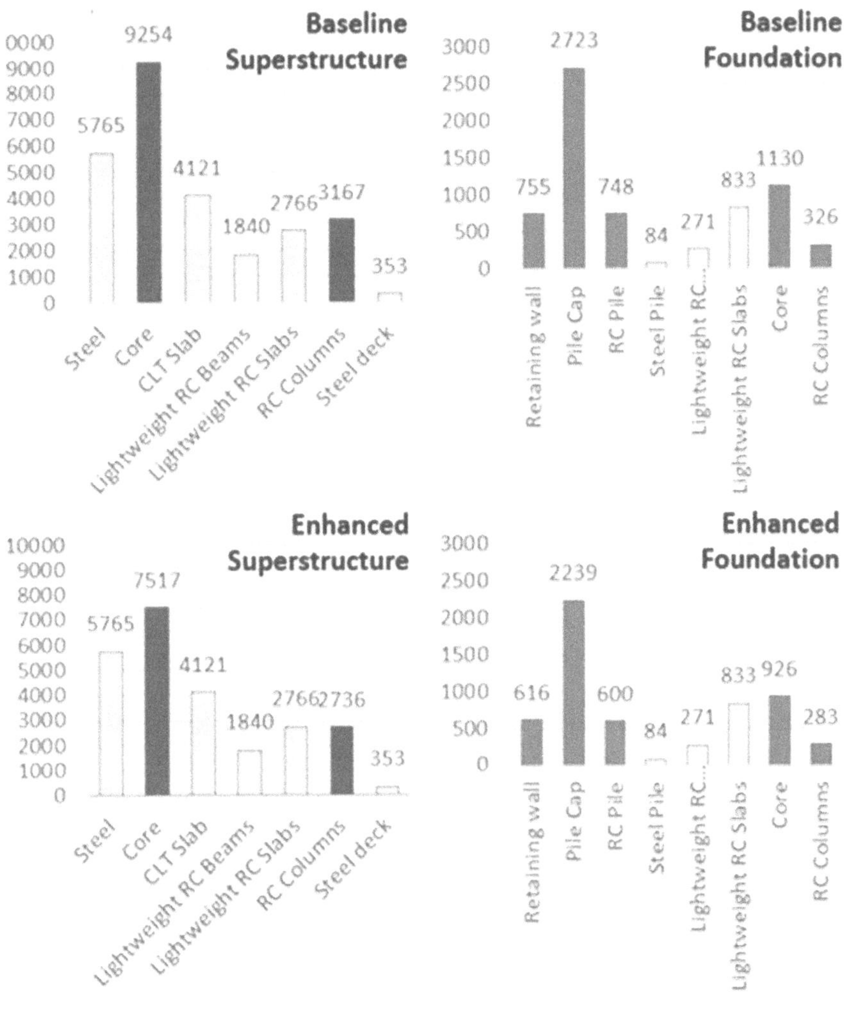

Figure 2.14 Consideration of low-carbon structural materials

2.5.1 Reversing Urban Heat Island

The Japanese CASBEE-Heat Island environmental assessment method evaluates how buildings can be designed to reduce UHI. The assessment focuses on the quality of the outdoor environment by evaluating natural ventilation, shading, ground surface, and facade materials and exhaust heat from buildings. The Treehouse team conducted outdoor climate engineering studies to understand the impact of the UHI effect and extreme temperatures on human well-being and the energy consumption of buildings. The results indicated that for every one degree in temperature rise, annual cooling energy use will increase by 3% for

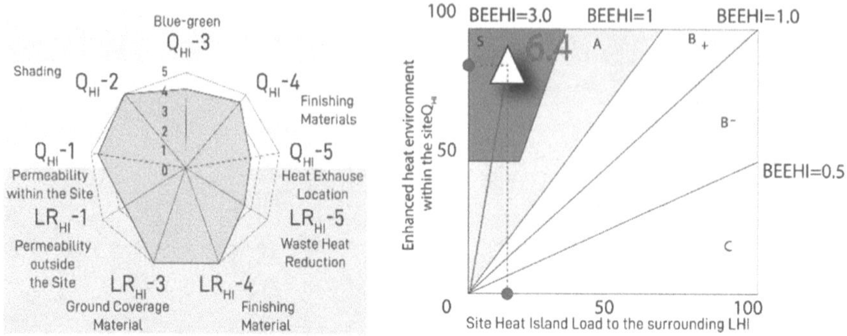

Figure 2.15 Summary performance of CASBEE-Heat Island (This evaluation result is based on self-assessment by Ronald Lu & Partners Ltd)

commercial buildings (Fung et al., 2006). Our research revealed that the three most effective strategies to alleviate the UHI effect were:

1 building permeability to channel wind to pedestrian spaces,
2 the provision of green spaces, water bodies, and cool materials; and
3 sufficient shade created by trees or canopies.

In a high-density high-rise urban context, Treehouse guides urban wind to pedestrian spaces, creates green spaces and water bodies for passive cooling, and provides sufficient shades by trees and an urban canopy to effectively alleviate the UHI. The design scores 6.4 and achieves the best performing Excellent Class under the assessment (Figure 2.15).

Trees and vegetation in parks help reduce CO_2 (a dominant GHG) by directly removing and storing CO_2 and indirectly by reducing air temperature and building energy use in and near parks. Tropical rainforests are by far the most important ecosystems for mitigating climate change. They collectively sequester more CO_2 from the atmosphere than temperate or boreal forests. Temperate winter deciduous species generally have higher levels of C reserves (Non-Structural Carbon concentrations) than evergreens (Vanderklein and Reich, 1999; Hoch et al., 2003). A study looking at the carbon storage of forests in southern China (Dunne, 2018) found that each additional tree species introduced to a plantation could add 6% to its total carbon stocks. The findings suggest that afforestation plans to sequester CO_2 should switch from using just one plant species to a more diverse mix (Dunne, 2018).

2.5.2 Design for Adaptive Comfort

The design team observed that conventional office designs in Hong Kong that use Variable Air Volume (VAV) systems have tightly controlled comfort parameters. Users often perceive the temperature to be too cold – this is mainly due to an emphasis on dehumidification. For Treehouse, a variety of ventilation, dehumidification, and cooling strategies were considered, including natural ventilation, radiant cooling, displacement ventilation and cooling, VAV, ceiling fans, and others. Radiant cooling was determined to be the most energy-efficient, with the associated risks of condensation overcome by high-temperature cooling, decoupling of ventilation, dehumidification, cooling, and airtight façades. Displacement cooling was considered, but when the integration of ceiling fans was carefully studied, the team found that the downward airflow from ceiling fans did not work well with displacement cooling.

The office spaces in Treehouse are fully air-conditioned. The perimeter zone (about 50% of the total office space) is conditioned to 29°C by a decentralised AHU located on every floor. These units also control humidity to 10 g/kg. Ceiling fans drive air movement in the perimeter zones, at an average speed of 0.8 m/s, to ensure high thermal comfort while allowing individual comfort control. This fan-induced air movement provides adequate thermal comfort (at a predicted mean value of ±0.5 for 90% occupancy) that is highly cost-effective, uses a low amount of energy, and avoids overusing cooling under extended temperature and humidity ranges (Figure 2.16). An EUI reduction of 5 kWh/m^2 is estimated from this adaptive comfort and hybrid ventilation strategy.

2.5.2.1 *Landlord-Managed On-Demand Shared Amenities*

In a post-pandemic era, people are making new choices about how they want to live and work. A hybrid of remote work mode from home, workplace, and satellite co-working spaces near home is envisaged. The conventional office model does

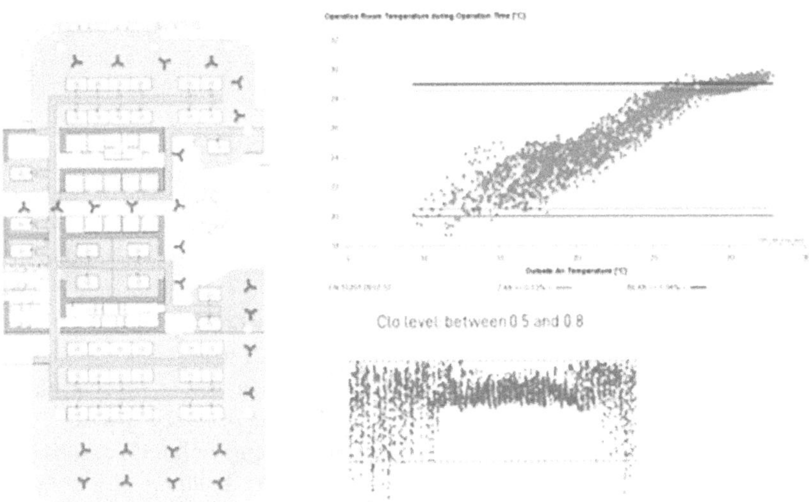

Figure 2.16 Schematic diagram of adaptive comfort and hybrid ventilation and thermal comfort analysis

not cater for such variation of workforce day-by-day. The Treehouse offers a new model of workplace of which the boundary of tenant-landlord-managed premises is flexible to cater for post-pandemic workforce.

Shared amenity rooms in Sky Commons are under landlord's control and are available to tenants on-demand. They provide quality multi-functional spaces for collaboration, focused work, physical or digital meetings, and workshops. The on-demand booking system is managed by AI and Internet of Things (IoT). The benefit of being under central management of the landlord is that state-of-the-art smart facility management and smart-active building systems, such as those discussed in Section 5.6 AI and big data analytics, can be integrated and be available to tenants across tenant- or landlord-managed spaces seamlessly and subject to the best available energy management practices.

2.5.2.2 Design for Adaptability

Providing occupants a diversity of workspaces that match their work style can lead to higher productivity and morale. The proposed design caters for activity-based working where occupants are encouraged to choose when, where, and how they work throughout the day. Besides open-planned desktop workspaces, quiet work/phone booths, formal and informal conference rooms, soft seating in common areas, cafes with hot desks, loops for walking meetings, semi-outdoor terraces in Sky Commons/gardens, and many other types of spaces are available to occupants. Occupants can check-in at Sky Commons physically or via smart phone app for their desired types of workspaces or request the virtual concierge to locate a colleague based on their needs for collaborative teamwork or match a space according to their personal comfort preferences. Such is powered by the IoT infrastructure, and the advanced indoor positioning technology deployed.

2.5.2.3 Virtual Desktop Infrastructure

The activity-based workplace concept is powered by an edge data centre and the virtual desktop infrastructure (VDI). VDI uses a client-server model of computing and separates the personal computing desktop environment from a physical machine in the workplace. Organisations benefit from cost savings, simplified management, enhanced security, and increased productivity. Occupants may check-in the VDI service at their Sky Commons and are free to work anywhere with superb and secure virtual connectivity in the building. The VDI works with the edge data centre located in the building which enjoys more efficient cooling of sensible heat from equipment, and thus a reduced cooling load within occupied spaces.

Conventional personal computer desktop arrangements will require substantial copper vertical riser ICT cabling throughout the entire building. It is estimated that this solution will substantially reduce the amount of copper ICT cabling to be deployed in the building by about 40 tonnes. Refining copper is extremely energy and carbon-intensive and the reduction in weight in the building would result in a lighter building structure with less embodied carbon.

2.5.3 Sky Common for Skyrise Active Biophilic Living

To enable hybrid working models at Treehouse, the design team proposed the concept of the "sky commons" – three-storey atria linking communal spaces and shared amenities, located on every third floor. Added to these are aesthetically pleasing and walkable 1.5 m-wide active stairwells, connected lifts, and shared spaces and amenities that promote socialising and physical activity. The shared amenities are available on-demand and could include regularly scheduled fitness opportunities for occupants such as yoga and stretching classes. The building's four sky gardens also provide naturally ventilated and vegetated semi-outdoor environments with physical activity areas such as jogging tracks, climbing walls, and gym equipment. Being naturally ventilated and surrounded by greenery, these spaces require lower cooling loads from the heating, ventilation, and air conditioning (HVAC) system. These adaptable "third spaces" provide occupants with a diversity of casual, formal, social, or focused spaces that match their individual needs, to enhance productivity and morale.

2.5.4 Granular User-Centric Environmental Control

The location of employees within an office impacts their performance due to air quality, thermal and acoustic comfort, visual acuity, perceived productivity, and access to biophilic elements. The Treehouse increases the occupants' control of thermal and lighting levels to maximise individual comfort and energy efficiency. High-precision real-time location solutions (RTLS) based on ultra-wideband (UWB) technology inform IoT-managed systems about the position, movement, and number of people with low power and high granularity. Smartphones integrated with UWB technology offer accuracy to within at least 10 cm.

Each user may adjust the temperature and lighting at their location based on their preferences and work zone. For example, lighting may automatically turn off or dim when a user moves away from their desk to save energy. The same process applies to air conditioning: the system can raise or lower the set temperature using personal identification (Figure 2.17).

Indoor environmental quality parameter sensors will continuously monitor data at five-minute intervals and provide real-time data analysis for each sensor's location. Parameter variations allow the facilities management team to assess changes in the environment and make appropriate building operation adjustments. The system will seek a balance between thermal comfort and energy consumption optimisation.

2.5.5 Holistic Passive Cooling System

We aim to reduce mechanical ventilation using the building height to tap cooler air at the top. According to the Hong Kong Observatory data, the air at 200 m above ground is around 2°C cooler than that at grade. We combine a high-rise-scale wind catcher, a wetland heat exchange, and a solar chimney, working together as a free-cooling system to produce 500 kW of cooling under Hong Kong's climatic conditions (Figure 2.18).

Figure 2.17 Visualisation of a thermal gradient approach in office spaces

2.5.5.1 Wind Catcher

The wind catcher's aerodynamically shaped inlet on the upper roof faces the easterly summer prevailing wind. It captures the stronger and cooler winds and channels it down through a 3 m by 2.5 m shaft to the basement level. Hygroscopic terracotta cladding lines the interior of the wind catcher to passively regulate the moisture content of the incoming air. The wind catcher enables an EUI reduction of 2 kWh/m^2.

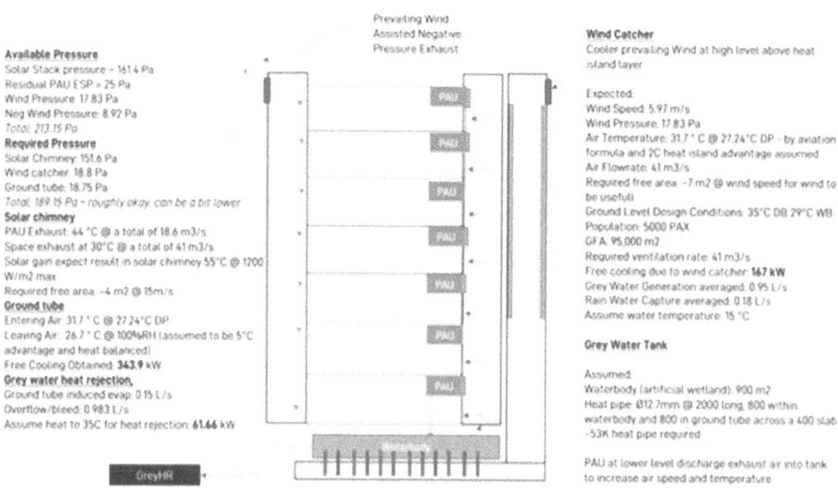

Figure 2.18 Holistic passive cooling by wind catcher, wetland heat exchange, and solar chimney

2.5.5.2 *Artificial Wetland for Natural Cooling*

The Zero Carbon Park at grade hosts a constructed wetland. The base of the wind catcher connects with a serpentine ductwork system underneath the wetland in the basement. Heat is transferred from the air in the ductwork to the wetland by heat pipes. The wetland releases excess heat by evapotranspiration, surface wind action, and plants water uptake. The wetland also serves as a stormwater management feature and enhancement of the urban ecology. Further heat recovery is achieved by a wastewater heat exchanger for the chiller plant prior to discharge. The wetland is expected to produce approximately 340 kW of cooling and the wastewater heat rejection contributes another 60 kW of heat rejection.

2.5.5.3 *Side Core with Solar Chimney*

A vertical air plenum connects the pre-cooled air to PAU on individual floors. A horizontal air distribution is organised by the ceiling-integrated ductwork. The central air exhaust on each floor takes the air to the solar chimney, which provides negative pressure by buoyancy and reduces use of mechanical ventilation. Before entering the solar chimney, the exhaust air will pass the heat recovery until there is an energy benefit. PAU at lower levels discharge exhaust air into a tank to increase air speed and temperature.

The outer glazing skin of the solar chimney is BIPV-clad, and the inner skin is insulated from the rest of the building. The chimney shields the western façade from direct solar gain. Solar heat gain from the glazed façade together with waste heat from BIPV, PAU desiccant regeneration and heat exchange under the wetland is used to create natural ventilation by buoyancy effect. The upper level of the chimney is protected from back pressure by a gooseneck stovepipe arrangement, which works complementarily with the widening of the chimney as it gains height. The prevailing wind assists negative pressure for the exhaust of the chimney. In summertime, the worst buoyancy period, the solar chimney is expected to produce approximately 40 m^3/s of air movement at 160 Pa. During winter, the pressure created would be more than double. This solar chimney is estimated to achieve a EUI reduction of 2 kWh/m^2.

2.5.6 **AI and Big Data Analytics for Building Energy Management**

We propose the use of AI and big data analytics for pattern recognition, performance tracking and optimisation, fault detection, and predictive operations such as pre-cooling.

The AI-equipped Building Management System (BMS) operates the building. It provides plant efficiency optimisation by collecting historical occupant behaviour and predicting the up-coming plant operating schedule to avoid any overprovision. For example, the AI counts the number of people using the lift and learns that usage is low in the afternoon, thus, it will turn off the fresh air supply accordingly. Smart meters record the real-time energy use per equipment and send the data to the BMS. The AI can detect the peak demand period and analyse the possibility of turning off certain equipment in advance and avoiding peak demand the next day.

Treehouse uses energy dashboards (water, electricity, gas, etc.) as it is easier to read and assess how energy is being used in the building. Algorithms and rules predict maintenance, including interfacing with room booking systems and interfacing with inventories. Real-time outdoor and indoor environmental sensors, together with recorded occupancy patterns, building usage and thermal comfort preferences, the environmental conditions for each interior zone and the building services' capacity for the next hour(s) can be predicted. Mobile apps and web browsers give access to occupants, visitors, and maintenance crew. They can access data, bookings spaces and services with certainty and ease. The maintenance crew's mobile app retrieves intuitive, interactive, and real-time information in a self-service and on-demand manner.

Traditionally, BMSs are designed and procured in silos. Siloed systems do not allow for easy transfer of data. With equipment and devices installed, having a proper automatic control system would offer the users faster and responsive control of the building. Control systems include the BMS, lighting control, security, vertical transportation, etc. Nowadays, BMS are a common platform for all these systems. The sequence to integrate them will involve the following:

- Firstly, connect and unite the disparate data produced by systems and devices. This is important from the initial design stage, to ensure all systems can communicate with the protocols which will enable the data from systems and devices to be the most useful.
- Secondly, the data needs to be processed to enable analytics, fault detection, and diagnostics. During the design stage the data is tagged appropriately.
- Thirdly, real-time and historical data is archived following a standard methodology. This methodology clearly defines the assets and entity relationships across all the systems. The BIM model is setup using appropriate tags and defines the methodology for the supply contractors to use.
- The last step provides the visualisation for the building-tenant engagement, and the fault detection, and diagnostics for the facilities maintenance team.

2.5.7 Automated Carpark

A parking system using AI and robotics offers a carbon-smart and enhanced car parking experience. Minimal lighting and ventilation for the unmanned automated carpark is estimated to consume 30% less energy. It features an electric vehicle (EV) charging system with auto-connectors between EV parking pallets and chargers. Smart load management using advanced control algorithms enables the system to distribute supply current according to the electricity demand of each EV charger, which optimises the use of limited power. User convenience and wellness is enhanced with less time-spent parking and pollutant exposure. Drivers can use their phone app to schedule car retrieval, check parking availability and car status.

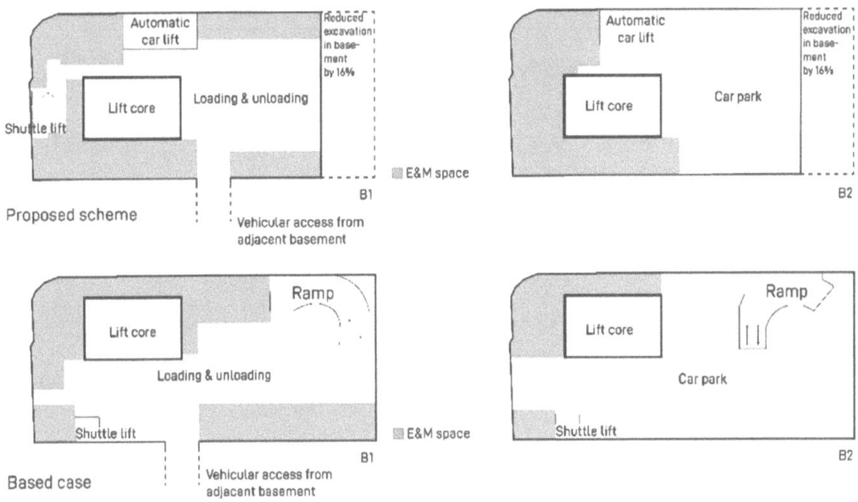

Figure 2.19 Automated carpark schematic layout study

From an embodied carbon perspective, comparing with the base case (Figure 2.19), an automated carpark saves 1/3 of the carparking spaces by enabling optimal parking in multiple rows and columns. Although additional plant room space is needed for the PCM thermal storage, we have saved the basement excavation extent by 16%.

Even with the increased space allocation for the PCM thermal storage, the Treehouse automated carpark has avoided ramps, circulation corridors, pedestrian access to other floors, or fire exits. The parking spaces are optimal since car doors do not to open or close. The automation of the parking process also eliminates the need for additional lighting, gas exhaust, and air filtering and conditioning. The car engines will be switched off and only maintenance crews will be allowed in the parking basements for inspections and repairs.

2.5.8 Lightweight Cross-Laminated Timber Floor Slabs

Timber is a sustainable and lightweight structural material and has a long service life. The CO_2 is temporarily stored within timber until it is released at end of life. The estimated sequestration value of CLT slabs is 125 tCo_2e/m^2. A CLT slab has a good load-bearing and thermal insulation capacity. Lightweight helps reduce the size and embodied carbon in supporting beams, columns, and foundations. We refer to the emission factor for CLT from the UK Inventory of Carbon & Energy (ICE) in the calculation.

Replacing post-tension (PT) concrete beams and slabs with steel beams and CLT floor (Figure 2.20) achieves a lower carbon footprint by replacing a carbon-intensive material (concrete) with a low-carbon alternative (timber). The proposed design also offers a greater floor to ceiling height, lightness of materials (reduced dead load), and ease of deconstruction at the end of life.

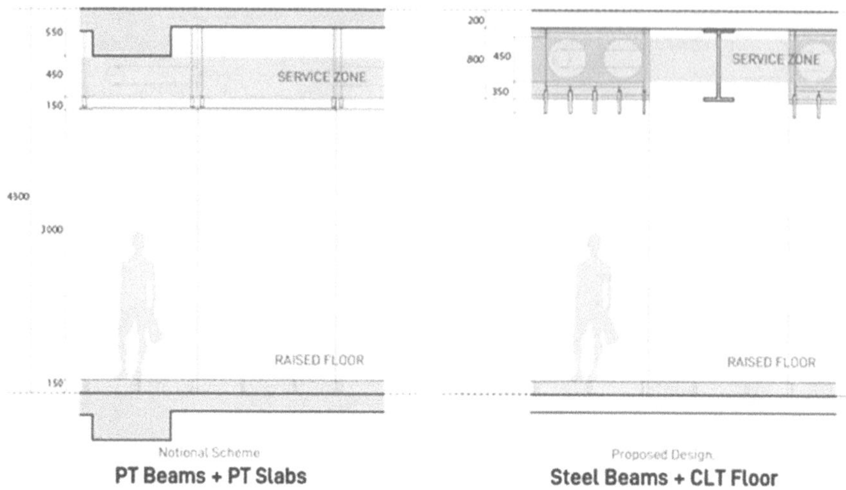

Figure 2.20 Comparison of notional base case and proposed case

2.5.9 *Lightweight High-Efficiency Perovskite PV*

Typical crystallised silicon PV is sensitive to the angle of solar incidence. PV panels installed on the façade vertically (solar intensity of about 0.5–0.8 Sun compared with 1 Sun on a flat roof) may not be operating effectively most the time. Perovskite PV is a thin film with a better generation efficiency under low-light intensity (<1 Sun). Research shows that a 750 nm thick Perovskite PV reaches 20% efficiency at 0.5–0.6 Sun or same efficiency of mono-crystallised silicon PV under STC (Figure 2.21). Future Perovskite PV panels are expected to be lightweight, flexible, and easy to install on vertical facades. Perovskite PV on façades will not increase structural loading or the embodied carbon.

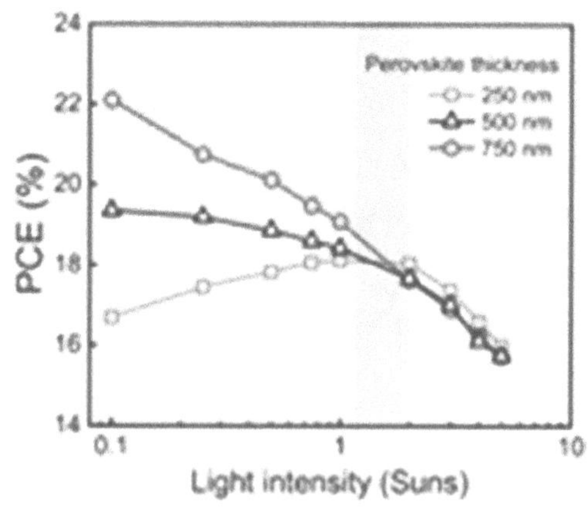

Figure 2.21 Efficiency of Perovskite PV under different light intensities

2.6 Conclusion

Achieving net zero is an exceptional challenge in hot and humid climates, high-density cities, and high-rise built environments. Subtropical cities like Hong Kong face this “triple challenge”. Currently, cost-effective carbon capture technologies remain out of reach for typical commercial projects. Emerging technologies will enable the decarbonisation of buildings but, in the meantime, market transformation, supply chain readiness, and professional capacity regarding net zero should all be accelerated.

Carbon-neutral design strategies should be evaluated for their potential positive environmental, social, and economic impacts versus any cost, technical, regulatory, and cultural challenges. Treehouse’s step-by-step trajectory to zero carbon is illustrated in Figure 2.22. It strives to deliver an adaptable design – something which the team believes is crucial for future-ready buildings. It is designed for both quick-win and exemplary carbon-neutral strategies during the whole life cycle of the building, from design, construction, and operation till end of use. It will deliver the cost-effective adaptations required by visionary strategies during the later stages of the building cycle.

Treehouse’s design brings harmony and balance with the natural world and celebrates the spirit of the place. Biophilic elements are weaved visually and spatially throughout: from green facades, a green roof, and Sky Commons to interior green walls, water features, and the patterns and colours of nature in the materials and furnishings. Daylight bathes the perimeter space, a sea breeze refreshes the common areas, and occupants find their own thermal comfort zone.

We use passive architecture to respond and adapt to the urban microclimate and its variations in different vertical zones. A self-shading built form, tilted glazing, larger and suitably angled BIPV-clad overhangs, and double-skin PV-clad and green roof canopy respond to the exposed upper zone of the urban microclimate

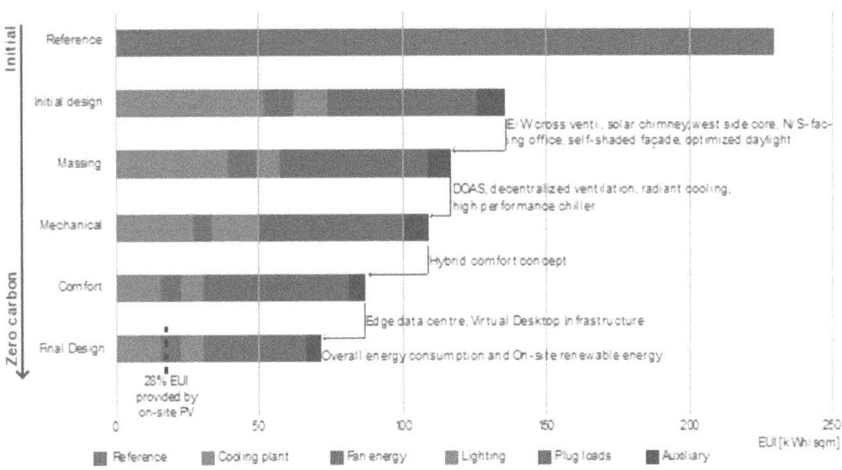

Figure 2.22 Step-by-step trajectory to zero carbon

with more shading, solar capture, and thermal insulation. High building permeability, reduced extent of overhangs and PVs, more vegetated cool surfaces on facades are designed to suit the weaker wind and daylight availability in the low zone of the tower. A reduced building footprint in the low zone also facilitates improved urban ventilation and results in better shading on streets. Operable façade permeability for communal areas adapts to varied wind and daylight environment at different heights and times in the day.

A holistic passive cooling strategy is designed with wind capture by a 200-m tall wind catcher facing easterly prevailing wind, free-cooling by heat exchange under an artificial wetland and the enhanced stack effect by a full-height solar chimney. Structural optimisation and use of super-low-carbon materials enable built form optimisation. The design achieves enhanced whole life carbon performance by applying solar energy harvesting, and high-capacity but light-weight RE on-site generation technologies.

Smart-active building systems recognise and address individual needs with integrated and granular design of the technologies. The IoT infrastructure and sensor-actuator-based systems optimise energy use and IEQ based on real-time data and occupants' needs. The holistic HVAC strategies, VDI, and adaptive workplace spatial planning engage occupants to adopt an adaptive approach in activity-based workplaces where occupants may work anywhere, meet collaborators, and find or adjust their preferred thermal zones. Landlord-managed shared amenities meet the on-demand needs of post-pandemic hybrid work patterns.

We used dynamic thermal building modelling with TRNSYS 18 to evaluate the building's energy use and to inform the design process. The model considers the building geometry and envelope performance, the internal layout, operational schedule based on BEAM Plus New Buildings Version 2.0, the designed mechanical and electrical systems, and the efficiency of the cooling plant. The summary of the estimated EUI is presented in Table 2.1.

The reference design, or base case, showed an annual EUI over 200 kWh/m^2. Optimising and adapting the massing by locating the offices facing north and south, reducing solar loads by tilting the south facade, introducing a solar chimney for cross ventilation, and optimising daylight availability, the energy use could be reduced to around 160 kWh/m^2. In the optimisation process of the mechanical system,

Table 2.1 Summary of estimated EUI, upfront embodied carbon, and on-site RE generation

Whole building EUI	71 kWh/m^2/yr (69% and 45% lower than the reference building's and targeted maximum EUIs of 230 and 130 kWh/m^2/yr, respectively)
Landlord EUI	29.1 kWh/m^2/yr (58.4% lower than the targeted maximum EUI of 70 kWh/m^2)
Upfront embodied carbon	342 kg CO_2-e per m^2 of CFA (40.5% lower than the targeted embodied carbon performance of 575 kg CO_2-e per m^2 of CFA)
On-site RE generation	20 kWh/m^2/yr (28% of the Whole Building EUI)

the design team decided to apply a dedicated outdoor air system in combination with hydronic radiant panels, reducing ductwork and fan energy by decentralised AHUs – instead of a traditional, centralised air conditioning system with high air change rates and a recirculation of air. The recent coronavirus pandemic showed the risks of a recirculating air system. In addition, using high-performance chillers with indirect sea heat rejection significantly improved the potential chiller COP.

The holistic passive cooling strategy using the wind catcher, wetland heat exchange, and solar chimney further reduced the annual EUI to about 100 kWh/m^2. An additional EUI reduction was achieved by applying thermal comfort strategies, which give occupants a choice. An EUI significantly below 100 kWh/m^2 is achievable by introducing a climate concept with different means for arriving at excellent thermal comfort conditions. These include ceiling fans, which generates high comfort with low energy use at the perimeter (with individual control). A system of virtual desktops with the computational power centralised reduces plug loads by about 20%.

Our design adopts low embodied carbon materials including CLT slabs, low-carbon concrete with cement replacement, and high-recycled contents material. Resulting in a total embodied carbon (A1–A5) of 342 kg CO_{2e}/m^2, or 40.5% lower than the base case (575 kg CO_{2e}/m^2).

The ideas and solutions proposed are practical, replicable, and scalable to be applied to other commercial buildings, project sites, high-density urban contexts. An innovative transdisciplinary design approach and close collaboration between members of the design team were crucial to arrive to such seamless integration of solutions. Transdisciplinary design is an effective approach to complex problems like designing for carbon neutrality. Treehouse has been recognised for its successful and innovation design, significantly reducing its embodied and operational carbon footprint in a high-rise, high-density, and hot-humid urban climate environment. With human behaviour being a strong influencing factor on consumption and carbon neutrality, the transdisciplinary approach creates a deeper understanding of the human social purpose; understanding which helped shape this holistic carbon-neutral building. Essential elements of success for the transdisciplinary design process in Treehouse included a complementary, trusted team with aligned values and goals, and the ability to execute the integrated design prioritising interactivity, flexibility, and user centricity.

Transdisciplinarity requires co-creation through truthful communication and open-minded learning among multiple disciplines working together towards a common goal. As the leaders of the transdisciplinary design team, the authors had the responsibility of forming a team of committed members with complementary skills and knowledge, the ability to adapt with agility to different situations, and the courage and drive to innovate. The outcome-based transdisciplinary design framework we used provided notable and innovative guiding principles for urban sustainability that may not have arisen if working in silos. Some of these innovations related to vertical microclimate-responsive planning, smart and adaptive comfort strategies, and an outcome-based biophilic approach for future-ready buildings.

The outcome-based transdisciplinary net-zero-ready building design framework (Figure 2.23) that we sought to implement considered a wide spectrum of

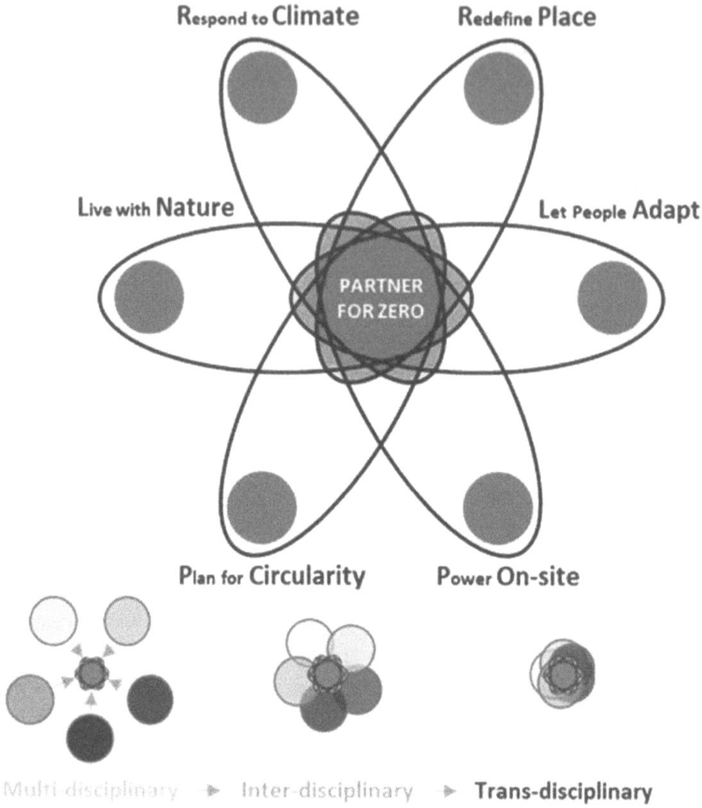

Figure 2.23 Outcome-based transdisciplinary design framework for net zero

carbon-neutral strategies for their potential positive environmental, social, and economic impacts versus cost, technical, regulatory, and cultural challenges.

An adaptive design approach should be adopted to take on board carbon neutrality quick-win and exemplary strategies at the outset, and eventually moon-shot strategies in later stages of a building life cycle. This will be done in synchrony with emerging technology development, net-zero-related market transformation, supply chain, and professional capacity advancement. Using the strategy mapping methodology (Figure 2.24), stakeholders can be engaged to review potential positive environmental, social, and economic impacts against significant cost, technical, regulatory, and cultural challenges of various design strategies. During the design process, we learned that transdisciplinarity evolves as an understanding of and a consensus about the design problems. It is built among the team members as values and goals become more closely aligned. We also learned that, particularly in the early stages of design, transdisciplinarity requires intensive design effort and more time to be spent on knowledge exchange and evidence-based evaluation of ideas.

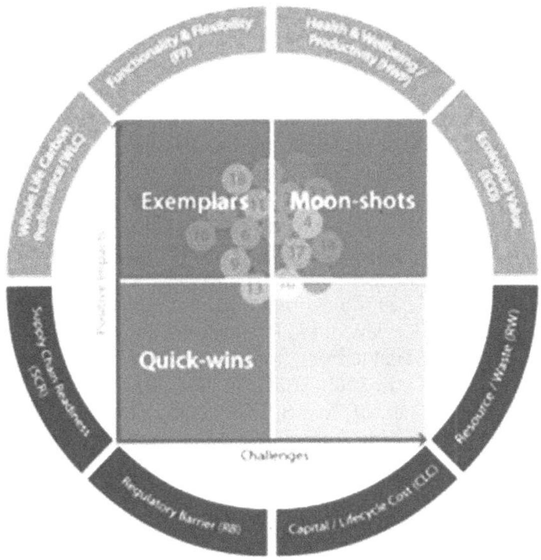

Figure 2.24 Net-zero-ready design strategies mapping

Notes

1 PET is the air temperature in a typical indoor setting (without wind and solar radiation), at which the heat budget of the human body is balanced with the same core and skin temperature as under the complex outdoor conditions assessed. PET allows us to compare the integral effects of complex thermal conditions outside with our own experience indoors (Höppe, 1999).

2 The adopted double glazing's key characteristics are: Uf – 1.7 W/m^2K, Ug – 1.0 W/m^2K, SHGC – 0.26, TVL – 0.5, and airtightness of <2 $l/s/m^2$.

3 sDA measures daylight illuminance in office areas over 300 lux for 10 hours per day per year. An sDA of around 70% is considered good, especially in a high-density compact urban context.

References List

British Standard (BS). 2011. *BS EN 15978:2011. Sustainability of Construction Works. Assessment of environmental performance of buildings. Calculation method. Available from:* https://standards.iteh.ai/catalog/standards/cen/62c22cef-5666-4719-91f9-c21cb6aa0ab3/en-15978-2011 [Accessed 10 June 2022]

Buildings Department (BD) of HKSAR. 2016. *Practice Note for Authorized Persons, Registered Structural Engineers and Registered Geotechnical Engineers (PNAP) APP-152: Sustainable Building Design Guidelines.* Available from: https://www.bd.gov.hk/doc/en/resources/codes-and-references/practice-notes-and-circular-letters/pnap/APP/APP152.pdf [Accessed 10 June 2022]

Bulkeley, H., Broto, V.C. and Maassen, A. 2010. Governing urban low carbon transitions. In *Cities and Low Carbon Transitions*. UK: Routledge. 45–57.

Campbell-Lendrum, D. and Corvalan, C. 2007. Climate Change and Developing-Country Cities: Implications for Environmental Health and Equity. *Journal of Urban Health: Bulletin of the New York Academy of Medicine*, 84(10), 109–117.

Chou, W.H. and Wong, J.J. 2015. From a Disciplinary to an Interdisciplinary Design Research: Developing an Integrative Approach for Design. *International Journal of Art and Design Education*, 34(2), 206–223.

Civic Exchange. 2020. *Decarbonising Hong Kong Buildings Policy Recommendations and Next Steps*. Available from: https://www.hkgreenfinance.org/wp-content/uploads/2020/12/Decarbonising-Hong-Kong-Buildings-Policy-Recommendations-and-Next-Steps.pdf [Accessed 28 March 2022].

Coutts, A., Beringer, J. and Tapper, N. 2008. Impact of increasing urban density on local climate: Spatial and temporal variations in the surface energy balance in Melbourne, Australia. *Journal of Applied Meteorology and Climatology*, 46(4), 477–493. Available from: https://journals.ametsoc.org/view/journals/apme/46/4/jam2462.1.xml [Accessed 14 July 2022].

Dodman, D. 2009. *Paper 1 Urban Density and Climate Change*. United Nations Population Fund (UNFPA) Analytical Review of the Interaction between Urban Growth Trends and Environmental Changes. Available from: https://www.uncclearn.org/wp-content/uploads/library/unfpa14.pdf [Accessed 14 July 2022].

Dunne, D. 2018. Planting a Mix of Tree Species 'could double' Forest Earbon Storage. Carbon Brief [Online]. Available from: https://www.carbonbrief.org/planting-a-mix-of-tree-species-could-double-forest-carbon-storage/ [Accessed 14 July 2022].

Fung, W. Y., Lam, K. S., Hung, W. T., Pang, S. W. and Lee, Y. L. 2006. Impact of Urban Temperature on Energy Consumption of Hong Kong. *Energy*, 31(14), 2623–2637.

Halawa, E., Ghaffarianhoseini, A., Ghaffarianhoseini, A., Trombley, J., Hassan, N., Baig, M., … and Azzam Ismail, M. 2018. A Review on Energy Conscious Designs of Building Façades in Hot and Humid Climates: Lessons for (and from) Kuala Lumpur and Darwin. *Renewable and Sustainable Energy Reviews*, 82(3), 2147–2161.

Hoch, G., Richter, A. and Körner, C. 2003. Non-Structural Carbon Compounds in Temperate Forest Trees. *Plant, Cell and Environment*, 26, 1067–1081.

Hong Kong Green Building Council (HKGBC). 2018. *Guidebook on Urban Microclimate Study*. Hong Kong. Available from: https://www.hkgbc.org.hk/eng/engagement/file/UMC_Guidebook_amended_reduced.pdf [Accessed 10 June 2022].

Hong Kong Green Building Council (HKGBC). 2021. *Advancing Net Zero (ANZ) Ideas Competition in Hong Kong*. Available from: https://anzideascompetition.hkgbc.org.hk/ [Accessed 10 June 2022].

Höppe, P. 1999. The Physiological Equivalent Temperature – A Universal Index for the Biometeorological Assessment of the Thermal Environment. *International Journal of Biometeorology*, 43(2), 71–75.

Intergovernmental Panel on Climate Change (IPCC). 2007. *Climate Change 2007: Synthesis Report*. Contribution of Working Groups I, II and III to the Fourth Assessment Report of the Intergovernmental Panel on Climate Change [Core Writing Team, Pachauri, R.K and Reisinger, A. (eds.)]. IPCC, Geneva, Switzerland, 104 pp. Available from: https://www.ipcc.ch/report/ar4/syr/ [Accessed 10 June 2022].

ISG. 2020. *The Power of Place: Defining the Future of the Workplace in a Post-COVID-19 World*. [Online]. Available from: https://www.isgltd.com/en/what-we-think/defining-the-future-of-the-workplace-in-a-post-covid19-world [Accessed 10 June 2022].

J. Love, J. Wingfield, A.Z.P. Smith, P. Biddulph, T. Oreszczyn, R. Lowe, C.A. Elwell. 2017. Hitting the target and missing the point': Analysis of air permeability data for new UK dwellings and what it reveals about the testing procedure, Energy and Buildings, Volume 155, Pages 88–97.

Kong, L., Lau, K. K.-L., Yuan, C., Chen, Y., Xu, Y., Ren, C. and Ng, E. 2017. Regulation of Outdoor Thermal Comfort by Trees in Hong Kong. *Sustainable Cities and Society*, 31, 12–25. Available from: https://www.sciencedirect.com/science/article/pii/S2210670716306576 [Accessed 10 June 2022].

Larsen, K., Pitt, H., Grant, M. and Houser, T. 2021. *China's Greenhouse Gas Emissions Exceeded the Developed World for the First Time in 2019*. Rhodium Group. Available from: https://rhg.com/research/chinas-emissions-surpass-developed-countries/#_ftn1 [Accessed 10 June 2022].

Leadership in Energy and Environment Design (LEED). 2020. *Reference Guide for Neighborhood Development*. United States Green Building Council. Available from: https://www.usgbc.org/resources/leed-reference-guide-neighborhood-development [Accessed 10 June 2022].

Lund, S., Madgavkar, A., Manyika, J., Smit, S., Ellingrud, K. and Robinson, O. 2021. *The Future of Work after COVID-19*. McKinse & Company. Available from: https://www.mckinsey.com/featured-insights/future-of-work/the-future-of-work-after-covid-19 [Accessed 10 June 2022].

Mugayar-Baldocchi, M., Schaninger, B. and Sharma, K. 2021. *The Future of the Workplace: Embracing Change and Fostering Connectivity*. McKinse & Company. Available from https://www.mckinsey.com/business-functions/people-and-organizational-performance/our-insights/the-organization-blog/the-future-of-the-workplace-embracing-change-and-fostering-connectivity [Accessed 10 June 2022].

Sijmons, D. 2014. *Landscape and Energy, Designing Transition*. Rotterdam, The Netherlands.

Steelcase. 2021. Changing Expectations and the Future of Work. Available from https://www.steelcase.com/content/uploads/2021/02/2021_AM_SC_Global-Report_Changing-Expectations-and-the-Future-of-Work-2.pdf [Accessed 10 June 2022].

Swett, B., Henn, M., Schantz, M. and Grayson, B. 2020. *Decarbonizing the Built Environment: 10 Principles for Climate Mitigation Policies*. Urban Land Institute. Available from: https://knowledge.uli.org/-/media/files/research-reports/2020/decarbonizing-the-built-environment_10-principles.pdf?rev=d97633017bf74893b4b3c65082132c0c&hash=536E299A2AA8AE5DFB643DACAD44742D [Accessed 10 June 2022].

Trenerry, B., Chng, S., Wang, Y., Suhaila, Z.S., Lim, S.S., Lu, H.Y. and Oh, P.H. 2021. Preparing Workplaces for Digital Transformation: An Integrative Review and Framework of Multi-Level Factors. *Frontiers in Psychology*, 12, 1664–1078.

United Nations Environment Programme (UNEP). 2020. *Global Status Report for Buildings and Construction: Towards a Zero-emission, Efficient and Resilient Buildings and Construction Sector*.

Vanderklein, D. and Reicht, P.B. 1999. The Effect of Defoliation Intensity and History on Photosynthesis, Growth and Carbon Reserves of Two Conifers with Contrasting Leaf Lifespans and Growth Habits. *New Phytologist*, 144(1), 121–132.

Yang, W. and Li, J. 2021. *What China's Carbon Neutrality Means for Its Building Stock*. Tianjin University, China. Buildings & Cities. Available from: https://www.buildingsandcities.org/insights/commentaries/cop26-china.html [Accessed 10 June 2022].

Yu, W. and Shaw, D. 2018. The Complexity of High-Density Neighbourhood Development in China: Intensification, Deregulation and Social Sustainability Challenges. *Sustainable Cities and Society*, 43, 578–586.

3 Retrofitting Solutions towards Net-Zero Buildings

Roberto Minunno, Matthew van Someren, and Karl Welsch

3.1 Introduction

This chapter explores pathways for energy efficiency retrofitting operations towards net-zero buildings. The net-zero concept was ideated to mitigate and prevent the catastrophic ramifications of anthropogenic climate change. Net zero has been proposed in high-profile reports and global agreements; however, its initial definitions were ambiguous and poorly understood. For example, although not explicitly stated, the concept of net zero was proposed in the Paris Agreement 2015 as 'a balance between anthropogenic emissions by sources and removals by sinks of greenhouse gases in the second half of this century' (Paris Agreement p. 4). Key details regarding kinds of greenhouse gas (GHG) emissions, sources and sinks, and timeframes are absent from this definition and have likely contributed to the varied understanding and applications from governments and businesses. Since the Paris Agreement, there has been a proliferation of net-zero commitments from organisations and governments alike. However, the required action to fulfil these commitments has not met the supposed ambition. In 2018, a further attempt promoted by the Intergovernmental Panel on Climate Change (IPCC) report pushed the boundaries of what should be achieved to avoid climate disasters. According to the IPCC (Nature Editorials, 2021), limiting global warming to 1.5°C can be achieved only by reaching net-zero-carbon emissions before 2050, coupled with considerable reductions in other planet-warming non-CO_2 GHGs. Global emissions have since continued to rise, as has the frequency and ferocity of climate change-driven extreme weather events, such as heatwaves, droughts, and heavy rainfall (Clarke et al., 2022), leading to some describing the planet as a state of climate emergency. What was wrong with the first definitions of the net-zero concept and its applications? It's complicated. One of the critical issues is the concept's global framing, which is too broad for individual entities to garner any definitive actions or responsibilities for reaching net zero. Consequently, governments and businesses have collectively failed to decrease their emissions, hiding behind incomplete and flawed applications of net zero. To put it in simple terms, the net-zero concept is broken. The literature consistently recognises that definitions of net zero are ambiguous and lack the necessary detail to effectively be implemented by individual actors (Black et al., 2021; Fankhauser et al., 2021; Hale et al., 2022). Still, net zero is a scientifically robust concept with extraordinary potential. As such,

DOI: 10.1201/9781003317890-4

it must be redefined to enable globally consistent understanding and application. It might sound ambitious, but re-routing the net-zero concept along a path aligned with its original scientific imperative is achievable. This goal will require the proposal of a new scientifically prescriptive and decision-useful definition and an orchestrated effort from all actors, particularly those in high-emitting industry sectors, to deliver on it.

Against this backdrop, the building sector is one of the largest emitters of carbon dioxide and other GHGs, thus representing a substantial opportunity to deliver progress towards net-zero societies (Zhong et al., 2021). Buildings are responsible for adverse environmental impacts during all life cycle stages, from production to operation to end of life. Many solutions are being researched to limit the impact of construction and demolition, including promising circular economy strategies that could cut embodied carbon emissions by nearly 90% (Minunno et al., 2020). However, the problem of high ongoing operational emissions still needs to be addressed. During several decades of operation, old and new buildings underperform (Gouveia & Palma, 2019; Miller et al., 2018). Homes and offices require vast amounts of energy to maintain optimal temperatures and ensure the health and comfort of their users. As a result, the operation stage of buildings globally impacts 28% of global GHG emissions (World Green Building Council, 2019).

In this context, effectively designed and implemented building retrofits could extend the life cycle of buildings while making them more operationally efficient. In other words, buildings' operations are a perfect fit for applications of the net-zero concept, as they involve a small number of actors (typically owners, building managers, users, builders, and sales agencies). Therefore, relatively limited – albeit complex – work would allow substantial reductions in carbon emissions. This chapter will first explain how stakeholder orchestration could move the built environment towards net zero through retrofitting operations. Then, the chapter will describe the many advantages of applying net-zero strategies to homes and offices, linking energy saving to increased liveability, a healthy indoor environment, and real estate economic value. Further, this chapter focuses on the context of Australian ageing residential buildings.

3.2 Net-Zero Potential for Australia's Ageing Residential Sector: Concepts and Challenges

This section defines some of the critical concepts required to appreciate the potential that the ageing Australian residential sector has to contribute towards net-zero societies. First, however, it is critical to explore the state of ageing Australian homes in the context of their GHG emissions and health outcomes for residents. This explanation is essential to clarify why residential buildings represent considerable opportunities to contribute to the net-zero transformation that cities will undoubtedly undertake. Second, it is critical to understand the complexities that stand in the way of making residential buildings net zero. These complexities arise from political, technical, and societal standpoints, further complicating large-scale retrofitting applications. Finally, this chapter will explore how novel definitions may influence future implementations of net-zero building initiatives.

3.2.1 The Status Quo *of Ageing Australian Homes*

Globally, the typical ageing residential home must meet the same thermal performance and energy efficiency standards that new buildings deliver (Collins & Curtis, 2018). Modern regulation further increases the distance in thermal performance between existing and new buildings. Typically, increased insulation is particularly efficient in reducing heating consumption, while, in some climates, it results in higher energy consumption for cooling (Lee et al., 2017). For these reasons, older homes with outdated insulation are linked with lower thermal resistance (R-value), resulting in inefficient heating and ventilation and higher energy costs. Similarly, these homes often have energy-inefficient appliances, such as heating and ventilation systems, operating at low efficiency compared to modern systems. These disparities illustrate how evolving standards primarily benefit new constructions while widening the performance gaps that existing buildings must overcome to achieve net-zero status (Daly et al., 2023). Double-glazing also represents a way to improve the energy efficiency of their old and ageing homes, allowing homeowners to renew their residences and save on their energy bills (Somasundaram et al., 2020). For these reasons, most countries are adopting increasingly strict energy efficiency rating systems. Still, these are typically only enforced for new buildings, leaving old ageing homes unchecked with occupants living uncomfortably and with substantial health risks (Gouveia & Palma, 2019; Howden-Chapman et al., 2012). The building stock is responsible for up to 90% of the carbon emissions from the residential building sector (Seo et al., 2018). These figures clarify why the residential sector, specifically existing and ageing Australian homes, represents a substantial opportunity to help reduce emissions towards net-zero goals (more on that in Section 3.2.3) while improving the quality of life of their millions of inhabitants.

Ageing Australian homes underperform compared to new homes and compare poorly to those in developed countries in the northern hemisphere. Northern European buildings, for example, are typically well-built and provide a stable and comfortable internal temperature and humidity, with minimal heating and cooling input (Oreszczyn et al., 2006). Conversely, Australian homes have minimal insulation and rely on gas or electric heating and cooling to maintain their occupants' comfort (Strengers, 2010; Zhuang, 2021). Interestingly, in many Australian states, single-glazed windows are still the *status quo*, partially because of the high cost of double-glazing and partially because Australian homes rely on active temperature regulation more than insulation. Window glazing retrofit is required in relatively modern homes built at the end of the 1990s and older homes (Willand et al., 2019). In Melbourne, for example, homes can be classified into three categories according to their thermal performance and based on the 10-star Nationwide House Energy Rating Scheme (NatHERS; 10 stars represent the best performance). Houses built before 1990 average 1.57 stars, residences completed between 1991 and 2005 average 3.14 stars, and houses built after 2005 should achieve 5 stars (a minimum

5-star rating requirement was introduced in 2005) (Seo et al., 2018). In South Australia, houses built before 1930 achieved only 1.1 stars, properties built before 1950 achieved 1.6 stars, and more houses built in 1996 averaged only a 1.2-star rating. Interestingly, ceiling insulation would bring the rating up to 2.8 stars at AU$230/m^2 (Whaley et al., 2017). These results are particularly concerning. Cold, damp, and insufficiently ventilated houses can host mould and foster poor indoor air quality, causing respiratory illnesses (Al-Rawi et al., 2021; Putnam & Brown, 2021) and mental illnesses (Putnam & Brown, 2021) for their inhabitants. Childhood respiratory diseases, including asthma, are related to thermal inefficiency, humidity, and discomfort and could be drastically reduced with appropriate retrofitting interventions (Sun et al., 2021).

From an economic standpoint, a case study from Spain shows that heating poorly insulated homes can result in owners spending up to 10% of their income on heating (Ortiz et al., 2019). Even though retrofitting options would be costly, they can potentially reduce healthcare costs by up to 34 times (Ortiz et al., 2019). In many instances, heating costs are so prohibitive that low-income families cannot afford to heat their dwellings (Delmastro et al., 2016).

Energy poverty is often prevalent in societies with low living standards and is characterised by households that cannot afford their energy bills. This can be compounded by poorly insulated energy-inefficient dwellings (González-Eguino, 2015). Energy poverty is a significant problem in Australia and is often associated with harmful health outcomes, including respiratory problems, mental health disorders, chronic illness, and cardiovascular diseases (Awaworyi Churchill & Smyth, 2021). Ageing residential buildings could be retrofitted to tackle these issues by improving indoor comfort and low energy consumption, particularly relevant to low-income brackets (Tardy & Lee, 2019).

3.2.2 Barriers to Home Retrofitting

On the one hand, the technology to effectively retrofit homes exists and is often cost-effective. On the other, many barriers prevent the retrofitting of millions of Australian homes that would see improved environmental and health outcomes. Table 3.1 provides a summary of the critical barriers to home retrofitting with selected vital references. Among these is the homeowner-tenant dilemma, where property owners may hesitate to invest in retrofits if they don't directly benefit from lower utility bills. Occupant behaviour also plays a role, as even the most advanced retrofits can underperform if occupants don't engage in energy-efficient practices. Additionally, there is often a lack of trust in the efficacy of retrofitting, fuelled by misinformation or past failures. Finally, regulatory and technical challenges, such as complex building codes and the need for specialised skills, can make retrofitting a daunting task that sometimes requires professional assistance. By providing the references for each source, this table can serve as a resource for further reading on the topic. To illustrate the owner-tenant dilemma, consider the example of a rental property in a major Australian city. The property is an older home with outdated insulation, inefficient heating and cooling systems, and drafty windows and doors.

Table 3.1 Identified barriers to retrofitting initiatives and critical references

Barrier to home retrofitting	*References*
Homeowner-tenant dilemma	• Ástmarsson et al. (2013) • Collins and Curtis (2018) • Greenough and Tosoratti (2014) • Kholodilin et al. (2016)
Occupant behaviour	• Xing et al. (2011a) • Risch and Salmon (2017) • Aranda et al. (2017)
Lack of trust in retrofitting	• Bergman and Foxon (2020) • Weiss et al. (2012)
Regulatory and technical challenges	• Friedman et al. (2018) • Galvin (2014) • ABS (2019)

The landlord hesitates to invest in energy-efficient retrofits because of the high upfront costs and the uncertainty of return on investment. Meanwhile, the tenants pay high energy bills and live in an uncomfortable environment. Additionally, they cannot afford to move to a better insulated property, as the demand for housing in the city is too high, and rents are continually rising. In this scenario, the interests of the homeowners and tenants are vastly misaligned, creating a situation where the homeowners prioritise their financial interests over energy-efficient retrofits. More specifically, according to the homeowner-tenant dilemma, homeowners are responsible for improving the energy efficiency of dwellings, whether they live in the house or if they are leasing them. Nearly a third of Australians are renting their homes. Their interests are vastly misaligned with the owners' interests, effectively creating what is known as the owner-tenant dilemma or the principal-agent problem (Ástmarsson et al., 2013; Collins & Curtis, 2018; Greenough & Tosoratti, 2014; Kholodilin et al., 2017). In these circumstances, the owners are interested in maximising their profit and minimising their expenses, which often means maintaining the *status quo* and avoiding retrofitting interventions. Tenants would prefer to live in a more comfortable, healthier house with cheaper bills. Still, due to the overwhelming demand for homes in Australian cities, tenants often cannot negotiate on their dwellings' quality or energy efficiency. Almost two-thirds of Australians own the buildings where they live, and for them, retrofitting would be a more viable option (ABS, 2019).

Another point of conflict in retrofitting is occupant behaviour, specifically, individuals' commitment to and knowledge of using their homes most energy-efficiently (Xing et al., 2011b). Research hasn't yet agreed on how much the human factor affects houses' energy use. Building panels, construction quality, retrofitting options, and appliances are vital parameters in decreasing buildings' energy use. Because these parameters can significantly affect how the building operates and how much energy is required to maintain thermal comfort, in many instances, the socio-demographics of the occupants – such as age, income, or educational level – are nearly uninfluential. This should not be confused with human factors and users' behaviour, which do have an impact. Specifically, socio-demographics often inform general trends but do not directly dictate individual actions within a building. In contrast, user behaviour, such as thermostat settings or lighting usage, has a direct and immediate impact on energy consumption and is discussed further in the section below (Risch & Salmon, 2017). Still, as a critical part of the home system, occupant behaviour can substantially affect buildings' thermal performance and energy use. One of the easiest and most efficient ways to save energy in the public housing sector, potentially up to two-thirds, is by educating occupants on how to use their homes more efficiently (Aranda et al., 2017). For instance, a household may have their thermostat set to 25°C all year round, regardless of the outside temperature or time of day, leading to unnecessary energy expenditure. However, if the occupants were informed of the most efficient temperature range for their location and the benefits of adjusting their heating, they could save a considerable amount of energy and money. Furthermore, some households may leave electronic devices and appliances on standby, leading to wasteful use of energy, even when the devices are not being used (Chen & Gou, 2022).

Another barrier to retrofitting is the lack of trust in the upgrade option. Retrofitting an ageing house can be expensive, and the owners investing often doubt the benefits and thus can be discouraged from embarking on such an endeavour (Bergman & Foxon, 2020). Relatedly, higher income households with higher levels of education are typically more likely to retrofit their homes (Weiss et al., 2012). In contrast, the lack of trust in retrofitting options manifests differently and is not solely dependent on education or awareness. For instance, homeowners may be sceptical due to conflicting reports about the long-term durability of certain sustainable materials, or they may have heard anecdotal accounts of retrofits failing to deliver the promised energy savings (Risholt & Berker, 2013). In some cases, the absence of widely accepted, third-party assessments confirming the effectiveness of newer, sustainable materials can also exacerbate this distrust. This scepticism, distinct from a simple lack of awareness, can be a significant barrier to adopting retrofitting options. From a regulatory and technical standpoint, retrofitting ageing houses can be complex and challenging. For instance, heritage status can impose strict limitations on the materials and methods used, often requiring specialised and expensive craftsmanship (Friedman et al., 2018). Zoning laws might also restrict alterations to building layouts or additions like solar panels. Similarly, integrating modern energy-efficient systems may clash with existing electrical or plumbing configurations, necessitating costly upgrades (Dadzie et al., 2018). These regulatory and technical hurdles can add complexity and cost, making it less straightforward for asset managers, engineers, and installers to implement retrofitting measures, even when residents are willing.

3.2.3 Net-Zero Buildings

The IPCC (IPCC, 2018) states, '*Net zero emissions are achieved when anthropogenic emissions of greenhouse gases to the atmosphere are balanced by anthropogenic removals over a specified period.*' This definition has been widely adopted in policy-making, industry operations, and research (IPCC, 2018; Loveday et al., 2022). Since the emergence of net zero, many applications of the concept have emerged. A quick search of academic databases shows that net zero is often associated with critical industry sectors. More specifically, the term is associated with waste, energy, and water management and, more frequently, with sustainable building operations, materials, and energy use. That is not surprising, as buildings consume about a third of global energy production and emit a quarter of global CO_2 emissions (Hong et al., 2020; Shen et al., 2019). Buildings have the potential to become positive contributors to the goal of global net zero through the minimisation of operational impacts, reduced energy demand, and enabling broader concepts such as net-zero cities (Alam & Lahdelma, 2019; Bhavsar et al., 2020; Englehardt et al., 2016; Kilkis, 2012). Despite the prominence of the term net-zero buildings, the definition of the concept is still vague. Undoubtedly, the concept of net-zero buildings is made complex by the compound nature of net-zero buildings. However, to avoid oversimplifying the problem at hand, we defined these two concepts separately and then blended the two concepts. In so doing, we defined the concept of net-zero buildings as '*buildings that are built and operate in such a way that their carbon emissions are in ongoing balance with their embodied and operational emissions.*' In other words, the carbon emissions associated with their construction and operation would eventually be nullified with a maintained state of balance through efficient operations. For example, a building whose embodied energy equals 100 tonnes of CO_2 equivalent and has a lifespan of 50 years would need to produce a yearly amount of renewable energy equivalent to 2 tonnes of CO_2 emission (i.e., 100 tonnes divided in 50 years) so that it would, in its lifetime, counterbalance its embodied emissions.

In the context of this chapter, carbon emission reduction is not considered through concepts such as carbon storage or air capture but through energy production and intelligent management. The following sections explain these concepts in more detail, providing recommendations for practitioners, homeowners, tenants, and policymakers on how they can positively contribute to net-zero buildings.

3.3 Towards Net-Zero Residential Buildings

So, how can the definition of net zero be applied to residential buildings? Net-zero residential buildings are structures optimised to reduce their carbon emissions in all seasons and climates. Emission reduction is critical because carbon removal is exceptionally complex; risks associated with carbon removal can be mitigated by simply producing fewer emissions. In a nutshell, net-zero buildings produce and help reduce the same amount of atmospheric GHG emissions. Residential buildings can help the built environment move towards net-zero operations. This section provides recommendations for practitioners, homeowners, tenants, and policymakers on how they can contribute positively to net-zero buildings.

3.3.1 How to Transform to Net-Zero Operations and Materials?

The net-zero concept is on global and local organisations' agendas, with timeframes that vary from 2030 to 2050. These organisations are contributing to transforming their industry sector through fundamentally changing their business models. Achieving this transformative goal necessitates a carefully strategised orchestration of a diverse array of stakeholders. This includes policymakers who set the legislative framework, industry leaders who drive technological innovation, technologists who develop and implement the solutions, and financial institutions that provide the economic backbone for these initiatives. Orchestration in this context means not just the mere involvement of these entities but a synchronised and adaptive collaboration among them. It involves aligning their distinct roles, responsibilities, and actions in a cohesive manner to ensure that policies, technological advancements, legislative changes, and financial mechanisms are not only mutually reinforcing but also flexible enough to adapt to changing circumstances. The aim is to create a unified, robust, and agile strategy that can fundamentally transform the prevailing business models within the residential building sector, making them more sustainable and aligned with net-zero objectives. Problematically, shifting entire industry sectors should be supported by standardised and monitored principles aligned with what key scientific bodies consider to be best practices to achieve net-zero goals (Pineda et al., 2020). Such a guideline for the residential building sector is still missing, but many initiatives can help shift the sector to achieve the first step towards net zero: reducing emissions. In this vein, climate laws established in countries, such as the European Union, the United Kingdom, and New Zealand, set targets and provide support for long-term planning (European Commission, 2021), allowing investors and planners a guiding framework to enforce a transformation to sustainable buildings.

Further, mechanisms, such as carbon prices, carbon taxes, and emission trading schemes, can be implemented to boost the uptake of clean technologies and materials. Carbon taxes and emission trading schemes create markets for carbon offsets, which have been implemented in several countries, including Australia, South Korea, the USA, China, and Finland. These countries adopted the initial or future iterations of markets based on emission trading schemes (Hammerle et al., 2021; Kim et al., 2023; Lou et al., 2022; Wei et al., 2022). Such schemes have yielded positive results towards the decarbonisation of various industry sectors, including the built environment (Jung & Song, 2023; Lou et al., 2022). However, it's crucial to note that these concepts are still underdefined, and their long-term outcomes are ambiguous for several reasons. For instance, the allocation of emission permits can vary widely between jurisdictions, leading to market inefficiencies. Additionally, the lack of a global standard for emission trading complicates cross-border cooperation and could lead to carbon leakage, where companies move operations to countries with less stringent regulations. Finally, short-term success in emission reduction doesn't necessarily translate to sustainable long-term environmental benefits, especially if these schemes do not account for other forms of environmental degradation (Isenhour & Feng, 2016). Simultaneously, countries, including the USA and China, are investing substantially in renewable energy technologies, batteries, heat pumps, hydrogen production, and networks (Statista, 2021). These technologies are critical factors that help residential buildings become more sustainable. Many global companies are committing to transforming their shipping fleets to zero-carbon fuels by 2040 through technologies such as green hydrogen (see Cullinane & Yang, 2022; Perčić et al., 2022). This approach could, in turn, help reduce the embodied carbon of construction materials such as steel (Thomas, 2021).

These interventions will help decarbonise the procurement and transport of building materials and improve building operational efficiency towards net-zero goals.

Drawing upon the specialised expertise of one of our authors in sustainable building practices, the term 'net zero' holds significant relevance in the Australian construction sector, particularly concerning home energy efficiency. This author has an academic background in sustainability and extensive practical experience in the construction industry, lending a nuanced understanding of how 'net zero' is conceptualised and applied. The concept is so relevant that it could be harmonised with the NatHERS, Australia's standardised rating system for residential energy efficiency, thereby providing a robust framework to certify net-zero homes. Counterintuitively, a house with a poor NatHERS star rating, which traditionally implies low energy efficiency, could still achieve a net-zero energy status. This seeming contradiction is reconciled through alternative strategies that go beyond the scope of traditional energy efficiency measures. For instance, a homeowner could install a high-capacity solar panel array that generates more energy than the home consumes, effectively offsetting its energy use (Dong et al., 2021).

3.3.2 Recommendations to Retrofit Ageing Homes Towards Net Zero

Achieving net zero in ageing homes is challenging and hindered by many barriers. For this reason, several recommendations can help orchestrate the various technologies, stakeholders, policies, and financing systems. Accordingly, recommendations can be grouped into three overarching themes: technical, social sustainability, political, and regulatory:

1 Recommendations within the technical theme

- Retrofitting with low-carbon materials ensures that the retrofit won't substantially increase the residents' carbon emissions.
- Evaluating the distance and means of transport of the retrofitting materials.
- Decreasing the operational energy with smart devices, building management systems (BMSs), or high-energy intensive appliances and lighting fixtures.
- Producing energy to offset energy use or removing carbon emissions with innovative materials.

2 Recommendations within the theme of social sustainability

- Educating households on the necessity of low-carbon living and net-zero goals.
- Considering more support for people with a low income as they often suffer more from poorly performing houses.

3 Recommendations within the political and regulatory theme

- Promoting reforms to incentivise low-carbon homes.
- Creating rating schemes to validate net-zero homes.

The recommendations listed here are further explained in the following subsections. Specifically, Section 3.3.2.1 provides an overview of how to apply recommendation (1) on the technical theme. Section 3.3.2.2 explains how social sustainability can be addressed in retrofitting buildings (recommendation (2)). Finally, recommendation (3) on the political and regulatory scheme is expanded in Section 3.3.2.3. These subsections provide several practical examples of how to move towards net-zero buildings through retrofitting options. It's crucial to underscore that these recommendations should not be undertaken in isolation or selected piecemeal. Instead, they are designed to function as a comprehensive, interrelated strategy, where each component amplifies the others. The interaction between technologies, stakeholders, policies, and financial mechanisms is vital for successfully transitioning ageing homes to net-zero status.

3.3.2.1 Technical Theme – Lowering Energy Use and Increasing Energy Production

Due to the complexity of the home system, it's helpful to divide the system into two life cycle stages derived from life cycle assessment methodologies of buildings, explicitly building production and building operation. The building production stage accounts for emissions from raw material extraction, the manufacturing of construction products, and the transportation of these materials to the construction site. Each of these activities has its own set of environmental impacts, such as the energy-intensive nature of manufacturing processes like steel and cement production (Cabeza et al., 2014). On the other hand, the building operation stage focuses on the emissions generated during the actual use of the building, including heating, cooling, lighting, and the use of appliances. This division into two life cycle stages allows for a more granular analysis, making it easier to identify opportunities for emission reduction in each phase (Minunno et al., 2020). Indeed, while other life cycle stages like building maintenance, renovation, and end-of-life management exist, their relative contribution to overall emissions is often significantly less than that of production and operation. In the context of building production, retrofitted buildings would include materials for retrofitting purposes, while concerning building operations, we explore operational energy required to heat and cool the house and its water, ventilation, and eventually air filtration, appliances, and lighting. Our approach aligns with elements discussed in Omrany et al. (2022), who also underscore the importance of considering operational and embodied energy in residential buildings. Specifically, their work illuminates how excluding embodied energy – energy embedded in the materials and construction processes – from energy efficiency regulations can skew efforts to reduce overall energy consumption. This is particularly relevant as we aim to provide a holistic view of energy use and emissions, making Omrany et al.'s findings pertinent to our discussion. With that in mind, the definition given in Section 3.2.3 for net-zero residential buildings finds its roots in the scientific underpinnings of the concept. Transforming the built environment, especially ageing residential homes, to net-zero buildings can primarily be achieved with technical solutions (Ohene et al., 2022). Firstly, retrofitting options must look into the materials used to retrofit (linked to the production life cycle stage). Low-impact materials should be prioritised, providing this choice does not negatively affect the increase in the energy efficiency of the building (Füchsl et al., 2022; Minunno et al., 2021). To this extent, technicians and professionals, such as designers, builders, and trade workers, should be trained to support a transition towards net-zero buildings (Godin et al., 2021). Further, materials should be sourced locally or with attention to the distances and carbon emissions related to transport (Arduin et al., 2022). In some instances, transport emissions derived from trucks can be ten times greater than those from ocean transport (Minunno et al., 2021). To illustrate, consider the transport of timber for a building project. If a construction site in Perth sources timber from a supplier in Sydney, the truck journey would cover approximately 4,000 km. In contrast, if the timber is sourced from a European supplier, it could be shipped a distance of approximately 16,000 km to Perth. Despite the much greater distance, the ocean transport's lower emissions profile could mean that the environmental impact of the two options is surprisingly similar. This counterintuitive example highlights how the mode of transport can significantly influence the overall carbon footprint of sourcing materials, even when the distances involved are vastly different.

When examining the operational life cycle stage of a building, it's crucial to break down the operational energy used. In ageing homes in Australia, about 40% of annual energy use goes towards heating and cooling (Department of Climate Change Energy, the Environment and Water, 2022). This sizable portion directly correlates with the insulation properties and window efficiencies of a building. Subpar insulation and windows necessitate more heating and cooling, thereby increasing energy consumption. For reference, underperforming appliances, such as fridges, washing machines, and dishwashers, consume about 30% of total household energy. The remaining portion is divided between heating water (25%) and lighting (5%) (Famiglietti et al., 2023; Sanyé Mengual & Sala, 2023), although these percentages can vary drastically depending on the use of the building and the efficiency of the used devices (Ferreira et al., 2023; He et al., 2023). Therefore, the primary requirement for reducing energy use in an ageing residential building centres on improving its thermal performance, which would address the most significant proportion of annual energy use and subsequently reduce the need for heating and cooling. That can be done by improving the building's insulation properties, eliminating draught, and upgrading to double-glazed windows. These strategies effectively reduce the 40% quota required for heating and cooling (Oldfield, 2022). Other recommendations include increasing the energy efficiency of home appliances when existing appliances need to be changed (i.e., we don't recommend changing still-functioning appliances to avoid increasing their embodied energy and creating more waste). When it comes to water heating, the main recommendation is to change the showerhead for a model that consumes less water, reducing the energy needed to heat it. Also, water heating devices such as heat-pump water heaters effectively cut a household's emissions; however, their cost remains prohibitive in many cases. To limit the energy used for lighting, it is recommended to substitute incandescent lightbulbs (ordinary in ageing homes) for LED lights, which can drastically reduce energy use (European Commission Joint Research Centre, 2011). With particular attention to lighting, technicians such as interior architects play a crucial role in advising homeowners on low-impacting technologies and optimisation towards energy use minimisation (Doulos & Tsangrassoulis, 2022).

We recommend producing renewable energy onsite where achievable to offset energy use and reduce carbon emissions. The adoption of solar panels and battery systems can be restricted for ageing residential homes depending on the local government authority and related regulations, roof type, and whether the home is an independent residence or an apartment. If solar panels can be installed, an innovative way to move closer to net-zero residences is virtual power plants (VPPs) (Kumar et al., 2022; Saboori et al., 2011). By aggregating multiple assets, a VPP allows homes to share energy, supply energy to the grid at times of peak demand, and limit solar curtailment from individual homes. In addition, homes can be equipped with smart devices and software such as BMSs. A BMS is software linked to sensors and robots integrated with blinds, windows, and switches to enable optimisation of energy use (Yamaguchi et al., 2023). Although BMSs are still in their infancy, they are becoming more efficient and popular.

Finally, sequestrating carbon at home is only possible in some circumstances. For example, if some works involve concrete or cement, the builder should ideally use only low-carbon cement, which reacts with the CO_2 in the air and absorbs it. A secondary strategy to decrease carbon emissions is by planting fast-growing trees. These two recommendations are limited in scope and effect. They can only be applied in limited cases (such as when masonry works are undertaken or in households with a large backyard).

3.3.2.2 Social Sustainability Theme – Educating Households and Considering Implementation Strategies Based on Equity and Fairness

While homes can theoretically become net-zero-ready by applying the recommendations listed in Section 3.3.2.1, these technological gains can be nullified by suboptimal occupant behaviour. Take, for instance, an energy-efficient air conditioning system: if it is cooling a room, but a window is left open, the system's energy-saving capabilities are undermined. Similarly, a highly efficient showerhead can promise water savings of up to 50%, but if occupants take excessively long showers, the intended benefits are negated. Addressing this disconnect requires a multi-faceted approach that goes beyond technology to include occupant education. Public awareness campaigns, advertising, or sensitisation initiatives can play pivotal roles in informing households about the importance of reducing emissions, both for environmental stewardship and financial well-being. Real-time feedback mechanisms can also be powerful influencers of behaviour. For example, BMSs can offer daily or weekly performance metrics, which not only inform but also motivate occupants to improve their consumption patterns. Electricity retailers might introduce neighbourhood challenges that leverage principles of social comparison and competition to nudge households into using less energy than the community average. Gamification strategies can also be employed to make the process engaging, offering rewards or recognition for households that achieve or surpass net-zero goals. Thus, a comprehensive strategy that combines technological innovation with behaviour-centric initiatives can significantly enhance the efficacy of efforts to make homes net-zero-ready. Besides educating households on how to use their homes in the most energy-efficient way, it is also fundamental to underline that all strategies proposed in this chapter would have to be carefully considered as to their achievability for low-income households. Recommendations explained in Section 3.3.2.1 (on the technical theme) are likely to be cost-prohibitive for some homeowners. Budget-constrained low-income households are unlikely to outlay considerable amounts of money to purchase low-emission appliances. Although energy-efficient appliances typically offer long-term cost savings, they often require a more significant initial investment compared to less efficient alternatives. This financial barrier is corroborated by Wang et al. (2021), who found that energy efficiency labelling policies do have an impact on sales, increasing the sales of refrigerators by 24.47%, for instance. However, the study also revealed a critical caveat: the higher cost associated with energy-efficient products can deter consumers, particularly those in the low-price range, from making green choices. This phenomenon, termed the 'energy efficiency paradox,' suggests that while energy efficiency labels can guide consumer behaviour, their efficacy is limited by the financial considerations that often prioritise short-term savings over long-term efficiency. People in low socioeconomic situations often suffer the most from poorly performing homes. Because of their situation, they might opt for cheap heating and cooling devices characterised by low energy efficiency. In short, low-income households might end up spending more on bills or living in unhealthy and uncomfortable conditions. For this reason, the recommendations proposed in this chapter related to the social sustainability theme are specifically designed to ensure social sustainability and fairness across all layers of society. Notably, we advocate for educating households on the importance of low-carbon living and net-zero goals. This education is crucial for all, but particularly for those in economically disadvantaged situations who may lack access to such information. Furthermore, we recommend additional support for low-income households, as they are often disproportionately affected by the issues associated with poorly performing homes. Such support could take the form of financial incentives for retrofitting, subsidised energy audits, or grants for purchasing energy-efficient appliances. These targeted recommendations aim to level the playing field, ensuring that the transition to net-zero living is inclusive and equitable. It is critical that policymakers at all levels, including municipalities, local governments, and state governments, support low-income households by investing in retrofitting schemes, as they are less exposed to financial risk (Fahlstedt et al., 2022). Low-income households are particularly vulnerable to such complexities and less capable of absorbing the financial risks of retrofitting. Therefore, targeted governmental support, such as subsidies or low-interest loans, can both improve living conditions for these households and contribute to broader goals of carbon reduction. Such policy initiatives can also serve as case studies to refine future retrofitting strategies, aligning with the call for more research in the cited study.

3.3.2.3 Political and Regulatory Theme – Incentives and Rating Schemes towards Net-Zero Homes

Economic incentives have proven to be a powerful tool in encouraging households to retrofit their ageing homes, as evidenced by multiple international programs. In Canada, the 'My Home Retrofit' initiative is particularly noteworthy. This program provides direct funding to homeowners undertaking projects like insulation, draught sealing, and the installation of high-efficiency appliances and upgraded windows (Government of Canada, 2021). This funding mechanism reduces the initial financial burden, making retrofitting more accessible to a broader demographic. Similarly, the French government has been proactive in promoting sustainable development through various financial incentives. Tax credits and zero-interest bank loans have been key drivers in this respect (Risch & Salmon, 2017). Notably, the French initiative has yielded tangible results, increasing renovation rates by 1.09% (Risch, 2020). These statistics indicate the efficacy of such financial tools: among homeowners who retrofitted their homes between 2009 and 2011, a significant majority utilised some form of financial aid – 52% used a tax credit, 28% took out a personal loan, and 5% availed of a zero-interest loan (Giraudet et al., 2021). These incentive programs serve dual goals: enhancing comfort and health conditions within homes while potentially increasing their market value. The latter is often certified through standardised rating schemes like Australia's NatHERS. To further amplify these benefits and align with our third set of recommendations within the political and regulatory theme, several strategies can be considered:

- ***Promoting Reforms to Incentivise Low-Carbon Homes***. Governments have a pivotal role in shaping the landscape for low-carbon homes through legislative reforms. Tax benefits or subsidies can be strategically designed to target specific aspects of low-carbon retrofitting. For instance, a tax credit could be offered for the purchase and installation of solar panels. This would make the technology more affordable and stimulate local markets for solar technology, creating jobs and boosting the economy. Additionally, subsidies could be designed to target energy-efficient HVAC systems. These systems often have a high upfront cost, acting as a barrier to adoption. By offsetting a portion of this cost through subsidies, governments can encourage rapid adoption among homeowners. Furthermore, these incentives could be structured in a tiered fashion, providing more extensive benefits to those who opt for more advanced and, consequently, more expensive systems. This would help drive technological innovation in the sector and lead to more significant reductions in carbon emissions.
- ***Creating Rating Schemes to Validate Net-Zero Homes***. Building on existing energy rating systems, like Australia's NatHERS, governments can introduce specialised certification tiers that exclusively recognise net-zero homes. This specialised tier would serve multiple functions. First, it would provide a credible means of validating the efforts of homeowners who have invested time, money, and resources into retrofitting their homes to achieve net-zero status. This could include a comprehensive evaluation of both operational and embodied energy in the home, perhaps drawing upon life cycle assessment methodologies for a more

holistic evaluation. Secondly, such a specialised rating would act as a market differentiator. Homes with a net-zero certification could command a premium in the real estate market, providing a tangible financial incentive for homeowners to engage in retrofitting projects. This market-driven approach could create a positive feedback loop; as more homeowners see the financial benefits of a higher rating, the demand for retrofitting services and technologies would likely increase, potentially driving down costs through economies of scale.

In summary, the role of political and regulatory frameworks in accelerating the retrofitting of ageing homes cannot be overstated. They have the power to mitigate upfront costs, alleviate financial risks, and speed up the transition to more sustainable housing forms. Given the proven efficacy of such approaches globally, it becomes imperative for policymakers to actively consider and implement these strategies in their broader agenda for social and environmental sustainability.

3.4 Discussion: Action Plan Towards Real-World Application of Comprehensive Net-Zero Retrofit Recommendations

The quest for achieving net zero in ageing homes is a multidimensional challenge that requires a harmonised approach encompassing technical, social sustainability, and political and regulatory elements. The recommendations proposed in the preceding sections offer a comprehensive blueprint for stakeholders to follow. However, to move from theory to practice, it is vital to discuss the implementation complexities and synergies among these three pillars of recommendations.

1 Interplay Between Technical and Social Sustainability Themes:

- *Knowledge Gap and Technology Adoption*. One of the critical hurdles in implementing technical solutions is the potential knowledge gap that exists among homeowners and even among professionals like builders and designers. Here, the social sustainability aspect of educating the households becomes pivotal. Technological interventions are effective only when the end-users are well-informed about their operation and benefits.
- *Economic Feasibility*. Technical solutions often come with a price tag that may not be affordable for all sections of society. The social sustainability theme's focus on supporting low-income households can play a crucial role in ensuring that cost is not a barrier to the adoption of green technologies.
- *Behavioural Economics*. Financial incentives alone may not be sufficient to nudge homeowners towards sustainable practices. Social sustainability efforts that use behavioural economics techniques like gamification or social comparisons can create a more compelling case for technical retrofits.

2 Political and Regulatory Framework as a Catalyst:

- *Legislation and Standardisation*. Political will can expedite the implementation of both technical and social sustainability recommendations. Legislation can set standards that mandate the use of energy-efficient materials

and low-carbon technologies. Rating schemes can further offer quantifiable metrics for households to aspire to, providing a direct link to market value and financial incentives.
- *Subsidies and Financial Instruments.* Governments have a crucial role in mitigating the financial risks associated with retrofitting, especially for low-income households. Particular low-interest loans, tax benefits, or direct subsidies can lower the entry barriers, driving faster adoption of technical recommendations.
- *Public-Private Partnerships.* Government bodies can collaborate with private organisations to facilitate bulk procurement of sustainable materials or technologies. This can drive down costs and encourage more households to retrofit their homes towards net zero.

3 Anticipating Future Challenges:

- *Technological Advancements.* As the technology landscape is continuously evolving, future-proofing the recommendations is essential. Modular and upgradable solutions should be preferred over static, one-time fixes.
- *Climate Change.* As climate conditions change, the effectiveness of retrofit solutions might vary. Thus, adaptive and resilient design principles should be incorporated.
- *Circular Economy.* As we move towards net zero, the concept of circular economy becomes increasingly relevant. Future policies should also focus on sustainable sourcing, reuse, and end-of-life management of materials used in retrofitting.
- *Global Lessons.* Learning from international case studies can provide valuable insights. Cross-border sharing of successful models can enrich the local implementation strategies.

The path to net-zero retrofitting is fraught with challenges that are technical, social, and political in nature. Addressing these in a siloed manner is a recipe for limited success. A systems-thinking approach that appreciates the complex interdependencies among these elements is crucial for the successful implementation of the proposed recommendations.

3.5 Conclusions

Urgent and decisive action is needed to limit global warming to 1.5°C (IPCC, 2022). As limiting global warming is now intrinsically connected to the concept of net zero, its reputation and validity as a global organising principle for decarbonisation are also at risk. Its success or failure as such will depend on the outcomes of its application. Fossil fuel-driven economies have appropriated environmental and atmospheric commons with little regard for principles of sustainable development. To counterbalance these actions, achieving global net zero is required and will mean fundamentally transforming all facets of economies and societies. The built environment can play a pivotal role in this transformation. Furthermore, one of its branches, the ageing residential sector, represents low-hanging fruit towards net zero and even beyond net-zero goals.

In this chapter, we proposed a set of recommendations divided into three themes: the technical theme, the theme of social sustainability, and the economic and regulatory theme. The recommendations illustrated in Sections 3.3.2.1–3.3.2.3 are not to be undertaken independently or selectively. Instead, these recommendations are meant to enable the orchestration of stakeholders towards retrofitting ageing homes in Australia in line with the proposed definition for net-zero buildings. Indeed, adopting these recommendations would change the built environment, but independently, they are not enough to transform the residential sector to achieve net zero. They should be considered a guiding framework more than a list of strategies of which only some may be chosen. Therefore, in assessing the feasibility and impact of our recommendations for net-zero homes, it becomes clear that an integrated action plan is imperative. Technicians will need to employ advanced tools for real-time monitoring, while builders must not only upskill their workforce but also validate these new competencies. Policymakers should go beyond financial allocation to include performance metrics that measure the effectiveness of their support. Homeowners, meanwhile, will benefit from ongoing education, potentially facilitated by intelligent building systems. The challenge lies not just in defining these roles but in orchestrating them in a manner that is both feasible and impactful, requiring a unified, data-driven strategy from all stakeholders.

Reference List

ABS, A. B. o. S. (2019). *Housing Occupancy and Costs*. Australian Bureau of Statistics. https://www.abs.gov.au/statistics/people/housing/housing-occupancy-and-costs/latest-release#housing-tenure

Alam, S., & Lahdelma, R. (2019). Towards net zero energy buildings: Building performance optimization, simulation and analysis. *IOP Conference Series. Materials Science and Engineering*, *609*(7). https://doi.org/http://dx.doi.org/10.1088/1757-899X/609/7/072061

Al-Rawi, M., Ikutegbe, C. A., Auckaili, A., & Farid, M. M. (2021). Sustainable technologies to improve indoor air quality in a residential house – A case study in Waikato, New Zealand. *Energy and Buildings*, *250*, 111283. https://doi.org/10.1016/j.enbuild.2021.111283

Aranda, J., Zabalza, I., Conserva, A., & Millan, G. (2017). Analysis of energy efficiency measures and retrofitting solutions for social housing buildings in Spain as a way to mitigate energy poverty. *Sustainability*, *9*(10), 22, Article 1869. https://doi.org/10.3390/su9101869

Arduin, D., Caldas, L. R., Paiva, R. d. L. M., & Rocha, F. (2022). Life cycle assessment (LCA) in Earth construction: A systematic literature review considering five construction techniques. *Sustainability*, *14*(20), 13228. https://www.mdpi.com/2071-1050/14/20/13228

Ástmarsson, B., Jensen, P. A., & Maslesa, E. (2013). Sustainable renovation of residential buildings and the landlord/tenant dilemma. *Energy Policy*, *63*, 355–362. https://doi.org/https://doi.org/10.1016/j.enpol.2013.08.046

Awaworyi Churchill, S., & Smyth, R. (2021). Energy poverty and health: Panel data evidence from Australia. *Energy Economics*, *97*, 105219. https://doi.org/https://doi.org/10.1016/j.eneco.2021.105219

Bergman, N., & Foxon, T. J. (2020). Reframing policy for the energy efficiency challenge: Insights from housing retrofits in the United Kingdom. *Energy Research & Social Science*, *63*, 101386. https://doi.org/https://doi.org/10.1016/j.erss.2019.101386

Bhavsar, R., Cupido, A., & Arriaga, M. (2020). Mohawk college's net zero energy and zero carbon building—A living lab for high efficiency and renewable energy technologies in buildings. *Journal of Green Building*, *15*(1), 185–214. https://doi.org/10.3992/1943-4618.15.1.185

Black, R., Cullen, K., Fay, B., Hale, T., Lang, J., & Smith, S. (2021). Taking Stock: A Global Assessment of Net Zero Targets. *Energy & Climate Intelligence Unit and Oxford Net Zero*. (March), 1–30. https://ec.europa.eu/clima/policies/strategies/2050_en

Cabeza, L. F., Rincón, L., Vilariño, V., Pérez, G., & Castell, A. (2014). Life cycle assessment (LCA) and life cycle energy analysis (LCEA) of buildings and the building sector: A review. *Renewable and Sustainable Energy Reviews*, *29*, 394–416. https://doi.org/https://doi.org/10.1016/j.rser.2013.08.037

Chen, X., & Gou, Z. (2022). Bridging the knowledge gap between energy-saving intentions and behaviours of young people in residential buildings. *Journal of Building Engineering*, *57*, 104932. https://doi.org/https://doi.org/10.1016/j.jobe.2022.104932

Clarke, B., Otto, F., Stuart-Smith, R., & Harrington, L. (2022). Extreme weather impacts of climate change: An attribution perspective. *Environmental Research: Climate*, *1*(1), 012001. https://doi.org/10.1088/2752-5295/ac6e7d

Collins, M., & Curtis, J. (2018). Rental tenants' willingness-to-pay for improved energy efficiency and payback periods for landlords. *Energy Efficiency*, *11*(8), 2033–2056. https://doi.org/10.1007/s12053-018-9668-y

Cullinane, K., & Yang, J. (2022). Evaluating the costs of decarbonizing the shipping industry: A review of the literature. *Journal of Marine Science and Engineering*, *10*(7), 946. https://www.mdpi.com/2077-1312/10/7/946

Dadzie, J., Runeson, G., Ding, G., & Bondinuba, F. K. (2018). Barriers to adoption of sustainable technologies for energy-efficient building upgrade—Semi-structured interviews. *Buildings*, *8*(4), 57.

Daly, D., Carr, C., Daly, M., McGuirk, P., Stanes, E., & Santala, I. (2023). Extending urban energy transitions to the mid-tier: Insights into energy efficiency from the management of HVAC maintenance in 'mid-tier' office buildings. *Energy Policy*, *174*, 113415. https://doi.org/https://doi.org/10.1016/j.enpol.2022.113415

Delmastro, C., Mutani, G., & Corgnati, S. P. (2016). A supporting method for selecting cost-optimal energy retrofit policies for residential buildings at the urban scale. *Energy Policy*, *99*, 42–56. https://doi.org/10.1016/j.enpol.2016.09.051

Department of Climate Change Energy the Environment and Water. (2022). *Quick Wins*. Retrieved 18/09/2022 from https://www.energy.gov.au/households/quick-wins

Dong, Z., Zhao, K., Liu, Y., & Ge, J. (2021). Performance investigation of a net-zero energy building in hot summer and cold winter zone. *Journal of Building Engineering*, *43*, 103192. https://doi.org/https://doi.org/10.1016/j.jobe.2021.103192

Doulos, L. T., & Tsangrassoulis, A. (2022). The future of interior lighting is here. *Sustainability*, *14*(12), 7044. https://www.mdpi.com/2071-1050/14/12/7044

Englehardt, J. D., Wu, T., Bloetscher, F., Deng, Y., Du Pisani, P., Eilert, S., Elmir, S., Guo, T., Jacangelo, J., Lechevallier, M., Leverenz, H., Mancha, E., Plater-Zyberk, E., Sheikh, B., Steinle-Darling, E., & Tchobanoglous, G. (2016). Net-zero water management: Achieving energy-positive municipal water supply. *Environmental Science: Water Research & Technology*, *2*(2), 250–260. https://doi.org/10.1039/c5ew00204d

European Commission. (2021). *European Climate Law*. European Commission. Retrieved October 2021 from https://ec.europa.eu/clima/eu-action/european-green-deal/european-climate-law_en

European Commission Joint Research Centre. (2011). *European LED Quality Charter*. Retrieved 18/09/2022 from https://e3p.jrc.ec.europa.eu/sites/default/files/files/documents/Lighting/eu_led_quality_charter.pdf

Fahlstedt, O., Temeljotov-Salaj, A., Lohne, J., & Bohne, R. A. (2022). Holistic assessment of carbon abatement strategies in building refurbishment literature—A scoping review. *Renewable and Sustainable Energy Reviews*, *167*, 112636. https://doi.org/https://doi.org/10.1016/j.rser.2022.112636

Famiglietti, J., Toppi, T., Bonalumi, D., & Motta, M. (2023). Heat pumps for space heating and domestic hot water production in residential buildings, an environmental comparison in a present and future scenario. *Energy Conversion and Management*, *276*, 116527. https://doi.org/https://doi.org/10.1016/j.enconman.2022.116527

Fankhauser, S., Smith, S. M., Allen, M., Axelsson, K., Hale, T., Hepburn, C., Kendall, J. M., Khosla, R., Lezaun, J., Mitchell-Larson, E., Obersteiner, M., Rajamani, L., Rickaby, R., Seddon, N., & Wetzer, T. (2021). The meaning of net zero and how to get it right. *Nature Climate Change*. https://doi.org/10.1038/s41558-021-01245-w

Ferreira, A., Pinheiro, M. D., de Brito, J., & Mateus, R. (2023). A critical analysis of LEED, BREEAM and DGNB as sustainability assessment methods for retail buildings. *Journal of Building Engineering*, *66*, 105825. https://doi.org/https://doi.org/10.1016/j.jobe.2023.105825

Friedman, C., Becker, N., & Erell, E. (2018). Retrofitting residential building envelopes for energy efficiency: Motivations of individual homeowners in Israel. *Journal of Environmental Planning and Management*, *61*(10), 1805–1827. https://doi.org/10.1080/09640568.2017.1372278

Füchsl, S., Rheude, F., & Röder, H. (2022). Life cycle assessment (LCA) of thermal insulation materials: A critical review. *Cleaner Materials*, *5*, 100119. https://doi.org/https://doi.org/10.1016/j.clema.2022.100119

Galvin, R. (2014). Why German homeowners are reluctant to retrofit. *Building Research & Information*, *42*(4), 398–408. https://doi.org/10.1080/09613218.2014.882738

Giraudet, L.-G., Petronevich, A., & Faucheux, L. (2021). Differentiated green loans. *Energy Policy*, *149*, 1. https://doi.org/http://dx.doi.org/10.1016/j.enpol.2020.111861

Godin, K., Sapinski, J. P., & Dupuis, S. (2021). The transition to net zero energy (NZE) housing: An integrated approach to market, state, and other barriers. *Cleaner and Responsible Consumption*, *3*, 100043. https://doi.org/https://doi.org/10.1016/j.clrc.2021.100043

González-Eguino, M. (2015). Energy poverty: An overview. *Renewable and Sustainable Energy Reviews*, *47*, 377–385. https://doi.org/https://doi.org/10.1016/j.rser.2015.03.013

Gouveia, J. P., & Palma, P. (2019). Harvesting big data from residential building energy performance certificates: Retrofitting and climate change mitigation insights at a regional scale. *Environmental Research Letters*, *14*(9), 23, Article 095007. https://doi.org/10.1088/1748-9326/ab3781

Government of Canada. (2021). *Eligible Grants for My Home Retrofit*. Government of Canada. https://www.nrcan.gc.ca/energy-efficiency/homes/canada-greener-homes-grant/start-your-energy-efficient-retrofits/plan-document-and-complete-your-home-retrofits/eligible-grants-for-my-home-retrofit/23504

Greenough, R., & Tosoratti, P. (2014). Low carbon buildings: A solution to landlord-tenant problems? *Journal of Property Investment & Finance*, *32*(4), 415–423. https://doi.org/10.1108/JPIF-09-2013-0060

Hale, T., Smith, S. M., Black, R., Cullen, K., Fay, B., Lang, J., & Mahmood, S. (2022). Assessing the rapidly-emerging landscape of net zero targets. *Climate Policy*, *22*(1), 18–29. https://doi.org/10.1080/14693062.2021.2013155

Hammerle, M., Best, R., & Crosby, P. (2021). Public acceptance of carbon taxes in Australia. *Energy Economics*, *101*, 105420. https://doi.org/https://doi.org/10.1016/j.eneco.2021.105420

He, R., Jin, J., Qiu, X., Zhang, C., & Yan, J. (2023). Rural residents' climate change perceptions, personal experiences, and purchase intention–behavior gap in energy-saving refrigeration appliances in Southwest China. *Environmental Impact Assessment Review*, *98*, 106967. https://doi.org/https://doi.org/10.1016/j.eiar.2022.106967

Hong, T., Chen, Y., Luo, X., Luo, N., & Lee, S. H. (2020). Ten questions on urban building energy modeling. *Building and Environment*, *168*, 106508. https://doi.org/https://doi.org/10.1016/j.buildenv.2019.106508

Howden-Chapman, P., Viggers, H., Chapman, R., O'Sullivan, K., Barnard, L. T., & Lloyd, B. (2012). Tackling cold housing and fuel poverty in New Zealand: A review of policies, research, and health impacts. *Energy Policy*, *49*, 134–142. https://doi.org/10.1016/j.enpol.2011.09.044

IPCC. (2018). Annex I: Glossary [Matthews, J.B.R. (ed.)]. In: *Global Warming of 1.5°C. An IPCC Special Report on the Impacts of Global Warming of 1.5°C above Pre-Industrial Levels and Related Global Greenhouse Gas Emission Pathways, in the Context of Strengthening the Global Response to the Threat of Climate Change, Sustainable Development, and Efforts to Eradicate Poverty*. Cambridge University Press. https://doi.org/https://doi.org/10.1017/9781009157940.008

IPCC. (2022). *Climate Change 2022: Mitigation of Climate Change*. Contribution of Working Group III to the Sixth Assessment Report of the Intergovernmental Panel on Climate Change. https://www.ipcc.ch/report/sixth-assessment-report-working-group-3/

Isenhour, C., & Feng, K. (2016). Decoupling and displaced emissions: On Swedish consumers, Chinese producers and policy to address the climate impact of consumption. *Journal of Cleaner Production*, *134*, 320–329. https://doi.org/https://doi.org/10.1016/j.jclepro.2014.12.037

Jung, H., & Song, C.-K. (2023). Effects of emission trading scheme (ETS) on change rate of carbon emission. *Scientific Reports*, *13*(1), 912. https://doi.org/10.1038/s41598-023-28154-6

Kholodilin, K. A., Mense, A., & Michelsen, C. (2016). The market value of energy efficiency in buildings and the mode of tenure. *Urban Studies*, *54*(14), 3218–3238. https://doi.org/10.1177/0042098016669464

Kholodilin, K. A., Mense, A., & Michelsen, C. (2017). The market value of energy efficiency in buildings and the mode of tenure. *Urban Studies (Edinburgh, Scotland)*, *54*(14), 3218–3238. https://doi.org/10.1177/0042098016669464

Kilkis, B. (2012). Net-zero energy cities: A hub for decentralized energy for better environment. *International Journal of Energy Research*, *36*(15), 1358–1365. https://doi.org/10.1002/er.1917

Kim, T. H., Lee, S. H., & Vourvachis, P. (2023). Accounting standard-setting for an emission trading scheme: The Korean case. *Journal of Business Ethics*, *182*(4), 1003–1024. https://doi.org/10.1007/s10551-022-05290-z

Kumar, L., Ahmed, S., Naidoo, R., & Bansal, R. C. (2022). Decentralized reinforce-control of renewable dynamic virtual power plant enabling it to offer ancillary services: An attempt towards net-zero targets. *IET Renewable Power Generation*. https://doi.org/https://doi.org/10.1049/rpg2.12545

Lee, J., Kim, J., Song, D., Kim, J., & Jang, C. (2017). Impact of external insulation and internal thermal density upon energy consumption of buildings in a temperate climate with four distinct seasons. *Renewable and Sustainable Energy Reviews*, *75*, 1081–1088. https://doi.org/https://doi.org/10.1016/j.rser.2016.11.087

Lou, Y., Yang, Y., Ye, Y., He, C., & Zuo, W. (2022). The economic impacts of carbon emission trading scheme on building retrofits: A case study with U.S. medium office buildings. *Building and Environment*, *221*, 109311. https://doi.org/https://doi.org/10.1016/j.buildenv.2022.109311

Loveday, J., Morrison, G. M., & Martin, D. A. (2022). Identifying knowledge and process gaps from a systematic literature review of net-zero definitions. *Sustainability*, *14*(5), 3057.

Miller, W., Liu, L. A., Amin, Z., & Gray, M. (2018). Involving occupants in net-zero-energy solar housing retrofits: An Australian sub-tropical case study. *Solar Energy*, *159*, 390–404. https://doi.org/https://doi.org/10.1016/j.solener.2017.10.008

Minunno, R., O'Grady, T., Morrison, G. M., & Gruner, R. L. (2020). Exploring environmental benefits of reuse and recycle practices: A circular economy case study of a modular building. *Resources, Conservation and Recycling*, *160*, 104855. https://doi.org/https://doi.org/10.1016/j.resconrec.2020.104855

Minunno, R., O'Grady, T., Morrison, G. M., & Gruner, R. L. (2021). Investigating the embodied energy and carbon of buildings: A systematic literature review and meta-analysis of life cycle assessments. *Renewable and Sustainable Energy Reviews*, *143*, 110935.

Nature Editorials. (2021). Net-zero carbon pledges must be meaningful to avert climate disaster. *Nature*, *592*(7852), 8–8. https://doi.org/10.1038/d41586-021-00864-9

Ohene, E., Chan, A. P. C., & Darko, A. (2022). Review of global research advances towards net-zero emissions buildings. *Energy and Buildings*, *266*, 112142. https://doi.org/https://doi.org/10.1016/j.enbuild.2022.112142

Oldfield, P. (2022). *Freezing Indoors? That's Because Australian Homes Are Closer to Tents than Insulated Eco-Buildings*. Retrieved 01/09/2022 from https://www.theguardian.com/commentisfree/2022/jun/13/freezing-indoors-thats-because-australian-homes-are-closer-to-tents-than-insulated-eco-buildings

Omrany, H., Soebarto, V., & Ghaffarianhoseini, A. (2022). Rethinking the concept of building energy rating system in Australia: A pathway to life-cycle net-zero energy building design. *Architectural Science Review*, *65*(1), 42–56. https://doi.org/10.1080/00038628.2021.1911783

Oreszczyn, T., Hong, S. H., Ridley, I., & Wilkinson, P. (2006). Determinants of winter indoor temperatures in low income households in England. *Energy and Buildings*, *38*(3), 245–252. https://doi.org/https://doi.org/10.1016/j.enbuild.2005.06.006

Ortiz, J., Casquero-Modrego, N., & Salom, J. (2019). Health and related economic effects of residential energy retrofitting in Spain. *Energy Policy*, *130*, 375–388. https://doi.org/10.1016/j.enpol.2019.04.013

Perčić, M., Vladimir, N., Jovanović, I., & Koričan, M. (2022). Application of fuel cells with zero-carbon fuels in short-sea shipping. *Applied Energy*, *309*, 118463. https://doi.org/https://doi.org/10.1016/j.apenergy.2021.118463

Pineda, A. C., Chang, A., & Faria, P. (2020). *Foundations for Science-Based Net-Zero Target Setting in the Corporate Sector*. (September), 53–53. https://sciencebasedtargets.org/resources/legacy/2020/09/foundations-for-net-zero-full-paper.pdf

Putnam, T., & Brown, D. (2021). Grassroots retrofit: Community governance and residential energy transitions in the United Kingdom. *Energy Research & Social Science*, *78*, 12, Article 102102. https://doi.org/10.1016/j.erss.2021.102102

Risch, A. (2020). Are environmental fiscal incentives effective in inducing energy-saving renovations? An econometric evaluation of the French energy tax credit. *Energy Economics*, *90*, 1. https://doi.org/http://dx.doi.org/10.1016/j.eneco.2020.104831

Risch, A., & Salmon, C. (2017). What matters in residential energy consumption: Evidence from France. *International Journal of Global Energy Issues*, *40*(1–2), 79–116. https://doi.org/https://doi.org/10.1504/IJGEI.2017.080767

Risholt, B., & Berker, T. (2013). Success for energy efficient renovation of dwellings—Learning from private homeowners. *Energy Policy*, *61*, 1022–1030. https://doi.org/https://doi.org/10.1016/j.enpol.2013.06.011

Saboori, H., Mohammadi, M., & Taghe, R. (2011, 25-28 March 2011). Virtual power plant (VPP), definition, concept, components and types. In: *2011 Asia-Pacific Power and Energy Engineering* Conference, Wuhan, China, 2011, pp. 1–4, https://doi.org/10.1109/APPEEC.2011.5749026.

Sanyé Mengual, E., & Sala, S. (2023). Consumption footprint and domestic footprint: Assessing the environmental impacts of EU consumption and production. In: *Life Cycle Assessment to Support the European Green Deal*. Luxembourg: Publications Office of the European Union, https://dx.doi.org/10.2760/218540.

Seo, S., Foliente, G., & Ren, Z. (2018). Energy and GHG reductions considering embodied impacts of retrofitting existing dwelling stock in Greater Melbourne. *Journal of Cleaner Production*, *170*, 1288–1304. https://doi.org/10.1016/j.jclepro.2017.09.206

Shen, P., Braham, W., & Yi, Y. (2019). The feasibility and importance of considering climate change impacts in building retrofit analysis. *Applied Energy*, *233-234*, 254–270. https://doi.org/https://doi.org/10.1016/j.apenergy.2018.10.041

Somasundaram, S., Chong, A., Wei, Z., & Thangavelu, S. R. (2020). Energy saving potential of low-e coating based retrofit double glazing for tropical climate. *Energy and Buildings*, *206*, 109570. https://doi.org/https://doi.org/10.1016/j.enbuild.2019.109570

Statista. (2021). *Investment in Clean Energy Globally in 2019, by Select Country*. Retrieved October 2021 from https://www.statista.com/statistics/799098/global-clean-energy-investment-by-country/

Strengers, Y. (2010). Air-conditioning Australian households: The impact of dynamic peak pricing. *Energy Policy*, *38*(11), 7312–7322. https://doi.org/https://doi.org/10.1016/j.enpol.2010.08.006

Sun, C., Cai, G., Liu, W., Zou, Z., & Huang, C. (2021). Thermal comfort in residences related to respiratory diseases among preschool children in Shanghai. *Energy and Buildings*, *236*, 110729. https://doi.org/10.1016/j.enbuild.2021.110729

Tardy, F., & Lee, B. (2019). Building related energy poverty in developed countries – Past, present, and future from a Canadian perspective. *Energy and Buildings*, *194*, 46–61. https://doi.org/https://doi.org/10.1016/j.enbuild.2019.04.013

Thomas, D. (2021). *Amazon, Ikea and Unilever Pledge Zero-Carbon Shipping by 2040*. Retrieved October 2021, from https://www.bbc.com/news/business-58970877

Wang, B., Deng, N., Liu, X., Sun, Q., & Wang, Z. (2021). Effect of energy efficiency labels on household appliance choice in China: Sustainable consumption or irrational intertemporal choice? *Resources, Conservation and Recycling*, *169*, 105458. https://doi.org/https://doi.org/10.1016/j.resconrec.2021.105458

Wei, Y., Zhu, R., & Tan, L. (2022). Emission trading scheme, technological innovation, and competitiveness: Evidence from China's thermal power enterprises. *Journal of Environmental Management*, *320*, 115874. https://doi.org/https://doi.org/10.1016/j.jenvman.2022.115874

Weiss, J., Dunkelberg, E., & Vogelpohl, T. (2012). Improving policy instruments to better tap into homeowner refurbishment potential: Lessons learned from a case study in Germany. *Energy Policy*, *44*, 406–415. https://doi.org/https://doi.org/10.1016/j.enpol.2012.02.006

Whaley, D. M., O'Leary, T., & Al-Saedi, B. (2017). *Cost Benefit Analysis of Simulated Thermal Energy Improvements Made to Existing Older South Australian Houses*.

Willand, N., Maller, C., & Ridley, I. (2019). Addressing health and equity in residential low carbon transitions – Insights from a pragmatic retrofit evaluation in Australia. *Energy Research & Social Science*, *53*, 68–84. https://doi.org/10.1016/j.erss.2019.02.017

World Green Building Council. (2019). *Bringing Embodied Carbon Upfront*. https://www.worldgbc.org/embodied-carbon

Xing, Y., Hewitt, N., & Griffiths, P. (2011a). Zero carbon buildings refurbishment—A hierarchical pathway. *Renewable and Sustainable Energy Reviews*, *15*(6), 3229–3236. https://doi.org/https://doi.org/10.1016/j.rser.2011.04.020

Xing, Y., Hewitt, N., & Griffiths, P. (2011b). Zero carbon buildings refurbishment—A hierarchical pathway. *Renewable & Sustainable Energy Reviews*, *15*(6), 3229–3236. https://doi.org/10.1016/j.rser.2011.04.020

Yamaguchi, Y., Shoda, Y., Yoshizawa, S., Imai, T., Perwez, U., Shimoda, Y., & Hayashi, Y. (2023). Feasibility assessment of net zero-energy transformation of building stock using integrated synthetic population, building stock, and power distribution network framework. *Applied Energy*, *333*, 120568. https://doi.org/https://doi.org/10.1016/j.apenergy.2022.120568

Zhong, X., Hu, M., Deetman, S., Steubing, B., Lin, H. X., Hernandez, G. A., Harpprecht, C., Zhang, C., Tukker, A., & Behrens, P. (2021). Global greenhouse gas emissions from residential and commercial building materials and mitigation strategies to 2060. *Nature Communications*, *12*(1), 6126. https://doi.org/10.1038/s41467-021-26212-z

Zhuang, Y. (2021). Why are Australian homes so cold? *New York Times*. Retrieved November 11 from https://www.nytimes.com/2021/06/25/world/australia/why-are-australian-homes-so-cold.html

4 Developers' Insights

Building-Integrated Photovoltaic (BIPV) Challenges and Potentials in Malaysia

Kai Chen Goh, Md Asrul Nasid Masrom, Sulzakimin Mohamed, Ta Wee Seow, and Hui Hwang Goh

4.1 Sustainable Development and Renewable Energy Technologies

The term "sustainable development" was first introduced in the Brundtland Report, also known as "Our Common Future," which was published in 1987 by the United Nations World Commission on Environment and Development. The report defined sustainable development as "development that meets the needs of the present without compromising the ability of future generations to meet their own needs." The definition of "Sustainable Development" is the development that meets the requirements of the present without jeopardising future generations' ability to meet their own needs (Farah & Nur, 2012). Therefore, this notion has been steadily modified to emphasise the quality of life that incorporates economic, social, and environmental dimensions (Yakob et al., 1999). The initiatives of Health Building, Green Building Congress, Sustainable Building International Conference and Sustainable City International Conference aim to make Sustainable Development a global development goal, with a focus on "healthiness" and "comfort" to find a balance between "sustainability," "green," and "healthy" environments (Chiang, 2005). According to Karrupannan and Sivam (2005), this phenomenon is a crucial contributor to local community development, local social justice and equality, and local economic prosperity. The projects are deemed effective when sustainability is translated into sustainable housing, in which many are aware of and concerned about the physical design part of a building by integrating or retrofitting as much green technology as possible, i.e. "Green Building" (Yakob et al., 1999).

"Green Construction" and "Green Building" are frequently used interchangeably. The green construction process begins with the early planning stage, from site to design, and continues through construction, operation and maintenance, renovation, and demolition. This green practice takes a more comprehensive approach to the conventional building construction process, enhancing and improving the building's economy, utility, durability, and comfort to achieve sustainable or high-performance green buildings that are more environmentally responsible and resource-efficient, thereby balancing the economic demands in creating a sustainable future (Isnin et al., 2013; Vatalis et al., 2013). In Malaysia, however, the

DOI: 10.1201/9781003317890-5

Green Building Index (GBI) is the overseeing and defining third-party organisation for green building. Their primary objective is to increase resource utilisation efficiency, such as electricity, water, and materials, while lowering the building's impact on human health and the environment through a greener construction process. However, GBI has been focused more on the application of renewable energy (RE) and energy efficiency in green buildings, which many countries have recently addressed (Chua et al., 2013).

The construction industry is a large contributor to global energy consumption and carbon emissions, making sustainable growth a crucial problem for the industry. Traditional building materials and energy sources in the construction industry contribute to environmental degradation, which undermines the industry's long-term viability. Thus, it is vital to support sustainable development in the building industry, with an emphasis on RE sources. RE sources play a crucial role in fostering sustainable development in the building industry. The construction industry is a significant energy consumer, and the use of RE sources such as solar, wind, and geothermal power can aid in reducing the sector's carbon footprint and fostering sustainable growth. According to the International Energy Agency (IEA), the construction industry is one of the most major contributors to climate change, accounting for approximately 39% of global energy-related CO_2 emissions.

In recent years, there has been an increasing tendency in the construction industry towards the usage of RE sources. BIPV systems, for instance, have been considered as a promising technology for fostering sustainable development in the sector. BIPV systems can generate electricity while also serving as windows, roofing, and facades, hence decreasing the demand for conventional building materials and energy sources. Several studies have emphasised the potential for RE sources to foster sustainable development in the building industry. Abanda et al. (2019), for instance, discovered that the utilisation of RE sources in the construction industry can reduce greenhouse gas emissions and support sustainable development. Another study by Akhavan et al. (2020) emphasised the ability of BIPV systems to reduce the environmental effect of buildings and promote sustainable development.

Malaysia's strategic location and physical topography circumstances permit four major sources of RE, namely hydro, solar, wind, and biomass, to be deemed perfect; thus, additional criteria must be applied to determine the most appropriate RE with the greatest capacity (Ahmad & Tahar, 2014). Several types of RE include hydro, geothermal, solar, marine, wind, and biomass (Chou & Ongkowijoyo, 2014). When utilised at total capacity, these kinds of RE have significant ramifications for the energy and construction industries (Table 4.1).

4.1.1 Hydro

The climate and topography of Malaysia are conducive to the abundant development of hydropower. There are approximately 189 named rivers in Malaysia's mountainous region, and the country receives an annual rainfall of 3549 mm, which exceeds the minimum requirement for hydropower generation. However,

Table 4.1 Summary of Malaysian Renewable Energy technologies in comparison

Types	*Hydro*	*Wind*	*Biomass*	*Solar power plant*	*Solar BIPV*
Source	Limited	Limited	Limited	Abundant	Abundant
Cost per kWh	Low	High	High	High	High
CAPEX	Extremely high	Low	High	High	High
OPEX	High	Average	Low	Low	Low
Land requirements	Extremely high	Low	High	High	Low
FiT Category	Inclusive	Not listed	Inclusive	Inclusive	Inclusive

Peninsular Malaysia's hydropower resources have been entirely mined, leaving an estimated 29,000 MW unused in East Malaysia. Considering the plant's lifespan and output, the cost per kilowatt-hour (kWh) generated is relatively cheap (Ahmad & Tahar, 2014). To provide a consistent capacity for hydropower plants, however, dams that hold water reservoirs must be constructed on a substantial quantity of land. This will result in significant social and environmental problems, including the displacement of native populations, soil erosion, loss of agricultural land, and ecosystem disruption (Rosnazri et al., 2012). The construction of this dam requires a high initial CAPEX involving billions of dollars and consistent operation expenditure (OPEX) to ensure the dam is in good condition and not subject to catastrophic failure, as well as mechanical maintenance of the turbines to increase their productivity lifespan (Ahmad & Tahar, 2014).

4.1.2 Wind

Due to the location of Malaysia, which has the 29th-longest coastline in the world at approximately 4675 km, the coastal wind has a low potential, with wind speeds predicted to be between 3.5 and 4.5 m/s. With a wind map, Malaysia offers wind potential in other areas, such as flexibility in equipment assembly, mobility in equipment movement, and relatively low CAPEX, despite wind power's modest potential. However, wind power could disrupt the flight path of migrating birds, cause electromagnetic interference for radio signals in the vicinity of large installations, generate noise from whirling blades, and deface the terrain (Ahmad & Tahar, 2014). In addition, mechanical parts are vulnerable to wear and tear, necessitating replacement to extend their lives, hence increasing OPEX.

4.1.3 Biomass

Biomass is used to generate electricity from organic waste. Due to Malaysia's position as the second-largest palm oil producer in the world, after Indonesia, with about 4.06 million ha of palm oil land in 2009, the biomass potential is estimated to be around 800 MW per year and has the potential to rise to 1300 MW per year. Palm oil waste is perfect for constant energy generation, with 1 ha of land providing 50–70 tonnes of trash; however, the fuel source of palm oil is unclear due to the competing pricing for palm oil waste, which has many applications in other industries. Other fuel sources for biomass include palm oil mill effluent and animal dung, which release biogas. Organic municipal solid waste, Malaysia produces between 17,000 and 28,500 tonnes per day, is an additional fuel source that releases methane. Biomass also necessitates a substantial quantity of land and water and may impact the surrounding ecosystem (Ahmad & Tahar, 2014).

4.1.4 Solar

Solar energy is the most environmentally friendly method of electricity generation. Malaysia's potential is considerable because of its strategic location near the equator, averaging 4.96 kWh/m^2 of annual solar irradiance and a range of 4.8–6.3 kWh/m^2 of monthly solar irradiance, except in December, when the monsoon season is in effect (Azhari et al., 2008). Malaysia's solar potential is reportedly four times the global fossil fuel reserve. However, the availability of materials for solar cells with higher efficiency is limited, and the production of solar cells can represent environmental risks if not properly managed (Oh, 2010). Solar panels are expensive due to their elaborate production, which incurs additional expenditures.

In contrast, the price of solar panels has reduced from MYR 31,410 per kilowatt peak (kWp) in 2005 to MYR 20,439 per kWp in 2009 (Ahmad & Tahar, 2014). The most recent prices announced in 2013 have decreased to an estimated MYR 10,000 per kWp. However, two-thirds of Malaysia's geographical area is tropical forest, so it is not environmentally feasible to construct a large solar power plant. Thus, Building-Integrated Photovoltaic (PV) (BIPV) is the ideal solution for rooftop solar power collection (Ahmad & Tahar, 2014).

4.2 Building-Integrated Photovoltaics (BIPV) Technologies and Development

BIPV is described as integrating the PV system into the building's skin and is considered in the project's conceptual phase before the building is constructed. Moreover, BIPV is one of the existing and mature green technologies that can be implemented as part of the project. Malaysia is the second-largest producer of solar PV modules in the world and has demonstrated expertise in the installation of large BIPV capacities in government and private buildings, including the Centre for Environment, Technology, and Development, the Green Energy Office (GEO) of Pusat Tenaga Malaysia (PTM), and Monash University (Sunway).

Silicon solar panels are the most prevalent solar PV panels since they have undergone extensive study and are now commercially viable. Currently, there are five types of solar PV panels. For example, monocrystalline has the highest efficiency but is the most expensive due to its high-tech lab-grown, sliced, doped, and etched growth process. The second type is polycrystalline, which has a lower price due to a less complicated manufacturing method, but a lower output efficiency. The cheapest is amorphous silicon, which is simple to make and inexpensive but has the lowest efficiency, between 5 and 10% (Foster et al., 2010). Cadmium Telluride

Table 4.2 Types of solar panels (silicon and non-silicon)

Types	*Solar panels (silicon)*			*Solar panels (non-silicon)*	
	Monocrystalline	*Polycrystalline*	*Amorphous silicon*	*Cadmium telluride (CdTe)*	*Copper indium gallium diselenide (CIGS)*
Material	Semiconductor-grade silicon from a single crystal ingot	Semiconductor-grade silicon from various crystal ingot	Amorphous silicon	Screen-printing from Cadmium liquid and Telluride liquid	Screen-printing from Cadmium liquid and Telluride liquid
Lifespan	20–25 years	20 years	10 years	–	–
Status	–	–	–	Under research	Under research
Efficiency	15–20%	13–15%	5–10%	Current: 6%	Current: 8% expected increment: 12–14%

and Copper Indium Gallium Diselenide are still being researched for non-silicon panels since the yield is still poor and the metals required for production are too rare and expensive. However, it is easier to manufacture solar PV panels from materials other than silicon (Markvart, 2000). Table 4.2 will outline both silicon and non-silicon solar panel types.

In addition, the thin film is the deposition of a thin layer of semiconductor onto a substrate, which is often made of glass or ceramic. This layer is constructed of amorphous silicon, CdTe, or CIGS and has a PV potential to generate electricity. It is typically utilised in BIPV facades because to its flexibility and suitability for use on tiles and glass windows. Silicon panels manufacture PV modules because of their durability and better efficiency yield. Due to their maturity and commercial availability, PV modules are extensively utilised in BIPV roofing arrays (Foster et al., 2010).

In the context of RE sources, solar energy is frequently emphasised due to its abundance, accessibility, and comparatively minimal environmental impact in comparison to traditional energy sources such as fossil fuels. In addition, solar energy has the ability to meet a substantial amount of the world's energy requirements in a sustainable and environmentally benign manner.

As a promising technology for fostering sustainable development, BIPV systems have gained considerable interest in the construction industry. BIPV systems can generate electricity while also serving as windows, roofing, and facades, hence decreasing the demand for conventional building materials and energy sources. BIPV systems have the ability to cut carbon emissions and support sustainable development.

While other solar energy technologies, like as solar panels, are available, BIPV systems are frequently favoured in the building industry because to their adaptability and aesthetic appeal. BIPV systems can be incorporated into building designs, making them a more desirable option than conventional solar panels, which are often erected separately from the building structure. Furthermore, it has been shown that BIPV systems have a lesser environmental impact than conventional solar panels, as they use fewer raw materials and less energy to generate.

Several studies have emphasised the capacity of BIPV systems to encourage sustainable development in the building industry. According to a study by Perez-Mesa et al. (2018), the installation of BIPV systems in buildings can cut energy consumption and greenhouse gas emissions while boosting the aesthetic appeal of the building. Wang et al. (2020) conducted a second study that demonstrated the economic viability of BIPV systems in building construction, demonstrating their potential to promote sustainable development.

Due to its abundance, accessibility, and minimal environmental effect, solar energy, particularly in the form of BIPV systems, offers considerable promise for encouraging sustainable development in the building industry. To attain sustainable development objectives, policymakers, industry stakeholders, and researchers should collaborate to encourage the implementation of BIPV systems in the building industry. The most recent push for BIPV was the Iskandar Regional Development Authority's proposal to build a community-based solar power project in the Iskandar Region Economic Corridor in Johor instead of keeping rooftops bare (IRDA). These results are from the property boom in the Iskandar Region brought about by the infusion of foreign investment, particularly from China (Mekhilefet et al., 2012). There is a good chance for developers to maximise their investments in BIPV for residential construction projects. Johor is still a developing economic state with a comparatively flat topography compared to other Malaysian state terrains, lessening concerns about natural shade from nearby mountains or hills. Johor has 706,089 stock houses, the second greatest among the states out of the total 4.7 million stock houses in Malaysia, accounting for around 15% of the total, providing an abundance of rooftops for BIPV applications (JPPH, 2013).

4.3 Government Policy and Subsidy on Building-Integrated Photovoltaics (BIPV)

According to Ekins (2004), no optimal model has developed, and probably none will do so in the contexts created by different histories and cultures. As a result, there are no government policies or subsidies that are ideal for all nations. Various nations will be selected based on their capacity and capability to invest in green technology, specifically BIPV. The United States, Germany, Australia, and Japan, among others, have utilised RE and BIPV as part of a government-created effort in addition to Malaysia.

4.3.1 Other Countries' Government Policy and Subsidy on Building-Integrated Photovoltaics (BIPV)

In the United States, approximately 50 MW of solar systems have been deployed, contributing approximately 8% of the nation's electricity. The United States government has implemented the Renewable Portfolio Standards (RPS), incentives, the Investment Tax Credit (ITC), the Production Tax Credit (PTC), and targets concerning RE. RPS is a framework that helps maximise energy production and RE responsibilities by exchanging RE credits or RE certificates by state utilities (Fayaz et al., 2011). This allows the other state utilities to invest in other states where the output of RE by the current state's utilities is greater.

Individual states have their own RE promotion programmes, although ITC is one of the most important factors in the rise of solar installation (Fayaz et al., 2011). In addition, the 1978 implementation of Public Utility Regulatory Policies (PURPA) compels utilities to purchase RE from authorised independent generators through long-term contracts, as opposed to the Feed-in-Tariff (FiT), which provides a premium for RE provided to the grid (Fayaz et al., 2011). This is because the utilities are state-based and have the option to negotiate contracts with state utilities and other state utilities that seek to invest to meet RPS. It is part of the government's incentive programme and the Production Tax Credit (PTC). In 1992, the United States government introduced the Production Tax Credit (PTC) for small-scale, large-scale wind turbine farms depending on the initial capital cost. For individuals and other entities like residential solar and wind projects, the Investment Tax Credit (ITC) will be available, which will help increase the RE investment by 30% and extend it for three years with the assistance of Congress for the tax extenders bill in their financial bailout plan in 2008, and is expected to generate an annual growth of 6000 MW for solar installations by 2016.

After the Fukushima disaster in Japan, countries with a high technology frontier, such as Germany and Australia, have adopted solar energy and the FiT policy with various incentives and abandoned nuclear energy (Fayaz et al., 2011). Australia began on RE around the same time as Malaysia, in 2001. Australia also implemented the FiT policy. However, it is selectively based on states alone, despite being considered in the RE Target. This action is intended to increase the commercialisation of mandated RE and FiT in the country since solar panels are too expensive. However, the Australian government is eager to promote RE to attain a national required RE target of 20% by 2020. Germany, on the other hand, implements FiT based on avoided costs. It indicates that a fee is based on the price of power displaced. This RE source has excellent priority access to the grid and will be accepted and delivered for up to 20 years, depending on the technology. Due to the attractive and convincing return on investment for RE, Germany is among the developed countries to implement FiT to promote RE sources (Abdullah et al., 2012; Fayaz et al., 2011).

Japan's New Energy and Industrial Technology Development Organisation (NEDO) has also used solar BIPV. They are more of a quasi-government organisation tasked with policy development, research, and design to innovate, solve problems, increase competitiveness, and promote the technology industry. They are Japan's largest organisation for managing research and development. Their focus is mainly on RE and its technology, as well as its application and efficiency, and they aim to support the progress of RE technology by recommending, developing, and demonstrating it (NEDO, 2013). Even Malaysia's neighbour, Singapore, is investigating the possibility of solar energy harnessing through BIPV, which is projected to supply 28% of the current electricity demand and reduce approximately 8% of the CO2 emissions with the 30% reserve margin above the peak that their government has mandated. By 2050, solar energy is anticipated to account for up to 65% of the morning's electricity generation (Wagner et al., 2012).

4.3.2 Malaysian Government Policy and Subsidy on Building-Integrated Photovoltaics (BIPV)

Although diversification was initiated in the 1980s under the National Energy Policy, National Depletion Policy, and the Four Fuel policy, which aimed to shift the reliance on petroleum to other sources for electricity generation, the reliance has been shifted from petroleum to natural gas, which is also a fossil fuel. Since the creation of these policies, more than 30 years have passed since the discovery of new petroleum resources, and the surviving fields are of poor quality, small size, and dispersed distribution. With a 10% annual decline in output and a dispersed distribution network, the outlook for natural gas is bleak. Through the 8th, 9th, and 10th Malaysian Plans, Malaysia shifted its focus to RE in 2000 (Cheng & Lalchand, 2014).

Under the 8th Malaysian Plan, the Four Fuel policy has been extended for a fifth time, resulting in a new Fifth Fuel strategy. This policy's purpose remains to diversify oil and gas dependence into alternative sources, if viable. Even though nuclear energy is not included in this Five-Fuel programme, a specific provision has been made to investigate the possibilities of using nuclear for power generation (Rosnazri et al., 2012). Special Committee on Renewable Energy (SCORE) has established the Small Renewable Energy Program (SREP) to accomplish the goals of the Third Outline Perspective Plan (2001–2010) and the Fifth Fuel strategy, which are to consolidate the development and exploitation of RE (Fayaz et al., 2011). With the founding of the Energy Commission or "Suruhanjaya Tenaga (ST)" under the Energy Commission Act of 2001, the Department of Electricity and Gas Supply was tasked with enhancing the supervision of the energy sector and promoting the efficient use of energy. The body is now responsible for regulating any energy-related issues (EPU, 2001). However, SREP developers must sign a Renewable Energy Electricity Purchase Agreement (REPPA) with Tenaga Nasional Berhad (TNB), the national power provider of Peninsular Malaysia. This resulted from four factors: a significant subsidy on fossil fuels, particularly for electricity generation, which made the RE power costs uncompetitive with fossil fuel energy pricing; high capital expenditures; and high-interest rates (CAPEX).

The 9th Malaysian Plan places a greater emphasis on energy efficiency, in line with Malaysia's sustainable development objective. It explains the approach for tackling environmental and resource management challenges in an integrated and holistic manner to ensure resilient and sustainable development (APERC, 2009). The formation of the Ministry of Energy, Communications and Multimedia to support a green economy that is clean and sustainable bolstered this objective (Fayaz et al., 2011). The establishment of National Green Technology in April 2009, with the goal of Green Technology development and energy efficiency, proves that Malaysia is pursuing sustainable development at a higher level (Islam et al., 2009). In addition, the establishment of the Malaysian Building-Integrated Photovoltaic Technology Application Project (MBIPV) via the United Nations Development Programme (UNDP) launched the BIPV application in Malaysia (Solangi et al., 2011). This project consisted of three categories: BIPV demonstration, the national

"SURIA 1000" programme, and the BIPV showcase, putting into action the establishment and participation of the private and public sectors in the BIPV market and its use in both residential and commercial structures (Mekhilef et al., 2012).

The 10th Malaysian Plan's Renewable Energy Policy and Action Plan indicate the government's continued interest in RE. The target increased to 985MW by 2015, or 5.5% of Malaysia's total power generation mix, from the previous 5%. When the Renewable Energy Act 2011 and Sustainable Energy Development Authority Act 2011 [Act 726] were approved to permanently replace the Special Committee on Renewable Energy (SCORE), the Sustainable Energy Development Authority (SEDA) was established (Fayaz et al., 2011). Under these two Acts, a new mechanism known as FiT has been implemented, allowing RE investors to enjoy a premium tariff payback rate per unit of energy generated when selling to Tenaga Nasional Berhad (TNB).

The Malaysian government has acknowledged the significance of RE, such as BIPV systems, in accomplishing sustainable development objectives. Both the 11th Malaysia Plan (2016–2020) and the 12th Malaysia Plan (2021–2025) underline the need to increase the usage of RE in the construction sector, including BIPV systems.

Under the 11th Malaysia Plan, the government introduced a number of regulations and efforts to encourage the use of RE sources, including BIPV systems. For instance, the government offered homeowners and companies who installed BIPV systems tax exemptions and Feed-in-Tariffs as subsidies and incentives. The government also developed the Green Technology Funding Plan to support enterprises that invested in green technologies, such as BIPV systems.

The government has continued to highlight the importance of RE, especially BIPV systems, in the building sector in the 12th Malaysia Plan. The strategy aims to increase the nation's RE capacity to 20% by 2025, with an emphasis on expanding solar energy use in the building sector. In addition, the government has launched the MySuria programme to stimulate the installation of solar panels, including BIPV systems, on public buildings such as schools, hospitals, and government structures.

In addition, the government of Malaysia has established the Sustainable Energy Development Authority (SEDA) to promote the growth of RE sources, such as BIPV systems. SEDA provides a number of financial incentives and programmes to promote the use of RE, such as the Net Energy Metering scheme, which allows households and businesses to sell excess solar energy back to the grid.

In conclusion, the Malaysian government has developed numerous laws and measures to encourage the usage of RE sources, such as BIPV systems, in the building industry. These policies and activities are crucial for accomplishing sustainable development objectives and decreasing the nation's reliance on fossil fuels.

4.4 Application of Building-Integrated Photovoltaics (BIPV) in Malaysia

In contrast to Australia, Germany, and Japan, Malaysia has implemented BIPV in individual, small-scale homes that get premium FiT rates from the government, as illustrated in the diagram below. Figure 4.1 illustrates the current Malaysia FiT

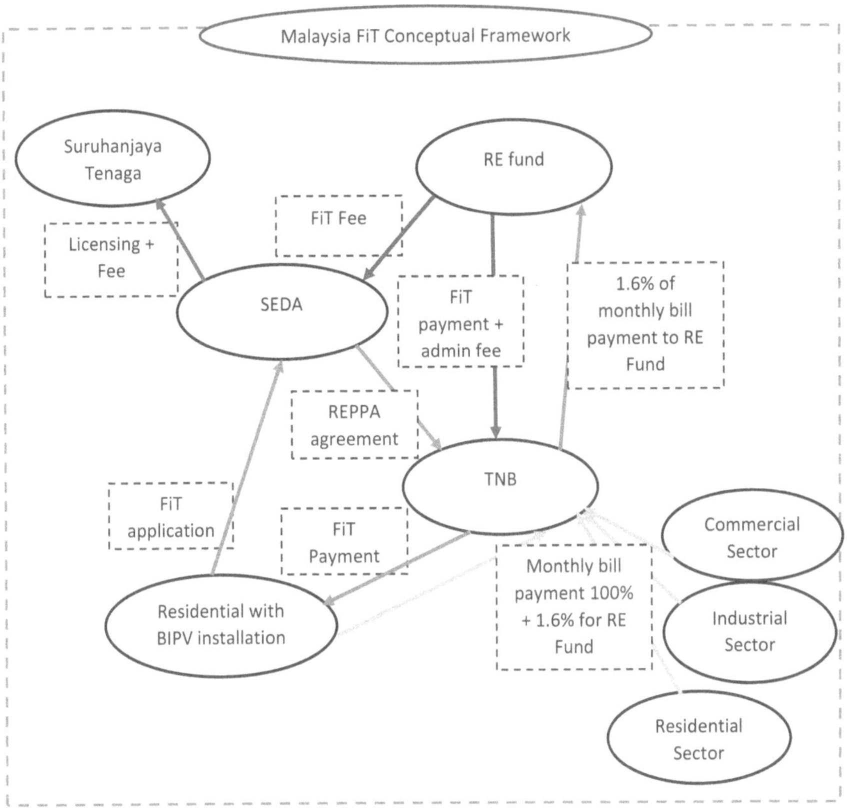

Figure 4.1 Malaysia FiT Conceptual Framework

Legend: SEDA – Sustainable Energy Development Authority
RE fund – Renewable Energy Fund
TNB – Tenaga National Berhad
REPPA – Renewable Energy Power Purchase Agreement
FiT – Feed-In-Tariff

Conceptual Framework, as Malaysia has not yet implemented a permanent, large-scale BIPV housing scheme. The Malaysian government has implemented this FiT Conceptual Framework to permit the installation of BIPV as an individual project. SEDA, the governing agency in charge of RE, manages all FiT scheme applications for BIPV. The application must then be authorised by Suruhanjaya Tenaga, the governing body responsible for energy-related matters under the government. Upon approval, SEDA must individually sign a "Renewable Energy Power Purchase Agreement (REPPA)" with Tenaga Nasional Berhad (TNB), which allows TNB to directly pay out the FiT premium to individual residents using the RE fund. Both residential and commercial sectors use monthly electricity bills for usage above 300 kWh per month (Fayaz et al., 2011). However, the approval procedure is lengthy due to the extensive documentation required, the quota cap on BIPV quotas, and the non-economic FiT incentives that can be profitable for the applicant, particularly developers (Solangi et al., 2011).

4.5 Challenges of Building-Integrated Photovoltaics (BIPV) Adoption

The deployment of BIPV is limited to single-family homes and has not been used in large-scale housing developments. The adoption preparedness, lack of marketing and incentive, and limitations in competence are the obstacles that prevent developers from integrating BIPV on a large scale in their applications. Barriers are the obstacles that impede BIPV deployment in the housing business that developers are currently facing. This is a necessary stage so that a solution may be devised to overcome situation to minimise its impact on the implementation.

4.5.1 Specialised Knowledge and Certification: Overcoming Obstacles to BIPV Implementation in the Construction Sector

In all professions, including the construction sector, specialised knowledge is indispensable. According to research, developers concur that limited experience is one of the obstacles to BIPV implementation in a project. Without expertise, there is no standard of excellence for the work performed. Developers agree that this obstacle may impede the spread of BIPV. This also demonstrates that a lack of certified and specialised BIPV contractors in Malaysia is one of the obstacles, as very few contractors in Malaysia are certified and specialise in BIPV. According to the study, the good response from developers indicates a promising outcome if the BIPV deployment were to reach full size in Iskandar, given their desire and openness to change and their ability to source locally and internationally, if necessary. This obstacle is considered from an economic perspective as a cost is associated with obtaining local and international expertise, although it is not as significant as the availability of specialised labour for sourcing.

4.5.2 The Need for Government and Developer Promotion of BIPV Implementation in Malaysia: A Two-Stage Challenge

This lack of promotion occurs in two stages: from the government to developers and from developers to homebuyers. One of the reasons for the delayed BIPV implementation in Malaysia since its inception is that the government is not currently promoting the technology to developers. This demonstrates the need for promotion in motivating BIPV implementation developers. According to Cheng, et al., BIPVs have been in Malaysia for approximately ten years, since their inception under the 10th Malaysian Plan (2014). Still, it is deficient in numerous aspects, making it an undesirable option among developers in the housing market. Xue et al. (2021) did emphasise a lack of cooperation and information flow between the commercial sector and the government. However, this lack of information flow and cooperation is attributable more to the government's reluctance to implement a private sector-wide promotion campaign for BIPV than to the developers' perception of this as an obstacle.

An obstacle preventing the complete application of BIPV in housing projects is the absence of advertising for homebuyers on the part of developers. According to research, some developers have attempted to promote BIPV to homebuyers independently, but it has been case-by-case and at the homebuyers' request. Using a retrofitted BIPV system, BIPV is typically an ad hoc addition to the structure. Aside from this, there are currently no developers who incorporate BIPV into a dwelling project. Some who attempted to promote BIPV ultimately opted to abandon their efforts owing to the failure of their efforts to promote BIPV for various reasons, including the difficulty to convey accurate information to house purchasers, among others effectively.

4.5.3 The Impact of Rising Global Energy Costs on the Prospects of BIPV Adoption in Malaysia

Subsidised power is one of the primary obstacles to the widespread adoption of BIPV. The conventional input energy source is typically subsidised and exempt from taxation to generate inexpensive electrical output. Currently, electricity in Malaysia is relatively inexpensive. This is because raw materials such as coal and natural gas are the usual fuel source for electricity generation. According to Cheng et al., this fuel receives substantial government subsidies (2014).

This subsidy drastically affects the competitiveness of green electricity, particularly BIPV, making it undesirable to both homebuyers and developers, despite the fact that BIPV has greater potential than conventional electricity and taps into an endless supply, as stated by Oh (2010). As a result, it becomes more difficult for developers to convince homebuyers to install BIPV in their homes, posing a barrier to the widespread application of BIPV. In addition, energy in Malaysia is tax-free for the user, thus cutting the price. However things has change since 2022, many causes, including the ongoing geopolitical war, a rapid global post-pandemic economic recovery, continued high reliance on fossil fuels, and a severe imbalance between energy demand and supply, have contributed to the rise in global energy costs (Grebski & Kuzior, 2022). The current crisis could expedite the deployment of cleaner, more sustainable RE sources, such as solar application in Malaysia due to their enormous potential.

4.5.4 Developers' Preparedness and Selective Adoption of BIPV

Adoption is a technique for reproducing an existing procedure in a new project, in this example, by developers attempting to apply BIPV in their housing development. The current status of developers is not prepared to incorporate BIPV in their housing developments, which can be a significant disincentive if BIPV is implemented fully. Current practice in Malaysia is to use FiT for BIPV, in which the utility company purchases the electricity at a premium price, but exclusively between the government and the homebuyers; the developers are not involved. SEDA established this following the Malaysia Fit Conceptual Framework (2011).

Nevertheless, the separation of the developers from the conceptual framework has consequences. The majority of developers are unprepared for the implementation of BIPV in their home projects. It impedes BIPV deployment, necessitating correct guidelines and forethought to cut costs and properly integrate BIPV in housing projects. Selective preparation is implemented on a case-by-case basis or at the customer's request. It is typically for a single property or a project segment when the house purchasers desire it, and the developers must provide a superior level of service. However, several businesses have been preparing to use BIPV as necessary. They are preparing selectively to install BIPV in their housing complexes partially. The result of the preparations may have been the same for both companies, namely the partial adoption of BIPV to make BIPV a particular feature, but the motivations behind the preparations that drive such anticipation may differ.

4.5.5 The Dependability of Technology

Technology might also be a barrier to BIPV implementation; in this situation, technology dependability. The purpose of technology has always been advancement; however, there are still concerns regarding the dependability of technology. Despite claims that BIPV's lack of mechanical components makes it simple to maintain and highly dependable, the developers disagree. The primary components of a BIPV system are the cabling, the panels, and the inverter, which converts direct current (DC) to alternating current (AC), the current residential voltage standard. Foster et al. (2010) suggest that the average lifespan of BIPV panels is 20 years, guaranteed by most BIPV producers. Cables, on the other hand, present no difficulty because standard cables are readily accessible on the market and are routinely used for electrical distribution within homes.

Consequently, the next suspect component is the DC/AC inverters. A malfunctioning inverter renders BIPV panels ineffective since electricity cannot flow through. If BIPV is used in housing development, this has financial and maintenance repercussions for the developers. Therefore, this is a hurdle to the long-term application of BIPV.

4.5.6 Financial Obstacles Faced by Developers

When adopting BIPV in their housing project, developers must overcome financial obstacles. Finance is crucial to any endeavour and is the project's lifeblood. There are numerous areas of finance to consider, but in this instance, there are only two: the "Bank" and the "Green material cost." Finances frequently impact a project and might be an obstacle. One of the things that can affect a project is banks. However, most projects require a bank or financial institution to act as a guarantee, and implementing BIPV might be an additional burden for developers working on large-scale housing projects. In this situation, banks play a crucial role. There are substantial ramifications when it is the widespread practice in the housing business for a financial institution to serve as a project's guarantee. When it comes to home loan procedures and regulations, the financing bank can assist developers with paperwork, which is a choice many developers make. In addition, the interest rates charged by the financing bank can be a significant barrier to installing BIPV in the housing project.

Green technologies, such as BIPV, have high starting costs, discouraging investment. The expense of green materials, in this case, the price of the BIPV system, is one of the financial obstacles developers confront when implementing BIPV. The BIPV materials are expensive for developers to acquire. This is due to the poor yield with efficiency ranging from 5 to 20%, as stated by Foster et al. (2010), and the high production cost, which makes BIPV expensive, despite Malaysia being the second-largest producer of solar PV modules, as stated by Kamaruzzaman et al. (2012). Cost of BIPV materials is a concern for developers because it is one factor determining the viability of BIPV deployment throughout the entire housing project. Although the price of BIPV has decreased over the years, as predicted by Ahmad and Tahar (2014), the competitiveness of BIPV is still insufficient for the homeowner to install BIPV, let alone a BIPV-implemented housing project.

4.5.7 Contribution

Section 4.5 discusses the challenges of widespread adoption of BIPV in Malaysia. The section identifies four main obstacles that limit the use of BIPV in housing projects. Firstly, there is a lack of expertise among developers to implement BIPV projects. Secondly, there is a lack of promotion of BIPV technology to developers and homebuyers. Thirdly, subsidised electricity prices, which make conventional energy cheaper, negatively affect the competitiveness of BIPV. Lastly, there is a lack of readiness among developers to adopt BIPV, which can be a significant disincentive if BIPV is implemented fully. The chapter concludes that overcoming these obstacles is crucial to enable the widespread adoption of BIPV technology in Malaysia.

4.6 Developers Perspective: Building-Integrated Photovoltaics (BIPV) Selection Process Criteria

Before the actual selection, the selection criteria must be established. The selection determines whether or not the option meets its criteria by determining the specific choice's efficacy. In this instance, the developers have never completed a housing project with a comprehensive BIPV deployment. Before deciding whether or not to invest in BIPV, the developer would consider six parameters.

4.6.1 Feasibility Analysis for BIPV Implementation: The Significance of Locality and Future Surrounding Developments

Every project commences with feasibility analysis. It examines whether the project is feasible for execution and implementation, in this case, the BIPV implementation for the housing industry. This level investigates two subthemes deemed significant by BIPV's developers: "Locality" and "Future surrounding developments." According to developers, location is crucial when selecting BIPV for housing development. Location is essential since it influences housing development. A house's marketability can be aided by its location, and the incorporation of BIPV into the home adds significant value to the concept of eco-housing complexes. The future growth of the area's environs is a second factor similar to the neighbourhood. A BIPV housing project is anticipated to last at least 20 years to optimise the lifespan of the BIPV panel to achieve a return on investment as predicted under the FiT rate. With such a high CAPEX, developers must be able to justify the usage of BIPV in a housing project and promote it to homebuyers.

BIPV requires direct sun. Therefore, future growth of the surrounding area is necessary to ensure that the BIPV's lifecycle cost is profitable. To initiate the BIPV project, it is necessary to conduct a feasibility study to determine the viability of BIPV in the area; this is the first criterion. Several concerns must be addressed within the scope of feasibility studies. One of them is the location or neighbourhood of the housing development. Shading or shadows cast by neighbouring tall buildings or structures prevent the housing industry from maximising the potential of BIPV. The issue caused by proximity is shadowing. The BIPV capacity can be reduced by natural settings, adjacent tall buildings, or smog released by nearby industrial sites that cast shadows. Therefore, prospective shading from higher buildings or industrial expansion that could produce smog in the future diminishes the return on investment and renders the BIPV system ineffective.

4.6.2 Factors to Consider in Designing the Architecture for BIPV Implementation in Residential Structures

Designing the architecture is the next step in establishing the viability of BIPV implementation to establish whether or not the home is suitable for BIPV installation. This theme contains four subthemes, namely "House design," "Roof capacity," "Roof size," and "BIPV design." The architecture of a home is essential since it impacts the BIPV's efficiency and capability. The second criterion that developers take into account is architecture. The design must optimise the BIPV's capacity to generate electricity. There are several factors to consider when designing architecture. Home design is the most fundamental element. Because the house's orientation and the roof's design impact the BIPV capacity, the developers consider house design to be significant. As the BIPV requires maximum sunshine to generate the most electricity, the house must have the optimal orientation.

The design of the house's roof takes into account the maximum load its structure can support. When constructing the roof, it is crucial to use BIPV as one of the design elements. In addition, the roof load-bearing capacity must be considered to implement the BIPV. The developers emphasised the significance of roof capacity because the roof must support the BIPV system's weight to prevent interior roof collapse accidents caused by the weight of the BIPV installation. The design of the house's roof takes into account the maximum load its structure can support. When constructing the roof, it is crucial to use BIPV as one of the design elements. Moreover, the size of the roof affects its weight-bearing capacity. Consequently, the more the roof size, the greater the reinforcements required to support the load.

Similarly, architecture design is considered in BIPV design alongside roof capacity and size requirements. Only two types of BIPV designs can be installed in residential structures: integrated and retrofitted. Most developers choose to incorporate it as part of the roof, saving money on roof tiles and expediting roof installation rather than duplicate work on the roof. However, there are issues regarding the quality of the specialist's handiwork to produce a watertight seal in the gaps between the panels, which has not yet been proven in Malaysia's tropical climate, unlike the solar water heater is well-established. Alternately, some developers believe that retrofitting may be a better option; nevertheless, additional costs are incurred when the roof tiles are installed as a second layer after the BIPV panels, thereby doubling the work on the roof and making it less efficient.

4.6.3 The Importance of Budgeting for BIPV

The expense of the project must be given specific consideration. The project cost dictates the type of BIPV and the financial sustainability of BIPV in home development. The initial consideration while deploying BIPV is that the cost of the BIPV system impacts the viability of the overall development project to put BIPV inside the homes. After architecture design, the project cost is the following criterion to consider. It determines whether the design is viable for developers to keep the required profit margin and pay off the bank loan with interest, assuming the developers borrow from the bank measures the house's marketability among homebuyers, indicating whether or not it is affordable and straightforward to sell. Therefore, the first expense to consider is the BIPV cost. The cost of the BIPV system to be purchased provides a rough estimate of the house's selling price. There are further aspects that can assist developers financially, such as bulk savings from large purchases and green material subsidies, if available.

The budget for a private housing project is vital, particularly for smaller or medium-sized developer firms. It was stated by one respondent as a justification for BIPV within the housing development. Aside from that, the housing project budget provides a more precise representation of the total cost of the project. In addition to the cost of materials, the budget also includes labour and overhead expenses. Despite the fact that Ahmad and Tahar (2014) have predicted that the price of BIPV will decrease over the years, developers obtain the BIPV system at a fraction of the amount paid by homebuyers.

4.6.4 Property Price and BIPV: Evaluating Affordability, Profit Margin, and Marketability

Property price is a significant criterion since it defines the marginal profit of the houses as well as the affordability of the houses for different segments of homebuyers. This property price was determined after estimating the project cost, budget, and BIPV pricing aids in identifying the properties' marketability. The cost of the property is a crucial factor in selecting a BIPV determines the property's affordability and the housing project's profit margin. However, BIPV can be a surplus if all homeowners are provided with a complete loan to cover the expense and also subscribe to premium FiT rates that can aid in the repayment of their mortgage. In this instance, BIPV enhances the home's aesthetic appeal and reduces worries regarding the increased cost of the home.

4.6.5 Selection Criteria for BIPV Installation: Importance of Tender Specification, Subcontractor Nomination, and Specialist Experience

An essential component of the selection process is the invitation for bids allows contractors to submit bids to provide the project's materials and labour. In this instance, BIPV is included in the selection process for the tender. This theme contains three subthemes: "Tender Specification," "Subcontractor Nomination," and "Experience (Specialist)." The bidding process is vital to the construction industry. To facilitate the construction process, tendering is used to choose who will construct what. The practice in the business is to choose specification-based tendering since it does not restrict contractors' ability to gauge their profit against price fluctuations in sourcing goods and labour.

The practice of nomination is very prevalent in the building business. Nomination is direct engagement of a specialised subcontractor to execute the installation of BIPV. This is a significant consideration when it serves as a substitute for tender specifications. On the other side, the nomination is a selection based on the reputation of the contracting company, but it is not as profitable for the primary contractor because pricing is typically non-negotiable. There is some evidence that nomination is utilised in BIPV since it is a specialised job for which it is uncommon to acquire materials and labour from the market. Regardless of the use of nomination or tender specification, the experience of a BIPV specialist is essential as selection criteria for either the nominated or the primary contractor.

Any construction project, including BIPV, requires a bid. Though it can be awarded directly to a BIPV specialist or subcontractor, tendering ensures that a competitive price can be offered to build the house and install the BIPV system that has been incorporated into the house. The two types of tender procedures are "tender specification" and "nomination of the subcontractor." Most developers choose tender specification since it is more convenient to establish a lump sum bill of quantities for the project being bid out instead of going by component, saving expenses and boosting the project's price competitiveness. However, some developers may consider the nomination of a subcontractor if the bidder for the prime contracting position cannot deliver the project's BIPV element. To ensure the highest level of quality and safety when the BIPV is installed and operational, the developers account for the expertise of BIPV experts.

4.6.6 The Importance of Maintenance Considerations for BIPV Systems: Reliability, Lifespan, and Cleaning

Maintenance is an essential consideration during the selection procedure. A BIPV system that requires minimum maintenance may be a significant consideration in its selection. The optimal system requires low maintenance, as regular maintenance is rather pricey. The manufacturer guarantees that BIPV panels have a long lifespan; however, some components require maintenance. Though Mekhilefet et al. (2012) remark that BIPV does not have any mechanical components that are susceptible to wear and tear, there are worries about the inverters' reliability and the cables' susceptibility to insect damage. In addition to covering the part for wear and tear, maintenance includes cleaning the BIPV panels to maximise sunshine exposure. It requires a second set of specialists in Malaysia to maintain the BIPV system, a field in which there is a severe dearth of competence. The competent cleaning of BIPV systems has not yet been tested, and researchers and the industry must fill this hole.

4.6.7 Contribution

Section 4.6 discusses the selection criteria for developers who want to incorporate BIPV into housing projects. The first criterion is feasibility, which involves examining the project's viability, locality, and future surrounding developments. The second criterion is architecture design, which considers house design, roof capacity, roof size, and BIPV design. Project cost is the third criterion and is crucial for developers to determine whether the design is financially sustainable. The fourth criterion is system efficiency, which measures the BIPV's capability to generate electricity. The fifth criterion is safety, which examines the safety aspects of the BIPV installation process and system operation. Lastly, maintenance is the sixth criterion and considers the BIPV's durability and maintenance requirements to ensure its longevity. Developers must consider these six criteria before deciding whether to invest in BIPV to ensure that the BIPV system is viable and profitable.

4.7 Conclusion

In conclusion, sustainable development is a critical challenge in the developing world, and the construction and energy sectors have been identified as primary areas for achieving sustainable growth. Malaysia has invested heavily in research and development for RE, particularly in the implementation of BIPV systems. Despite limited implementation due to developers' focus on short-term earnings, the future of BIPV in Malaysia is promising due to recent economic investments. This study aims to involve housing developers in driving the adoption of BIPV in Malaysia's housing construction sector. The study's findings indicate that developers are generally receptive to the adoption of BIPV, highlighting the role of developers in driving the implementation of sustainable technologies in the construction industry. The study also identifies the barriers and opportunities for BIPV adoption in Malaysia, which can inform policy and decision-making processes related to sustainable development in the country.

References

Abanda, F.H., Egondo-Ndeh, E., Tening, A.S., Tambe, M.A. 2019. Sustainable development and greenhouse gas emissions reduction in the construction sector: The role of renewable energy. Renewable and Sustainable Energy Reviews, 105, 238–247.

Abdullah, A.S., Abdullah, M.P., Hassan, M.Y., Hussin, F. 2012. Renewable Energy cost-benefit analysis under Malaysian Feed-in-Tariff, *IEEE Student Conference on Research and Development*. pp. 160–165.

Ahmad, S., Tahar, R.M., 2014. Selection of renewable energy sources for sustainable development of electricity generation system using analytic hierarchy process: A case of Malaysia. Renewable Energy 47, 458–466. Retrieved January 03, 2014, from http://www.sciencedirect.com/science/article/pii/S0960148113005314

Akhavan, M., Golzari, A., Omidvar, A., Jamali, A. 2020. The potential of building-integrated photovoltaic systems in reducing environmental impacts of buildings. Journal of Cleaner Production, 255, 120294.

Asia Pacific Energy Research Center (APERC). 2009. APEC Energy Demand and Supply Outlook: 4th Edition (Volume 1&2). Retrieved December 01, 2013, from http://publications.apec.org/publication-detail.php?pub_id=946

Azhari, A.W., Sopian, K., Zaharim, A., Ghoul, M.A. 2008. A new approach for predicting solar radiation in tropical environment using satellite images: Case study in Malaysia. WSEAS Transactions on Environment and Development, 4(4), 373–378.

Cheng, K.S., Lalchand, G., 2014. A review on sustainable power generation in Malaysia to 2030, historical perspective, current assessment and future strategies. Renewable and Sustainable Energy Reviews, 29, 952–960. Retrieved January 03, 2014, from http://www.sciencedirect.com/science/article/pii/S1364032113005509

Chiang, C.M. 2005. Earth sustainability & human health concept design. *International Conference on Sustainable Culture and Creation*.

Chou, J.S., Ongkowijoyo, C.S. 2014. Risk-based group decision making regarding renewable energy schemes using a stochastic graphical matrix model. Journal of Automation in Construction, 37, 98–109.

Chua, F.Y., Yaacob, N.M., Hussein, H. 2013 Achieving Sustainable Development: Accessibility of green buildings in Malaysia, *AMER International Conference on Quality of Life 2013*, pp 120–129.

Economy Planning Unit (EPU) of Malaysia. 2001. *Rancangan Malaysia Ke-8*. Retrived May 1, 2012 from http://www.pmo.gov.my/dokumenattached/RMK/RM8.pdf

Ekins, P., 2004. Steps for decarbonising the energy system: Research needs for renewable, energy efficiency and nuclear power. Energy Policy, 32, 1891–1940. Retrieved January 03, 2014, from http://www.sciencedirect.com/science/article/pii/S0301421504000692

Farah, I., Nur Emma, M. 2012. Sustainable Housing Development: The Way Forward for Hillside Areas. *Proceeding of 3rd International Conference on Business and Economic Research (ICBER 2012)* pp272–298.

Fayaz, H., Rahim, N.A., Saidur, R., Solangi, K.H., Naiz, H., Hossain, M.S. 2011. Solar Energy Policy: Malaysia vs Developed Countries. *2011 IEEE Conference on Clean Energy and Technology CET Retrieved March 12, 2012* from http://ieeexplore.ieee.org/xpl/freeabs_all.jsp?arnumber=6041512

Foster, R., Ghassemi, M., Cota, A. 2010. *Solar Energy: Renewable Energy and the Environment*. United States, CRC Press.

Grebski, W., Kuzior, A. 2022. Opportunities and challenges in addressing the global energy crisis. Rynek Energii.

Islam, M.R., Saidur, R., Rahim, N.A., Solangi, K.H. 2009. Renewable energy in Malaysia. Engineering e-Transaction, 4(2), 69–72, Retrieved on February 27, 2012, from http://ejum.fsktm.um.edu.my

Isnin, Z., Ahmad, S.A., Yahya, Z. 2013 Challenges of the Unknown Building Material Substances for Greener Adaptation Projects, *ASIA Pacific*.

Jabatan Pernilaian dan Perkhidmatan Harta (JPPH). 2013. *NAPIC Housing Stock* Retrieved January 03, 2014, from http://www.napic.jpph.gov.my/portal

Kamaruzzaman, S.N., Abdul-Rahman, H., Wang, C., Karim, S.B., Lee, T.Y., 2012. Solar technology and building implementation in Malaysia: A national paradigm shift. Maejo International Journal of Science and Technology, Energy Procedia, 6(02), 196–215.

Retrieved January 10, 2014, from http://www.academia.edu/2714220/Solar_technology_and_building_impleme%20ntation_in_Malaysia_A_national_paradigm_shift
Karrupannan, S., Sivam, A. 2005. Sustainable Development and Housing Affordability. pp 1–7.
Markvart, T., 2000. *Solar Electricity*: Second Edition. England, John Wiley & Sons Ltd.
Mekhilef, S., Safari, A., Mustaffa, W.E.S., Saidur, R., Omar, R., Younis, M.A.A., 2012. Solar energy in Malaysia: Current state and prospects. Renewable and Sustainable Energy Reviews, 16, 386–396. Retrieved January 05, 2014, from http://www.sciencedirect.com/science/article/pii/S1364032111004138
New Energy and Industrial Technology Development Organisation (NEDO). 2013. *NEDO Project* Retrieved January 03, 2014, from http://www.nedo.go.jp/english
Oh, T.H., 2010. Carbon capture and storage in potential coal-fired plant in Malaysia. Renewable and Sustainable Energy Reviews, 14, 2697–2709. Retrieved January 03, 2014, from http://www.sciencedirect.com/science/article/pii/S1364032110001504
Perez-Mesa, J.C., Navarro-Esbrí, J., Molina-García, A. 2018. Building-integrated photovoltaics: A review of the state of the art. Renewable and Sustainable Energy Reviews, 82, 3585–3601.
Rosnazri, A., Ismail, D., Taib, S., 2012. A review on existing and future energy sources for electrical power generation in Malaysia. Renewable and Sustainable Energy Reviews, 16, 4047–4055. Retrieved January 04, 2014, from http://www.sciencedirect.com/science/article/pii/S1364032112001839
Solangi, K.H., Lwin, T.N.W., Rahim, N.A., Hossain, M.S., Saidur, R., Fayaz, H. 2011. Development of Solar Energy and Present Policies in Malaysia, in *IEEE Conference on Clean Energy and Technology CET*. pp. 115–120.
Vatalis, K.I., Manoliadis, O., Charalampides, G., Platias, S., Savvidis, S. 2013. Sustainability components affecting decisions for green building projects. *International Conference on Applied Economics (ICOAE) 2013*, pp 747–756.
Wagner, M., Schönsteiner, K., Hamacher, T., 2012. Impacts of photovoltaics and electromobility on the Singaporean energy sector. PV Asia Pacific Conference, Energy Procedia, 25, 126–134. Retrieved January 04, 2014, from http://www.sciencedirect.com/science/article/pii/S1876610212011794
Wang, Y., Liu, Z., Li, X., Wei, Q., Li, C. 2020. Economic feasibility of building-integrated photovoltaic systems in building construction. Journal of Cleaner Production, 250, 119525.
Xue, Y., Lindkvist, C.M., Temeljotov-Salaj, A. 2021. Barriers and potential solutions to the diffusion of solar photovoltaics from the public-private-people partnership perspective—Case study of Norway. Renewable and Sustainable Energy Reviews, 137, 110636.
Yakob, H., Yusof, F., Hamdan, H. 1999. Land Use Regulations towards a Sustainable Urban Housing: Klang Valley conurbation. *ASIA Pacific International Conference on Environment-Behaviour Studies*.

Part II

Planning, Design, Construction, Operation and Demolition of Sustainable Built Environment

5 Carbon Source/Sink Scenarios in Passive Buildings

Bao-Jie He, Xin Dong, and Mingqiang Yin

5.1 Introduction

Human activities and greenhouse gas (GHG) emissions are accelerating global climate change. Compared to pre-industrial times, the average global temperature (January–October) increased by approximately 1.1°C in 2019, leading to frequent natural hazards such as heatwaves, droughts, floods, heavy precipitation, and wildfires. According to the current warming rates, global warming is projected to be about 1.5°C by 2030–2040, at which time natural hazards will exceed levels that the environment and society can withstand (UNEP, 2022).

Carbon neutrality is a broad and systematic response to climate change. Carbon neutrality is a significant conceptual shift from energy conservation and efficiency toward comprehensive carbon reduction and decarbonization. Carbon neutrality is also a common goal of all sectors of society, such as buildings, industry, transport, electricity, agriculture, and forestry. Moreover, carbon neutrality necessitates the reduction of carbon emissions or the attainment of "zero carbon" given the high carbon intensity in the current era.

The building and construction sector has high energy and carbon emissions intensities. In 2019, the building sector accounted for 35% and 38% of global energy use and carbon emissions (IPCC, 2020), respectively. Accordingly, the building and construction sector becomes a focus for carbon neutrality. Previous actions on energy-saving have resulted in a significant reduction in building energy use and carbon emission intensity. However, the total building energy use and carbon emissions are still rising along with urbanization and population growth. While promoting the realization of the global "carbon neutrality" goal, the global demand for new buildings is 2.30×10^{10} m^2, and the average annual CO_2 emissions of building materials will also increase from 3.5 Gt in 2020 to 4.6 Gt in 2060 (RICS, 2014). Furthermore, both global and local warming lead to an increase in cooling electricity. Global warming will lead to an increase in average annual CO_2 emissions for heating/cooling in residential buildings, from 0.8 Gt in 2000 to 2.2 Gt in 2100 (Isaac and Van Vuuren, 2009). In addition, many buildings approach the end of their lifespan around 2060, and building demolition, renovation, or repair will also increase building material use and carbon emissions.

DOI: 10.1201/9781003317890-7

The key to the carbon neutrality goal is to radically reduce carbon emissions from the building sector by upgrading decarbonization technologies and strengthening policy regulations. Carbon emissions from buildings consist of operational carbon (OC) and embodied carbon (EC), while carbon sinks (CSs) play an important role in decarbonization. Reducing OC and EC and enhancing CS at the building stages of design, construction, and use are important ways to decarbonize the built environment. Reducing energy use, developing renewable energy grids, electrifying operations, developing low-carbon fuels, and implementing carbon capture and sequestration should be the five key strategies for systematically regulating OC, EC, and CS. However, in practice, industries still mainly focus on OC reduction, neglecting the EC and CS from production, transport, construction, maintenance, and other stages across the whole building life cycle.

Passive design is an important solution toward built environment decarbonization. On the one hand, passive buildings have good climate adaptability and can make proper use of regional resources and energy advantages to create good indoor physical environments while reducing or even eliminating OC. On the other hand, some passive design strategies, such as green roofs and fountains, can regulate the microclimate, and the implementation of passive strategies on a large scale will enhance the localized climate quality and reduce the energy load of buildings (He, 2019). However, passive technologies lack focus on EC and CS. For example, passive buildings require a high-performance envelope, which possibly leads to an increase in EC and potentially increases the whole life cycle carbon emissions of the building (Seto et al., 2021). Tree shading and wind protection are key strategies, but their CS potential has not been fully understood. In addition, the carbon performance of passive technologies can be affected by the climate, and their energy-saving potential varies with the climate zone in which they are located.

Overall, decarbonizing the building and construction sector is an urgent mission to address climate change challenges. This chapter offers a critical thinking on passive building technologies for built environment decarbonization. In particular, this chapter first introduces fundamental sciences in aspects of life cycle assessment (LCA), building climate adaptation, urban climatology, and urban morphology. Next, this chapter proposes two basic principles of carbon neutrality and climate adaptation for passive building design. Moreover, a comprehensive analysis of carbon source/sink scenarios of site space and form, building type, envelope properties, ventilation performance, lighting, renewable energy use, and other passive technologies was conducted. This chapter is expected to provide a basis for understanding energy-saving benefits and implementation costs of passive buildings for a dynamic "trade-off" among OC, EC, and CS.

5.2 Science of Built Environment Decarbonization

5.2.1 Life Cycle Assessment

A whole life cycle carbon assessment is important to regulate the OC and EC relationships and to control the total carbon emissions of buildings. Built environment decarbonization requires the reduction of carbon emissions throughout the building life cycle (OC + EC). OC emissions can be reduced by implementing increasingly stringent energy efficiency regulations, improving equipment efficiency, and decarbonizing energy supply systems, but EC emissions have received less attention. At the same time, high-performance buildings, through passive design, tend to have greater potential for emission reductions in terms of EC than code-compliant buildings. Therefore, EC reduction should be a focus of high-performance buildings.

LCA is adopted to assess the environmental impact of a product's production, use, and other aspects. The International Organization for Standardization (ISO) and European Committee for Standardization (CEN) developed a series of supporting standards (CEN, 2011, 2013; ISO, 2006a, 2006b). These standards suggest a standardized assessment step consisting of LCA goal and scope definition, inventory analysis, life cycle impact assessment, and interpretation of results. The CEN standard (CEN, 2011) specifies the life cycle stage of a building LCA into various phases and modules. The scope of the carbon assessment (e.g., cradle to gate/site/grave) can be defined according to the purpose or operationalization, while a full LCA should at least have a "cradle to grave" scope. Inventory analysis provides a carbon quantification process, requiring support from a building carbon inventory database covering the carbon emission factor for each of the relevant processes, e.g., raw material extraction, manufacturing, transportation, repair, and disposal. Finally, the impact analysis and interpretation of results is the post-processing phase, which aims to present and interpret the results in either "midpoints" (providing true quantitative values for different environmental impact categories) or "endpoints" (providing normalized assessment results for different environmental impact categories) method.

5.2.2 Building Climate Adaptation

Climate shapes the built environment in various aspects such as outdoor environment and indoor environmental quality so that building and construction activities have to adapt to the climate throughout planning, design, operation, maintenance, etc. Climate adaptation science reflects the requirements, techniques, and strategies of climate adaptation, beyond which it also includes the thought of climate response (Mao et al., 2018). The climate adaptation science consists of a "mitigation-adaptation" mechanism based on passive operation. This makes the building itself a regulator of the climate and an energy management medium through natural energy manipulation methods such as buffering, storage, and controlled release while meeting the comfort requirements of residents.

5.2.3 Urban Climatology

Urban modifications in building construction and operation have altered the original ground cover and, together with anthropogenic heat emissions, led to changes in near-surface atmospheric thermal circulation. Urban energy balance equations (Eq. 5.1) and (Eq. 5.2) explain the urban climate formation mechanism (He, 2019), where solar radiation (Q^*) is the most important heat input (Figure 5.1). Similar to

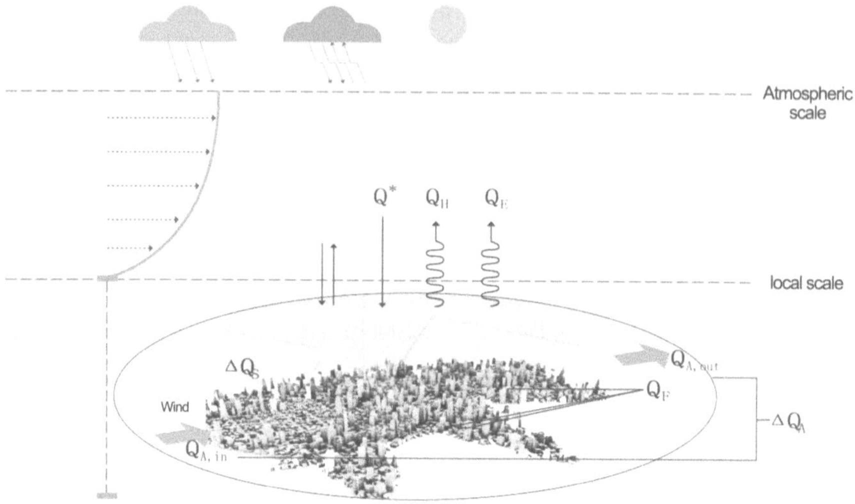

Figure 5.1 Urban climate model and energy transfer process

natural landscapes, incident short-wave radiation (K ↓) and long-wave radiant flux (L ↓) are always constant, whereas decreasing material albedo results in much less short-wave radiation (K ↑) than in natural landscapes. Although building materials have higher surface long-wave radiation (L ↑), most of the emitted long-wave radiation is captured by the street valley, and only a small amount is spilled into the atmosphere.

$$Q^* + Q_F = Q_H + Q_E + Q_S + \Delta Q_A \tag{5.1}$$

$$Q^* = K^* + L^* = K\downarrow - K\uparrow + L\downarrow - L\uparrow \tag{5.2}$$

where Q* represents solar radiation, including short-wave radiation K* and long-wave heat flux L*; K* includes incoming solar short-wave radiation K ↓ and reflected short-wave radiation K ↑ by the urban surface; L* includes atmospheric particles generated long-wave energy flux L ↓ and surface emitted long-wave radiation L ↑; Q_F is the heat from anthropogenic activities; Q_H represents turbulent sensible heat flux; Q_E is dissipated heat via evaporation and transpiration; Q_S is the heat stored in buildings or road systems; and ΔQ_A is the net horizontal heat transfer under the action of the airflow.

In built-up areas, dense buildings lead to an increase in net horizontal heat transfer (ΔQ_A) owing to airflow and an increase in anthropogenic heat (Q_F), forming a heat island. The heat island phenomenon leads to various negative impacts, such as air pollution, energy use increase, human thermal comfort deterioration, and ecological quality reduction. Therefore, in addition to macroclimatic characteristics, architectural design should focus on local climatic features.

5.2.4 Urban Morphology

Impacts of urban form on energy use and thermal comfort can be considered at both city and block (neighborhood) scales. At the city scale, urban complexity, compactness, and layout are the main indicators. Complexity represents the irregularity of urban plots, compactness reflects urban building density, and layout indicates urban development patterns (e.g., single-core and multi-core). Note that there is often a correlation between these indicators. For instance, compact cities generally have higher transport efficiencies and lower carbon emissions, but they are prone to environmental problems such as the heat island effect, reduced natural ventilation potential, and air pollution, thereby resulting in energy use increase in buildings. For microclimate regulation, a multinuclear urban development pattern can help mitigate heat islands and connect urban and suburban areas to form air-directed corridors (Figure 5.2), facilitating urban heat dissipation and reducing heat loads on buildings (Sanaieian et al., 2014). However, such a development pattern may increase travel distances and transport-related carbon emissions (Cai et al., 2021).

At the block (neighborhood) scale, street and block forms and configurations have a decisive influence on microclimates (Xiong et al., 2022), which in turn affect building energy use. Street form (e.g., street aspect ratio, vegetation configuration, and facade form) influences not only incoming solar radiation but also airflow patterns and wind speeds (Figure 5.3a). Compared to deep north-south streets, deep east-west streets are more susceptible to solar radiation intrusion during the daytime in summer and are therefore hotter and more uncomfortable. The building

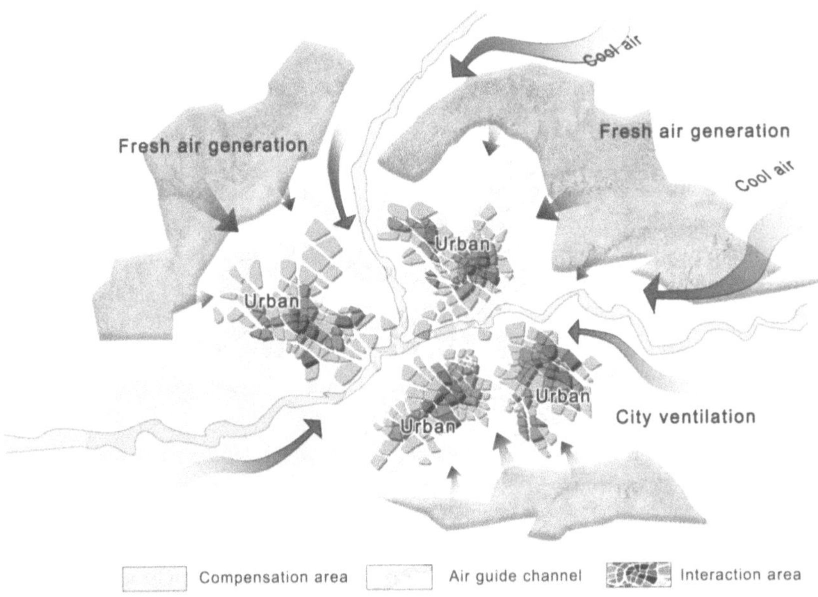

Figure 5.2 Urban ventilation under the multi-core development mode.

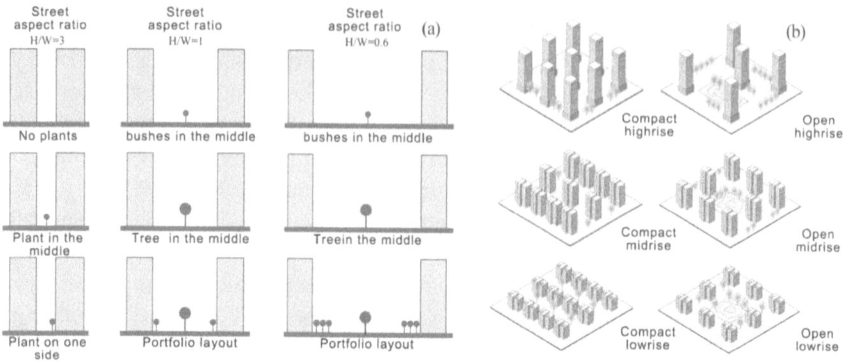

Figure 5.3 Street form (a) and block form (b).

characteristics and layout patterns within the site can strongly influence the microclimate as well (Figure 5.3b). Compared to low-rise blocks, the wind-thermal environment and air quality in compact high-rise blocks are poorer, limiting the feasibility of passive strategies. Additionally, the location of the blocks determines the passive strategy potential. For example, blocks located on the waterfront, near the sea, near large green areas, or forests are affected by local thermal circulation and have better ventilation and cooling performances in summer.

5.3 Principles of Passive Building Design

5.3.1 Carbon Neutral Principle

Given the spatiotemporal variations of climate resources, the selection of proper passive technologies that can effectively regulate indoor environmental quality and reduce building energy use is important (Schnieders et al., 2015). Moreover, passive technologies exhibit differences in both energy efficiency (OC) and implementation costs (EC) associated with local climatic characteristics. In cold climates, it is a common practice to increase envelope thickness to reduce heat losses and improve building energy efficiency. This helps to reduce OC but significantly increases building material use and EC. The association between OC and EC is more complex in tropical climates. For example, changing wall thickness can help insulate during the day but cannot effectively dissipate heat at night. The setting of the window-wall ratio is also crucial for balancing OC and EC. If the window-wall ratio is too high, the load on the air-conditioning system (OC) will be high; conversely, it will result in a high EC.

An assessment of the decarbonization potential of passive technologies should concern the "trade-off" between OC and EC to achieve the unified goal of carbon neutrality. Figure 5.4 suggests that the "trade-off" does not mean that OC reduction from the passive strategy significantly offsets EC but rather that carbon emissions are considered in all building stages to minimize whole-life carbon emissions. In addition, passive strategies associated with vegetation incorporate CS benefits,

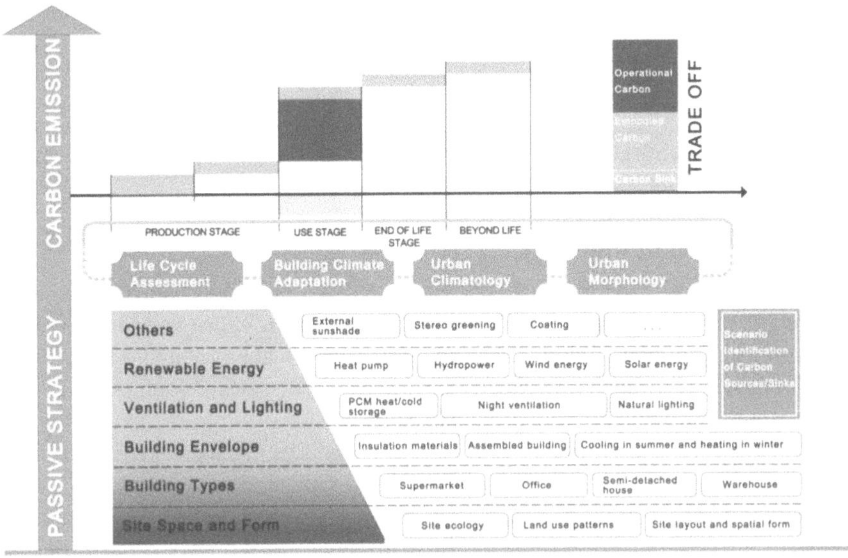

Figure 5.4 A passive building design framework based on principles of carbon neutrality and climate adaptation.

which can offset carbon emissions throughout the building life cycle. Overall, the decarbonization potential of passive technologies consists of OC reduction, EC increase, and CS inclusion.

5.3.2 Climate Adaptation Principle

Since the energy efficiency of passive buildings is constrained by macroclimatic, local, and microclimatic contexts, a prerequisite for using passive strategies is to fully understand the association between climatic characteristics and passive technologies. For example, in a similar climatic context, the local wind environment is influenced by urban spatial form (i.e., central zone location and breezeway planning), and local light resources are influenced by building density and street alignment. Figure 5.4 shows that the adoption of passive technologies needs to consider:

i passive design priorities in response to macroclimatic characteristics;
ii the combination of local heat island effects and natural ventilation effects to analyze natural ventilation and envelope insulation design potential;
iii analysis of the potential of various renewable energy sources, with emphasis on light, wind, and water resources, and local urban form; and
iv the selection and design of other passive technologies such as vertical/roof greening and architectural coatings, considering macroclimatic, local, and microclimatic characteristics.

Accordingly, building groups, individual buildings, and building elements can be adjusted and optimized, and building functions can be combined to maximize passive energy efficiency.

5.4 Passive Technology Analysis and Carbon source/sink Scenario Identification

5.4.1 Site and Urban Form

Land use patterns with high-density features (i.e., high building heights, high building densities) directly affect greenspace ratio, solar gains, and local ventilation performance. Accordingly, land-use patterns dominate local and microclimate conditions and the "spatial passive potential" of local spaces (Xiong et al., 2022). Urban cooling techniques and strategies, such as site shading, sprinkler cooling, cold materials, and permeable pavement, can significantly increase the "passive potential of space" (Figure 5.5). Particularly for typical urban centers, the use of shading strategies, cooling materials, and ventilation strategies on a large scale can alleviate outdoor thermal environments in summer and improve passive design potential.

Site configuration and spatial form affect EC in residential transport, climate use, resource allocation and utilization, etc. A compact urban layout can characterize a lower intensity of EC emissions. For instance, the per capita EC of low-density neighborhoods could be 2.0–2.5 times higher than that of high-density neighborhoods (Sanaieian et al., 2014). However, the "spatial passive potential" of buildings in dense spaces is generally low due to poor natural ventilation in summer and low daylight in winter. This is a major challenge for the application of passive technologies in high-density areas and a major conflict between compact urban design and carbon reduction.

Site ecology plays an integral role in reducing carbon emissions from buildings. There are two CS pathways. The first one is carbon sequestration and storage through plants and soil, and the second one is the reduction of air conditioning

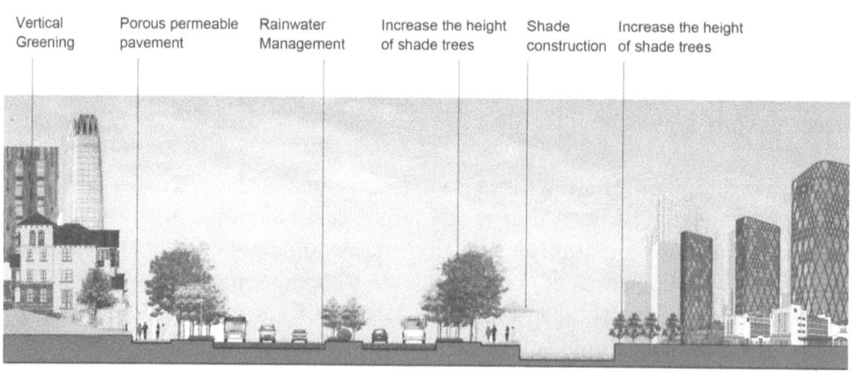

Figure 5.5 Urban space structure and site cooling technology.

and cooling energy use through environmental cooling. Both CS pathways can be achieved by increasing urban green infrastructure and green space surrounding the site and buildings.

Carbon source/sink scenario identification:

OC scenario: Land-use pattern influences the "spatial passive potential" of the site, further affecting the cooling and heating loads of buildings. Site configuration and spatial form influence local and microclimate, which in turn affects buildings' OC emissions. The vegetation can potentially reduce cooling loads through environmental regulation, indirectly reducing OC emissions. Renewable energy sources (light and wind) can compensate for building energy use in conjunction with site configuration and building form, providing another approach to OC reduction.

EC scenarios: Site configuration and spatial form can lead to differences in passive resource distribution and utilization, making a difference to EC in the construction and use stages of passive buildings. This is a key conflict between the compact city design and carbon reduction.

CS scenarios CS benefits can be achieved through carbon sequestration and storage in plants and soils as a compensatory approach in passive building design.

5.4.2 Building Types

Large commercial complexes typically exhibit high energy use and carbon emission intensities (Guo, 2022). These complexes are usually located in city centers, where the "spatial passive potential" of the site is usually low, and the high foot traffic leads to a high dependence on active equipment, making it difficult to operate passively during the summer and winter months. However, in transitional seasons, these complexes can be naturally ventilated or mechanically ventilated intermittently. Building size reduction is conducive to improving the effectiveness of mechanical and natural ventilation and reducing OC emissions. Compared with large commercial complexes, other types of buildings can potentially promote passive operation, with a more ventilated environment. For example, medical buildings can implement passive design in rehabilitation wards and outpatient clinics.

EC varies with building materials, structural design, and construction techniques. RICS (2014) analyzed the carbon footprints of different types of buildings across different stages over building life cycles (i.e., a 30-year life cycle). Their results show that the carbon footprints of different types of buildings vary considerably. However, there is still a lack of studies on EC emissions of different types of buildings. The development of an EC database that reflects parameters such as building type, structural selection, and material choice is highly needed.

Carbon source/sink scenario identification:

OC scenarios Differences in building types and functions can lead to different levels of OC emissions. The "spatial passive potential" of central urban areas is poor, making it difficult for large commercial complexes to achieve passive operation. Small commercial buildings have the potential to achieve passive operation in transitional seasons, and thereby OC reduction. Residential buildings, medical buildings, and some others have the potential to achieve passive operation with reasonable passive design and seasonal conditions.

EC scenarios Differences in building types and functions affect the EC emissions of buildings. Structural design and construction techniques should be considered to reduce EC levels in new and retrofitted buildings.

5.4.3 Envelope Properties

The building envelope typically consists of the main structure (i.e., walls, slabs) and insulation. Concrete, clay, and wood are common materials used for the main structure. Wood is known to be a carbon-negative material, and light or heavy timber construction will have a CS effect. Well-mixed concrete mainly contains calcium hydroxide [$Ca(OH)_2$], calcium silicate hydrate (CSH), and calcium aluminate hydrate (CAH), which can undergo carbonation reaction with CO_2; on the contrary, firing of bricks through clay will produce a large amount of carbon emissions. Insulation materials consist of conventional and environment-friendly types, and their adoption for insulation is a main carbon source scenario. Conventional materials mainly include organic insulation materials (e.g., expanded polystyrene boards, extruded polystyrene boards) and inorganic insulation materials (e.g., rock wool). However, conventional materials generally have a high environmental impact (e.g., GWP). First, the carbon emission intensity for producing such insulation materials is high. Second, these insulation materials have an environmental impact during waste disposal after demolition.

Climate conditions must be traded off against the adoption of insulation materials. Walls with low insulation performance are more energy-efficient in regions where summer cooling is needed, and walls with high insulation performance are more energy-efficient in regions where winter heating is needed (Ibn-Mohammed et al., 2013). In subtropical areas, since the need for winter heating is low, the integrated trade-off between wall insulation design and energy-saving potential should be highly concerned.

Prefabrication (or assembly) is a key technique in passive construction. Its advantage is industrial production and processing of building components (e.g., walls, floor slabs) which are then transported to the site and mechanically assembled through reliable connections. Compared with traditional construction methods, assembled buildings have the advantages of short construction periods, material savings, energy efficiency, and environmental protection.

Carbon source/sink scenario identification:

OC scenarios Thermal performance of building envelope is crucial for OC reduction of passive buildings, with an energy-saving potential of up to 40%. The energy-saving scenario is twofold: reducing cooling loads by reducing high outdoor thermal impacts in summer and reducing heating loads by reducing indoor heat losses in winter.

EC scenarios Wall construction and the extensive adoption of materials for insulation lead to strong EC emissions in passive buildings. If the building is not heated in winter, installing excessively thick wall insulation panels will increase the EC. The low-EC construction process through prefabricated buildings is possible in a passive construction. In addition, the use of timber walls is an effective way to reduce EC.

CS scenarios The use of concrete for buildings will have a CS effect.

5.4.4 Ventilation Performance

Natural ventilation systems can achieve an energy-saving potential of 20%–50%, depending on ventilation types and external climatic conditions (Zhang et al., 2021). There are three main passive building ventilation strategies: wind pressure ventilation, thermal pressure ventilation, and night ventilation. Wind-pressure ventilation is less feasible in areas of inland, basins, and equatorial windless zones owing to weak background wind field. Furthermore, heat islands can result in high wind temperatures, making daytime wind-pressure ventilation unsuitable.

Thermal pressure ventilation relies on temperature differences between the indoor and outdoor air in summer, and it requires the existence of some large cavity spaces (i.e., atrium and side atrium), which are often accompanied by curtain walls or large glass roofs. The adoption of thermal pressure ventilation then often leads to energy-saving trade-offs between winter and summer.

The use of night ventilation for cooling is a common approach in passive design, as the wind temperature of the site drops at night. High-thermal-mass building structures or the addition of high-thermal-storage materials are necessary for night ventilation cooling. On the phase change materials (PCMs), there is an increasing number of studies and practices on passive nighttime ventilation based on phase change thermal storage (Solgi et al., 2018). However, the adoption of PCMs may lead to a reduction in daytime air-conditioning cooling efficiency (Arumugam et al., 2022). In recent years, moreover, heat islands and heat waves have become increasingly prominent and serious, thereby decreasing the cooling storage potential of night ventilation. Whether night ventilation is necessary for passive design is dependent. PCMs can also be used in sunrooms in winter, but the same PCM cannot be used for both summer cooling and winter thermal storage (Kylili and Fokaides, 2016). In addition, PCMs also suffer from durability issues, and carbon emissions of their production processes should be studied.

Overall, ventilation breaks the indoor-outdoor climate boundary. Prolonged night ventilation makes the interior more susceptible to natural erosion, such as condensation and drifting rain, which can reduce the life of the envelope.

Carbon source/sink scenario identification:

OC scenarios Energy-saving potential of natural ventilation can be compromised by the predominance of weak winds and high temperatures. A passive nighttime ventilation strategy based on PCMs was designed. However, there are two key issues: (i) the increase in PCM may lead to low efficiency of air conditioning during the day, and (ii) PCM cannot be taken into account for winter and summer heat storage.

EC scenarios EC sources in night ventilation cooling are related to the material production process. In addition, humid climates at low latitudes make the envelope vulnerable to natural erosion, such as mold and condensation, leading to service life reduction.

5.4.5 Lighting

Natural light systems consist of side and top lighting. Central urban areas generally have high building densities, characterized by small building spacing and large building heights, both of which limit the use of these two types of natural lighting strategies. Moreover, energy used for lighting is eventually converted into heat, resulting in additional cooling loads in proportion to the total energy use.

The International Energy Agency indicates that artificial lighting accounts for 14% of the total electricity use in the EU and 19% globally. Traditional office buildings use up to 35% of the total electrical load for lighting, and residential buildings use up to 25%–35% (Zhen et al., 2019). Replacing electric lighting with natural lighting could reduce the amount of energy consumed by 25%–30% (building energy composition is mainly electricity- and fossil-based) (Gago et al., 2015). This demonstrates the large potential of natural lighting in carbon emission reduction.

When designing natural lighting, it is necessary to consider the contradiction between the energy efficiency of lighting and heating. Large windows and curtain walls enhance light but lead to increased energy use for cooling in summer and excessive heat loss in winter.

Carbon source/sink scenario identification:

OC scenario Natural lighting can effectively reduce energy use in buildings. However, in high-density urban areas, natural light is limited by dense urban spaces. The heat from lighting systems may increase the energy for building cooling, and excessive windows and doors can increase cooling energy use in summer and heat losses in winter.

EC scenarios The increase in windows and doors can save building materials to some extent. However, the carbon emission factor of making glass is greater (Arbelaez Perez et al., 2022), so an increase in window and door area may lead to an increase in EC.

5.4.6 Renewable Energy

Renewable energy technologies consist of technologies for hydropower, wind power, water-, ground-, air-source heat pumps, and photovoltaic light/heat. Maximizing on-site renewable energy supply and off-site renewable energy supply should be prioritized in built environment decarbonization.

Solar energy is an important renewable energy source, and building-integrated photovoltaics (BIPV) are often adopted in passive building design. The effectiveness of BIPV depends on factors such as latitude, building orientation, building form, shading of surrounding buildings, availability of roof/façade space, and PV efficiency. Therefore, the design of BIPV systems should be tailored to specific contexts.

Most small-to-medium residential buildings and standalone houses have the ability to reduce or even offset the energy required to run the building, given the large roof area available for BIPV implementation. For other buildings with high energy requirements, it is difficult for on-site production to meet their energy needs, and regional-level renewables must be considered.

Carbon source/sink scenario identification:

OC scenarios Renewable energy sources can help reduce operational energy needs and then reduce OC emissions. Solar power combined with BIPV technology can be the most reliable on-site renewable energy supply; regional-level (off-site) renewable energy sources are selected based on local resources.

EC scenarios Renewable energy consists of on-site renewable energy and off-site energy, where the latter relies on sound infrastructure construction, which may increase EC. Various types of heat pump technologies may also increase EC, owing to their high costs and low energy quality.

5.4.7 Other Passive Technologies

Flexible and convenient shading systems are widely used in areas with strong solar radiation, particularly in summer. Shading elements are generally inexpensive, and the EC for producing and setting up elements is low.

Surface coating is also a passive strategy. Materials with high emissivity and reflectivity release energy into the environment (increasing external infrared radiation and short-wave reflection) and reduce the cooling loads. In winter, materials with low emissivity and reflectivity conserve energy (reduce external infrared radiation and daylight reflection) and reduce heating loads.

Vertical greening has similar benefits to surface coating, but its decarbonization mechanism is more complex. First, its implementation generates EC. Second, it shows a CS effect, that is, plants can absorb CO_2 via photosynthesis or sequester it into soil. Third, it reduces urban energy use by regulating the microclimate and lowering indoor temperatures.

Carbon source/sink scenario identification:

OC scenarios Solar radiation gain can be regulated by flexible external shading and surface coatings, which reduce OC emissions. Vertical greening increases the thermal performance of envelopes and contributes to the microclimate cooling of the site, reducing OC emissions as well.

EC scenarios A small amount of EC was generated during shading device production and installation. However, EC in the production and maintenance of surface coatings is unknown.

CS scenarios CS benefits are achieved through carbon sequestration in plants and soils.

5.5 Exploring the Potential of Passive Building Design in the Context of Climate Change

A holistic analysis of passive design and the identification of carbon source/sink scenarios are conducted in this chapter (Figure 5.6). First, climate and site characteristics are key factors. High density, low ventilation performance, and strong heat island effects in tropical or subtropical regions reduce the benefits of passive design. Second, the process of reducing carbon emissions in passive buildings is highly complex, and the specific carbon reduction potential needs to be analyzed in building type, envelope, ventilation, lighting, renewable energy use, etc. In addition, attention must be paid to climate change, as it will significantly affect passive design.

Passive design can reduce heating and cooling demands through a range of strategies during the use stage. Moreover, unlike active facilities such as air conditioners, passive buildings do not release strong waste heat when regulating indoor environments, thereby avoiding vicious cycles due to outdoor thermal environment deterioration. However, passive construction raises requirements for building materials, leading to a significant increase in EC proportion. The carbon reduction potential of passive buildings needs to be clarified given various carbon source/sink scenarios throughout the whole life cycle.

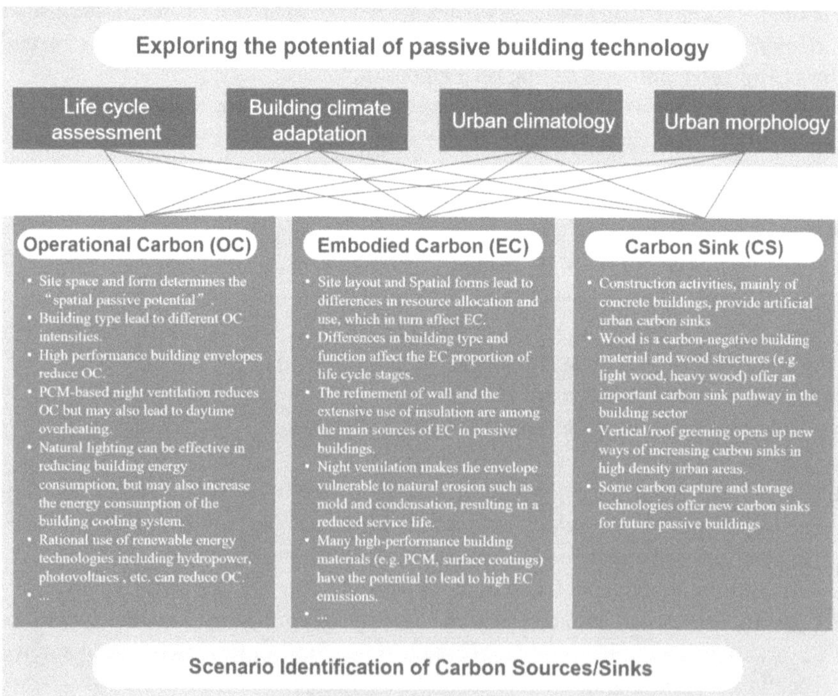

Figure 5.6 Identification of carbon source and sink scenarios in passive buildings.

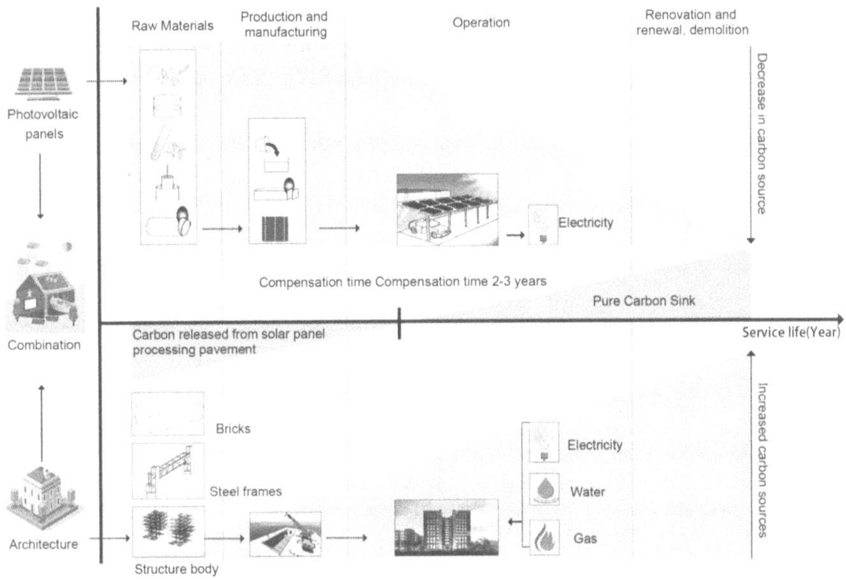

Figure 5.7 Carbon payback time for implementing passive strategies (Taking solar photovoltaic panels as an example).

Decarbonization potential of passive design should be assessed based on climate adaptability and their life cycle. The latter has not been well explored in practice but is on the track. Various international and societal actions are pushing for more systematic LCA applications. The net zero carbon built environment proposed by the World Green Building Council, Global Alliance for Buildings and Construction, and many national action plans (e.g., *UK-The RIBA 2030 Climate Challenge Plan*, *UK-The London Energy Transformation Initiative*, *USA-Architecture 2030 Challenge*) have been developed using the LCA approach. Both OC and EC were included to define a net zero carbon built environment. The quantitative indicator of "carbon payback time" (C-PT), an indicator to characterize the implementation of a structure, equipment, or entire building in terms of carbon offset time, should be emphasized for EC assessment in materials or components. Figure 5.7 presents the assessment C-PT of solar photovoltaic panels. Overall, it is important to embed an all-encompassing and deeper consideration of EC and OC with a whole life cycle perspective to form and refine a framework for assessing the carbon reduction potential of passive building design.

5.6 Concluding Remarks

Passive building design is a key approach enabling the building and construction sector to save energy and reduce carbon and promoting a net zero carbon built environment. This chapter provides a comprehensive analysis of passive design technologies, in aspects of site space and form, building types, envelope properties, ventilation performance, lighting, renewable energies, and others, and identifies their carbon source/sink scenarios. This study provides a reference for a comprehensive assessment of the energy efficiency benefits and implementation costs of passive buildings to achieve a dynamic "trade-off" between OC and EC emissions. Overall, this chapter is conducive to a better understanding of carbon emission measurements and the achievement of built environment decarbonization.

References

Arbelaez Perez, O. F., Florez, D. R., Zapata Vergara, L. M. & Hernández Benavides, K. V. 2022. Innovative use of agro-waste cane bagasse ash and waste glass as cement replacement for green concrete. Cost analysis and carbon dioxide emissions. *Journal of Cleaner Production*, 379, 134822.

Arumugam, P., Ramalingam, V. & Vellaichamy, P. 2022. Effective PCM, insulation, natural and/or night ventilation techniques to enhance the thermal performance of buildings located in various climates – A review. *Energy and Buildings*, 258, 111–132.

Cai, M., Shi, Y., Ren, C., Yoshida, T., Yamagata, Y., Ding, C. & Zhou, N. 2021. The need for urban form data in spatial modeling of urban carbon emissions in China: A critical review. *Journal of Cleaner Production*, 319, 12–45.

Cen 2011. EN 15978: Sustainability of Construction Works—Assessment of Environmental Performance of Buildings—Calculation Method.

Cen 2013. EN 15804: Standards Publication Sustainability of construction works—Environmental product declarations—Core rules for the product category of construction products.

Gago, E. J., Muneer, T., Knez, M. & Köster, H. 2015. Natural light controls and guides in buildings. Energy saving for electrical lighting, reduction of cooling load. *Renewable and Sustainable Energy Reviews*, 41, 1–13.

Guo, Y.-Y. 2022. Revisiting the building energy consumption in China: Insights from a large-scale national survey. *Energy for Sustainable Development*, 68, 76–93.

He, B.-J. 2019. Towards the next generation of green building for urban heat island mitigation: Zero UHI impact building. *Sustainable Cities and Society*, 50, 101647.

Ibn-Mohammed, T., Greenough, R., Taylor, S., Ozawa-Meida, L. & Acquaye, A. 2013. Operational vs. embodied emissions in buildings—A review of current trends. *Energy and Buildings*, 66, 232–245.

IPCC 2020. Climate change 2021: The physical science basis.

Isaac, M. & Van Vuuren, D. P. 2009. Modeling global residential sector energy demand for heating and air conditioning in the context of climate change. *Energy Policy*, 37, 507–521.

Iso 2006a. ISO 14040: environmental management—Life cycle assessment—Principles and framework.

Iso 2006b. ISO 14044: Environmental management—Life cycle assessment—Requirements and guidelines. International Organization for Standardization.

Kylili, A. & Fokaides, P. A. 2016. Life Cycle Assessment (LCA) of Phase Change Materials (PCMs) for building applications: A review. *Journal of Building Engineering*, 6, 133–143.

Mao, P., Li, J., Tan, Y., Qi, J. & Xiong, L. 2018. Regional suitability of climate-responsive technologies for buildings based on expert knowledge: A China study. *Journal of Cleaner Production*, 204, 158–168.

Rics 2014. *Methodology to calculate embodied carbon*, London, Royal Institution of Chartered Surveyors (RICS).

Sanaieian, H., Tenpierik, M., Linden, K. V. D., Mehdizadeh Seraj, F. & Mofidi Shemrani, S. M. 2014. Review of the impact of urban block form on thermal performance, solar access and ventilation. *Renewable and Sustainable Energy Reviews*, 38, 551–560.

Schnieders, J., Feist, W. & Rongen, L. 2015. Passive houses for different climate zones. *Energy and Buildings*, 105, 71–87.
Seto, K. C., Churkina, G., Hsu, A., Keller, M., Newman, P. W., Qin, B., & Ramaswami, A. 2021. From low-to net-zero carbon cities: The next global agenda. Annual review of environment and resources, 46(1), 377–415.
Solgi, E., Hamedani, Z., Fernando, R., Skates, H. & Orji, N. E. 2018. A literature review of night ventilation strategies in buildings. *Energy and Buildings*, 173, 337–352.
Unep. 2022. *Facts about the climate emergency* [Online]. Available: https://www.unep.org/zh-hans/explore-topics/climate-change/guanyuqihoujinjiqingkuangdeshishi [Accessed].
Xiong, K., Yang, Z. & He, B.-J. 2022. Spatiotemporal heterogeneity of street thermal environments and development of an optimised method to improve field measurement accuracy. *Urban Climate*, 42, 101–121.
Zhang, H., Yang, D., Tam, V. W. Y., Tao, Y., Zhang, G., Setunge, S. & Shi, L. 2021. A critical review of combined natural ventilation techniques in sustainable buildings. *Renewable and Sustainable Energy Reviews*, 141, 110–127.
Zhen, M., Du, Y., Hong, F. & Bian, G. 2019. Simulation analysis of natural lighting of residential buildings in Xi'an, China. *Science of The Total Environment*, 690, 197–208.

6 Home Retrofitting Initiatives towards a Sustainable Built Environment in the Australian Context

Roberto Minunno and Karla Fox-Reynolds

6.1 Introduction

Buildings contribute more than 30% of global energy consumption and 25% of greenhouse gas emissions, and building energy use is expected to increase by nearly 2% annually (Hong et al., 2020; Shen et al., 2019). Existing buildings represent a substantial opportunity to decrease societal adverse environmental impact and strive towards a sustainable built environment. In this chapter, the authors propose an analysis of retrofitting initiatives that could be implemented in Australia to improve homes' liveability, health, and thermal and energy efficiency. Australia's minimum building energy efficiency standards were first legislated in 2005, meaning that older buildings lack features (Bulut et al., 2022). As a result, about 86% of Australian homes typically rank overwhelmingly poorly regarding energy efficiency (Kumar et al., 2021).

Many of the 10 million Australian homes are extremely cold in winter and unbearably hot in summer, mainly because of their poor build quality (Daniel et al., 2019), with striking ramifications on household occupants' health. In New Zealand, healthier homes could help minimise the yearly 1,600 deaths from unhealthy and icy homes (Ade & Rehm, 2020; Wimalasena et al., 2022). Australian homes are reportedly considered extremely cold, much below the threshold established by the World Health Organisation for internal temperature (Chalmers, 2022; Oldfield, 2022; Southwell, 2022; World Health Organization, 2018). Retrofitting could essentially solve these issues by improving Australian homes' thermal efficiency. However, retrofitting solutions can be expensive, and therefore, households prefer to create a comfortable space by using active systems such as gas heaters and obsolete cooling devices. These solutions are not sustainable and are more expensive than a retrofit in the long run, but they are widely available. In short, a properly retrofitted home is critical for households to live in healthier and more comfortable spaces, improving their well-being (Fisk et al., 2020).

Australia is slowly moving towards zero-energy building initiatives, as outlined by the Trajectory for Low-Energy Buildings (Department of Climate Change, Energy, the Environment and Water, 2018). The Trajectory recommends a list of mandatory and voluntary initiatives, including following the National Construction Code and home rating through the NatHERS (Nationwide House Energy Rating Scheme).

DOI: 10.1201/9781003317890-8

The adoption of NatHERS is supported mainly by the Australian building industry, which considers it a reliable tool for evaluating a home's thermal and energy efficiency (Fox-Reynolds et al., 2021). As these Government initiatives show, there is substantial activity in the housing sector. The Australian housing context studied in this chapter is addressed explicitly towards single dwellings – a category of buildings falling into the description of Class 1a (Australian Building Codes Board, 2022b) – that are owner-occupied or tenanted. These building types represent the overwhelming majority of Australian homes (86%; for reference, apartment buildings represent only 12% of Australian dwellings); it's also critical to point out that not all of these would require retrofit, nor might not be suitable for it.

To support the Australian housing sector to become more energy-efficient and provide healthier and more sustainable homes, we propose a retrofitting initiative based on four hierarchical upgrades to transform existing Australian homes whilst aligned with the Trajectory and helping to provide a higher NatHERS rating. To help practitioners gain a better understanding of the proposed retrofitting options, we outline them in the following Section 6.2. Then, we explore the fundamental limitations of our proposed upgrades and present the concluding remarks of this study. This chapter was written as part of the RACE for 2030 Collaborative Research Centre project Pathways to Scale: Retrofitting One Million+ Homes (Fox-Reynolds et al., 2021).

6.2 Retrofitting Options

The hierarchical set of upgrades is rooted in the concept that home thermal and energy efficiency requires actions beyond installing solar panels. Ironically, however, installing solar panels seems to be the quick solution most households are implementing to improve their home's sustainability features. A more holistic approach is needed if we want to ensure sustainable operations. This chapter outlines four retrofitting upgrades to improve Australian homes' thermal efficiency and comfort. More specifically, the first upgrade involves drastically improving homes' insulation. Each home's whole envelope should be revised, ensuring a satisfying level of insulation and limited air leakage. That is because improved thermal efficiency starts with slowing the heat flow between the interior and exterior environments (Hardy et al., 2018; Muddu et al., 2021). Second, ceiling fans are an inexpensive and effective solution to decrease the energy consumption of air conditioning systems. Simultaneously, traditional air conditioning systems should be upgraded to reverse-cycle systems, while the windows should strategically be upgraded to double-glazing units. The third upgrade brings efficient lighting and water use. LED lighting would replace traditional and highly inefficient bulbs in this upgrade, while improved showerheads and taps would be installed to limit water use. Relatively inexpensive solutions like improved lighting and water fittings would ensure energy and water-saving with minimal cost. Finally, a complete conversion to an electric home is proposed in the fourth upgrade, including installing solar panels and battery systems, a hot water heat pump, and complete home electrification with an induction stovetop and electric oven. These targeted solutions and upgrades will deliver a more sustainable built environment and help households and tenants live in healthier spaces whilst reducing their emissions and reliance on fossil fuels.

These retrofitting options are explored in more detail in the following sections. Table 6.1 summarises the various proposed interventions in terms of their hierarchical importance, with Retrofitting upgrade 1 being the most impactful. Retrofitting upgrade 4 is the least impactful but still provides significant environmental and economic benefits in the long run (where the required capital costs measure impact). It is widely accepted within the industry that the suggested upgrade items can provide significant improvements without always requiring the user to possess particular skills or abilities. They also consider the importance of the building fabric on the thermal comfort of the home and in mitigating heat and cold stress. The building also has a relatively long lifespan, especially compared to technology-based installations. They would improve the quality of the existing Australian housing stock; the results of these improved buildings would be reflected in the houses' NatHERS rating, the home's comfort, and the occupants' health, and their energy bills may be reduced. On the other hand, the interventions above can be costly. To provide a complete picture, an estimation of these costs will be outlined relating to each intervention (all figures are in Australian dollars and were up to date in 2022). Specifically, the costs reported below are indicative and largely depend on the housing market – these were obtained through discussions with experts in the sector and, therefore, were not referenced in this chapter.

6.2.1 Retrofitting Upgrade 1 – Insulation, Pipe Lagging, and Draught Proofing

Most buildings in Australia need to be more well-insulated, wasting a substantial portion of energy through their walls, ceilings, floors, and pipes. An under-insulated building allows heat to flow through, so it is important to consider improving insulation and sealing draughts. Insulation is typically inexpensive, which was only often adopted in Australian buildings in recent years. For reference, the National Construction Code 2022 (the Australian building code serving as a regulatory framework for the design, construction, and performance of buildings across the country) was released when the authors wrote this chapter in 2022. This version finally places a significant emphasis on sustainability and energy efficiency, aligning with global trends and local initiatives aimed at reducing the environmental impact of the built environment (Australian Building Codes Board, 2022a). Thermal insulation affects the building envelope and the hot water system. Further, in poorly built homes, the energy flows easily through window and ceiling leakages. Table 6.2 summarises the proposed interventions, including the baseline parameter these will improve upon and approximate cost.

6.2.1.1 Roof, Wall, and Floor Insulation

Insulation materials prevent heat from transferring from across spaces. Specifically, buildings partially inhibit the heat from moving outside the building when

Table 6.1 Summary of retrofitting upgrades in Australian buildings with an explanation of their hierarchical importance

Upgrade	*Summary*	*Hierarchical importance*	*Impact level*	*Cost level*
Retrofitting upgrade 1 – Insulation, Pipe Lagging, and Draught Proofing	Focuses on insulation, pipe lagging, and draught proofing to improve energy efficiency and reduce heat loss in poorly insulated buildings.	This most impactful upgrade provides significant energy savings and comfort improvement. Buildings in Australia are typically poorly insulated, so improving insulation and draught proofing is crucial in reducing energy loss.	Very Significant	Moderate
Retrofitting upgrade 2 – Ventilation, Air Conditioning, and Double Glazing	Includes the installation of ventilation, air conditioning, and double-glazing windows to improve comfort, health, and energy savings.	This upgrade provides substantial benefits in terms of energy savings, cost, and health benefits. While it is hierarchically less impactful than upgrade 1, it can still significantly improve comfort and energy efficiency.	Significant	High
Retrofitting upgrade 3 – Appliances and Lighting	Involves upgrading appliances and lighting to energy-efficient models, adopting a clothesline, and installing water-efficient showerheads to achieve marginal energy savings.	This upgrade is less expensive and provides more marginal energy savings than Upgrades 1 and 2. It is still essential, however, as it promotes responsible purchasing and energy-efficient choices that can reduce overall energy consumption.	Moderately Significant	Moderate
Retrofitting upgrade 4 – Solar System, Electrification, and Hot Water	Requires significant capital costs but provides long-term environmental and economic benefits by installing solar panel systems, home batteries, hot water heat pumps, and induction cooktops to reduce reliance on fossil fuels.	While upgrade 4 requires more substantial capital costs, it provides significant environmental and economic benefits in the long run. It promotes the adoption of renewable energy sources and reduces reliance on fossil fuels, which is crucial in achieving sustainable development.	Significant	Very High

Table 6.2 Retrofitting interventions included in upgrade 1 with details on their baseline and improved parameters and cost

Intervention	*Description*	*Baseline parameter*	*Improved parameter*	*Cost per unit*	*Reference*
Roof insulation	Improvement of U-values of roof, wall, and floor insulation. (Roof U-values collected for light-coloured roofs.)	<1 W/m²/K	0.217 W/m²/K	$24 installed per m²	Australian Building Codes Board (2022b)
Wall insulation		0.5–1 W/m²/K	0.357 W/m²/K	$23 installed per m²	
Floor insulation		1 W/m²/K	0.444 W/m²/K (0.364 W/m²/K in Victoria)	$24 installed per m²	
Pipe lagging	Thermal insulation of the hot water pipes.	–	4% increased efficiency	$30 plus labour	Marini et al. (2021)
Draught sealing/proofing	Overall, it improved air tightness.	–	15% improved airtightness after retrofitting	$1,020 per house	Wills et al. (2021)

the external temperature is less than the temperature inside, and *vice versa* prevents the external heat from flowing inside in hot weather. In so doing, the heat generated indoors in winter is retained inside, resulting in less energy required to generate the heat dispersed through the envelope. As shown in Table 6.2, the baseline parameter for insulation in the roof, walls, and floor of Australian homes must be included. Its baseline parameter (the heat transfer coefficient) is typically 1 W/m²/K (Commonwealth Scientific and Industrial Research Organisation, 2016).

However, the Australian Building Codes Board (2022b) suggests that houses' envelope should grant a heat transfer coefficient between 0.444 W/m²/K on the floor and 0.217 W/m²/K on the roof. Typical building products that can be installed to improve the envelope insulation properties include earth wool, rock wool, glass wool, and rigid boards (such as expanded polystyrene, EPS or extruded polystyrene, XPS). Depending on the building structure, installing insulation can be easy and unintrusive.

6.2.1.2 Pipe Lagging

Domestic hot water systems are typically energy-intensive and require up to 50% of a house's energy use (Hadengue et al., 2022). As the water flows through cold and under-insulated pipes, heat is wasted as it transfers to the external environment. If the pipes are visible (i.e., not walled in), pipe lagging or thermal pipe insulation can be inexpensive and retain about 4% of the wasted energy (Marini et al., 2021).

6.2.1.3 Draught Sealing

Draught, or air leakage, happens when air flows through the building envelope. As the air flows through the unsealed envelope, it results in either internal heat loss or an increase, similar to what would happen, leaving an open window in winter when trying to keep the room warm or in summer when trying to keep it fresh. The envelope elements that are more prone to drought include (Fernández-Agüera et al., 2019):

- Suspended ceilings;
- Windows;
- Doors;
- Damaged or poor seals around windows and doors;
- Openings for light fixtures;
- Penetrations for piping;
- Cracks.

To contrast the air leakages, air tightness should be improved by identifying the gaps and sealing them with silicon or expandable foam, depending on the nature of the opening. These measures can be time-consuming and costly, amounting to about 1,000 Australian dollars per household. Still, they can reduce air leakages by about 15% (Wills et al., 2021), allowing for increased comfort and substantial energy savings.

6.2.2 Retrofitting Upgrade 2 – Ventilation, Air Conditioning, and Double Glazing

The retrofitting upgrade 2 (Table 6.3) includes hierarchically less impactful features than those seen in upgrade 1 but can result in substantially improved comfort, health, and energy savings. Upgrade 2 includes the installation of ceiling fans to help mitigate the energy used for cooling, reverse-cycle air conditioning systems, which promote additional savings for cooling with the added benefit of a highly efficient heating system, and the change of window panels from the more common single-glazing to double-glazing windows. The subsections below highlight the benefits of these upgrades regarding energy savings, comfort, and health benefits.

6.2.2.1 Ceiling Fans

Ceiling fans are already famous in rural Australia, whereas air conditioning systems are more common in urban areas (Gui & Gou, 2022). Ceiling fans are simple and affordable tools to help responsible households cool down their homes in summer (Omer, 2008). If installed in all house rooms (assuming two bedrooms, the kitchen, and the lunchroom), they can cost approximately $2,000. They can create a cooling sensation of up to 4°C, which makes an internal temperature of 30°C comfortable (Beng Ooi et al., 2017). Although the efficiency of the mechanical ventilation induced by ceiling fans depends on the ceiling height, these devices can help save energy when their fan power usage is less than 15 W. On the other hand, devices with power over 60 W might have higher energy usage than traditional cooling systems, such as a perfectly efficient HVAC system (Schiavon & Melikov, 2008). In addition to their low energy usage, the airflow generated by ceiling fans is considered to increase the occupants' comfort (Hasama et al., 2022). Although more research is needed to explore the health benefits of adopting ceiling fans

Table 6.3 Retrofitting interventions included in upgrade 2 with details on their baseline and improved parameters and cost

Intervention	*Description*	*Baseline parameter*	*Improved parameter*	*Cost per unit*	*Reference*
Ceiling fans	Addition of ceiling fans to minimise the need for air conditioning.	–	Air conditioning was reduced to zero.	$500 per unit	–
Reverse-cycle aircon	Substitution of the traditional heating or cooling system for a reverse-cycle split system.	–	Improved coefficient of performance (COP) by 11%	$2,000 per fixture installed	Department of Climate Change, Energy, the Environment and Water (2022a)
Double glazing	Double glazing in place of single glazing to improve the thermal insulation of the windows (decreased windows' U-value)	U-Value: 5–6 SHGC: 0.5–0.6	U-value 3.1 SHGC 0.63	$500 per window	Australian Building Codes Board (2022b)

versus other cooling systems, this ventilation method is not associated with any concern for human health (Beng Ooi et al., 2017).

6.2.2.2 *Reverse-Cycle Air Conditioning System*

Realistically, ceiling fans are often coupled with reverse-cycle air conditioning (or split) systems (Gui & Gou, 2022). Split systems include an external compressor and an indoor unit, and they use electricity to regulate the internal temperature (Department of Climate Change, Energy, the Environment and Water, 2022a). While ceiling fans move the air around the occupants without affecting its temperature, split systems transfer heat from one source and move it to another source to provide internal heating or cooling. These systems are extremely energy-efficient, as their efficiency varies between 300 and 1,000%, meaning that they transform a unit of energy used into three to ten times heating or cooling energy. For this reason, splits are considered the most energy-efficient heating and cooling systems on the market for residential buildings (Australian Government – Your Home, 2021), and they can help improve the performance of air temperature systems by 11%. Advanced air conditioning systems are also linked to health benefits, especially concerning improved indoor air quality, which helps mitigate symptoms of diseases such as asthma; they can filter particulate matter and total volatile organic compounds and regulate indoor carbon dioxide levels (Shriram & Ramamurthy, 2019; Xu et al., 2010). Our market research shows that the most affordable yet highly efficient reverse-cycle air conditioning model costs about $2,000 per installed unit.

6.2.2.3 Double Glazing

In a reasonably sealed house, windows are the portion of the envelope which allows for substantial indoor-outdoor unwanted heat transfer, and old windows can cause up to 18% heat loss (Safe Style Windows, 2022). Multi-layer glazing windows are widely considered an efficient alternative when retrofitting a home to contrast the energy lost through windows. Although triple-glazing windows are commercially available, they are required in harsh weather, such as in northern Europe and America (Jelle et al., 2012). In the more moderate Australian climate, double-glazed windows are considered the best option for a cost-energy efficiency balance (Tam et al., 2018). The most standard double-glazing technologies are transparent, reflective, and low-emissivity double glazing (Florides et al., 2002). Typically, existing Australian houses employ single-glazing windows, which must be retrofitted to double glazing to improve the homes' energy efficiency (Simko & Moore, 2021). Problematically, however, in some Australian states, old and underperforming windows are cemented into the brick wall panels, thereby increasing the cost of the upgrade to double-glazing systems (Horne & Hayles, 2008; Lawania & Biswas, 2016). To overcome this issue, modern window systems allow manufacturing a bespoke glazing portion and swapping it without changing the entire window frame for a considerable economic saving – upgrading a window to double glazing could cost about $500 (Jason Windows, 2022). These windows dramatically reduce the thermal transmittance (U-value) from 6.0 to 3.1 W/m^2/K and increase the window's solar heat gain coefficient (SHGC) to 0.63.

6.2.3 Retrofitting Upgrade 3 – Appliances and Lighting

The retrofitting upgrade 3 (Table 6.4) includes strategies that, compared to the previous upgrades 1 and 2, are less expensive and help achieve more marginal energy savings. Specifically, upgrade 3 proposes four interventions. First, we promote the responsible purchase of dishwashers, washing machines, and fridges, preferring the highly energy-efficient models. Second, we suggest substituting obsolete incandescent lightbulbs with LED lights. Third, where feasible, we promote the adoption and use of a clothesline, which, in the temperate weather of some Australian cities, could help eliminate the dryer. Finally, upgrade 3 suggests preferring water-efficient shower heads to reduce water consumption.

6.2.3.1 Energy-efficient Appliances

In Australia, appliances are responsible for about 30% of energy use per household (Department of Climate Change, Energy, the Environment and Water, 2022c). The most commonly used home appliances are dishwashers, washing machines, and a combined fridge and freezer. Switching to high-energy-efficient devices can help reduce the overall power density by up to 55%, reducing the households' total energy use by 16.5%. Dishwashers, washing machines, and fridges are expensive devices. Their most energy-efficient models cost over $2,000, $1,500, and $1,400,

Table 6.4 Retrofitting interventions included in upgrade 3 with details on their baseline and improved parameters and cost

Intervention	*Description*	*Baseline parameter*	*Improved parameter*	*Cost per unit*	*Reference*
Energy-efficient appliances	Dishwasher Washing machine Fridge	–	Overall reduced power density by 55% (W/m²)	\$2,000 \$1,500 \$1,400	Greenhouse Energy Minimum Standards Regulator (2022)
LED lighting	Substitution of the incandescent lightbulbs with LED or similarly high-efficiency lights	–	Reduced light power density by 85% (W/m²)	\$50 per light	European Commission Joint Research Centre (2011), Mills and Schleich (2014)
Clothesline	Elimination of dryer	–	Reduced equipment power density by 15% (W/m²)	\$200 per fixture	Greenhouse Energy Minimum Standards Regulator (2022)
Showerhead	Water-efficient showerheads can help reduce water	12.9 L/min	6.7 L/min		Willis et al. (2013)

respectively. Still, the Australian Government's Greenhouse Energy Minimum Standard Regulator provides a free-to-use energy rating calculator to help consumers make an informed decision on the cost of the appliances and the economic savings that the higher energy efficiency allows over ten years (Greenhouse Energy Minimum Standards Regulator, 2022). For example, compared to a 2.5-star rating 7-kg washing machine used three times per week, a five-star device helps save about \$250 over ten years. In another instance, if a family prefers a more expensive \$1,500 five-star rating fridge over a cheaper \$1,000 device, their total expenditure over ten years would be the same, considering that the best option helps them save about \$500 over the ten-year usage period. This proves that, even though the more expensive appliances require a higher initial investment, they grant long-term economic savings and help decrease the household's environmental impact. Fridges, washing machines, dishwashers, and similar goods embody substantial energy and carbon, which should be preserved. For this reason, substituting existing appliances for newer devices should be done responsibly. We do not suggest their substitution if the existing appliances are still in good working condition and their consumption is reasonably low.

6.2.3.2 LED Lighting

A well-lit home brings many benefits to people. Among these, good indoor lighting is linked to health benefits to the occupants, such as mood stabilisation (which, in turn, helps prevent diseases such as depression and metabolic syndrome; Veitch and Gifford [1996], Grimaldi et al. [2009]), but it can also help households save a substantial amount of energy. Australian homes consume between 8 and 15% of their energy from obsolete, inefficient lighting (Department of Climate Change, Energy, the Environment and Water, 2022c). LED lighting can help a household save up to 85% of its power density and 60% of energy use, especially if linked with smart technologies such as automatic dimming (Chew et al., 2017; European Commission Joint Research Centre, 2011; Mills & Schleich, 2014). Additionally, LED lighting is highly versatile in changing hue and intensity to help follow a circadian rhythm, which further helps users' sleep patterns and overall health (Stefani et al., 2021). LED lighting bulbs and fixtures cost about $50 per device, which may result in an about $500 investment when substituting ten fixtures.

6.2.3.3 Clothesline

About 80% of households in the US are fitted with clothes dryers (Bansal et al., 2011), and they are also common in many Australian homes (Romanach et al., 2017). Even though these devices are functional, they require substantial energy to remove the humidity from clean clothes (Ahmadi et al., 2021). Importantly, clothes dryers can be run at higher temperatures and require additional energy to ensure more efficiency against infectious pathogens that might otherwise form in their humid environments (Reynolds et al., 2022). Installing a clothesline could be an effective strategy to shift the use of clothes dryers, especially in a covered portion of the backyard, translating into a yearly saving of about 150 kWh for the cold cycle and 300 kWh for the warm cycle, or $37 and $74 of yearly cost saving, respectively (see the Australian Government Energy Rating Calculator [2023] for more details). Clotheslines are reasonably inexpensive, starting at $200. hanging clothes is more time-consuming than clothes dryers, and they require a backyard or a balcony in apartment buildings with a suitable climate.

6.2.3.4 Showerhead

The water consumed for showering largely depends on socio-economic indicators and country. For a five-minute shower, it can reach up to 51% of the household's water use (Dalmédico Ioris & Ghisi, 2022), or about 64.7 L per person daily (Willis et al., 2013). Water-efficient showerheads can help save almost half of the water used in each shower, resulting in substantial economic savings and beneficial environmental impact. Specifically, substituting the showerheads can result in a water bill reduction of up to $315 per family and reduce energy use, as less water needs to be heated (Department of Climate Change, Energy, the Environment and Water, 2022c). The cost of these devices varies substantially depending on the model and type of shower head (e.g., fixed, overhead, column), but the price of the most simple models starts from about $80.

6.2.4 Retrofitting Upgrade 4 – Solar and Hot Water

The final retrofitting upgrade requires more substantial capital cost but returns significant environmental and economic benefits in the long run. Upgrade 4 entails installing two distinct but complementary systems to reduce the home's reliance on

Table 6.5 Retrofitting interventions included in upgrade 4 with details on their baseline and improved parameters and cost

Intervention	*Description*	*Details*	*Cost per unit*	*Reference*
Solar panel system	Installation of the solar panel system, including battery.	Efficiency = 15% Tilt angle = 30° Panel area (5 kW) = 25 m^2 Storage capacity = 5.8 kWh	\$5,000 (solar panel system) \$5,000 (Battery)	Solar Calculator (2022)
Hot water heat pump	High-efficiency heat water pump (pipe lagging included)	30% less energy compared to traditional hot water systems	\$3,100 installed	Department of Climate Change, Energy, the Environment and Water (2022b)
Induction Cooktop	Higher efficiency compared to LPG tops	The efficiency of induction cooktops is 84%, whilst LPG is 39.9%	\$2,900 installed	Tama (2022)

fossil fuels significantly. The first is a solar panel system, which also incorporates a home battery for energy storage where suitable. The second is a separate hot water heat pump system. Together, these systems form a comprehensive approach to energy efficiency and sustainability. Swapping a gas cooktop for an induction cooktop, alongside these recommended upgrades, would enable a fully electric, efficient, healthy home (Table 6.5).

6.2.4.1 Solar Panel System

The production and installation cost of solar panel systems decrease steeply, and simultaneously, their efficiency increases at the same rate. For these reasons, solar panel systems are becoming extremely popular, helping societies decrease their environmental impact, promoting energy independence, stabilising electricity grids, and, if completed with home batteries, transforming houses into active parts of the energy production industry (Abdalazeem et al., 2022; Ruostetsaari, 2020). The efficiency of the modern commercially available panels varies between 15 and 22% (Solar Calculator, 2022). In these circumstances, a typical system of an Australian home could include a 5-kW solar panel system designed to cover about 25 m^2 of a 30° tilted roof, costing about \$5,000. Solar energy is produced during daylight hours and heavily around midday, but household energy is typically consumed early in the morning and late afternoon. The battery helps shift energy production and use. For Australian households, the battery should have a capacity of at least 5.8 kWh (which also costs about \$5,000) (Phadnis et al., 2019).

6.2.4.2 Hot Water Heat Pump

Heating the households' water for domestic use accounts for about a quarter of the total energy use (Department of Climate Change, Energy, the Environment and Water,

2022c). Costing about $3,100, heat pump water heaters save up to 30% of energy when compared with conventional electric hot water systems. This system can also utilise midday peak solar energy. It is essential to mention that modern systems promise an even higher efficiency by implementing direct-expansion solar-assisted heat pumps (Chow et al., 2010). Still, this technology is not the focus of this study.

6.2.4.3 *Induction Cooktop*

Installing an induction cooktop could cost about $2,900, but it would conclude the electrification process of the house and help households remove endangering gases such as liquified petroleum gas (LPG). At the same time, induction cooktops are safer to use and have an efficiency of 84% compared to the 39.9% efficiency of gas cooktops (Tama, 2022). These technologies, paired with solar panel systems and batteries, could help households save substantial energy and cost on bills (Zhang et al., 2020).

6.3 Retrofitting Initiatives

Australia needs a targeted and coordinated effort to retrofit millions of existing homes. It is an initiative to retrofit homes to support Australia's current and future comfort and energy needs and facilitate the transition to renewable energy. Government, community housing providers, landlords, and homeowners require an informed knowledge base to direct the housing sector towards providing the retrofitting solutions needed to improve Australia's housing stock quality.

This section will explore some barriers and opportunities for home retrofits in Australia.

6.3.1 Scale and Complexity of Retrofitting and Breakdown of Australian Homes

Australia is home to a substantial number of single dwellings, which sums up 86% of all households in the nation. To break this down into ownership schemes, only 29% of single dwellings are owned outright, 36% have a mortgage, and 20% are rented (Kumar et al., 2021). Most of these homes, particularly those built before the 2000s, urgently need retrofitting to bring them up to the modern standard of construction established by NatHERS. However, two core complexities stand in the way. First, the geographical diversity of Australia means various climate zones that necessitate a range of retrofitting techniques and materials (Dowling et al., 2014). Second, the construction industry's lack of specific retrofitting skills and standardised processes makes it exponentially challenging to plan for a unified approach to large-scale retrofitting (Pasichnyi et al., 2019; Späth & Rohracher, 2015). The challenges extend into the financial domain due to the diverse ownership status of these single dwellings. This problem is increased by the typically sizeable initial investment necessary to retrofit a home. Also, financing strategies will differ depending on whether the home is owned outright, mortgaged, or rented, which might require a nuanced approach to funding. The available financing options and incentives for retrofitting are currently limited, and the high upfront costs might often deter potential retrofit projects. Technological advancements could help solve these issues. Emerging materials and construction methods, currently in the research and development phase, could potentially revolutionise the retrofitting landscape by offering cost-effective and easily implementable solutions (Seghier et al., 2022). Problematically, despite these advances, Australian homeowners' attitudes towards sustainability and initial investment costs are historically significant behavioural factors affecting the adoption rate of these new technologies (Bond, 2011). Thus, the scale and complexity of retrofitting Australian homes are influenced by numerical or logistical factors and socio-economic and behavioural dimensions. Meeting this challenge requires a multi-pronged approach that addresses the physical and financial aspects and the human elements involved, as homeowners often fail to see the health, economic, and ecological advantages of retrofitting their homes.

6.3.2 Consumer Preferences, Financial and Non-Financial Barriers in Home Retrofitting

Insights were gathered from an industry reference group of focus groups involving government experts from most Australian states and key stakeholders in the building retrofitting sector. These discussions aimed to assess consumer preferences in the context of home retrofitting. These focus groups suggest that energy efficiency is not a high priority for homeowners and that they are "*Not really interested*" in energy efficiency retrofits. An initiative must understand homeowners' drivers and the trigger where they are most open to improving their home's livability. Would health and comfort be a higher priority for occupants and homeowners? At present, comfort is deemed a luxury in Australian homes. A home retrofit initiative should transform the market to demand comfort as a standard. Using a values-based approach to understanding customer motivations is an effective way to overcome intervention barriers.

The Green Finance Institute (2020) identified various financial and non-financial barriers that owner-occupiers could experience. Similar barriers exist for other sections of the market. Understanding and overcoming these barriers is critical to improving the Australian housing stock. A summary of these barriers is provided in Table 6.6. To make the problem of large-scale retrofitting more complex, these barriers are sometimes interlinked. For example, a lack of access to capital means that homeowners cannot afford upfront costs for improvements. This might also be compounded by their need for more awareness about the cost savings related to such improvements. Similarly, the duration of tenancy can create uncertainty in the financial availability, especially when the owners lack initial access to capital, as it would reflect on the low confidence in seeking full repayment via energy savings and tenants' rent charges.

6.3.3 Initiative Design Principles

A large-scale home retrofit initiative is required to support and enable market transformation. Combined with policy and regulatory alignment and a broader suite of

Table 6.6 Financial and non-financial barriers to retrofit projects in owner-occupied households

Financial barriers	*Non-financial barriers*
High upfront costs for improvements. **Lack of access to capital.** **Low confidence in energy bill savings: A barrier for homeowners seeking full repayment via energy savings.** **Duration of tenancy: Energy bill savings may not accrue to the original homeowner if they move property.** **Property value-add: Efficiency improvements not considered to increase and protect property values.** **Availability and accessibility of products: Low penetration and availability of attractive financial offers for efficiency measures.**	Homeowners' genuine concern for climate change is disconnected from the energy efficiency of their property. A lack of educational efforts from professional influencers means information is not widespread among homeowners. Lack of good quality information and support on products, choices, and suppliers to embark on a "sustainable renovation journey". Lengthy timelines can deter homeowners from initiating retrofit projects. Coordination and permitting requirements add layers of inconvenience. Navigating material choices, installation, and finances can overwhelm homeowners. Concerns about the reliability and quality of materials and services are linked to fragmented and inefficient supply chains (i.e., from materials to skilled labour to transportation). Leaseholders gaining permission: Getting collective agreement amongst groups of share-of-freeholders

actions, this will create the environment and mechanism for the thermal and energy efficiency retrofit of millions of homes across Australia. The initiative must have clear and demonstrable goals, as set out below in the following six fundamental principles.

1 *Creating future-ready homes.* A home retrofit initiative in Australia should aim to create future-ready homes that are comfortable, energy-efficient, and have a path towards electrification. The initiative should also improve the NatHERS rating of each home, help towards market transformation, and promote a culture shift towards demanding energy-efficient homes. To ensure fairness across the Nation, the delivery of home retrofits should be on a large scale across all states and regions in Australia.
2 *Focusing on single-dwelling homes.* A home retrofit initiative should primarily target NCC Class 1a single dwellings, irrespective of whether they are owner-occupied, tenanted, mortgaged, or owned outright. Particular attention should be given to those among them that are of poor built quality.
3 *Building and maintaining trust with homeowners.* A home retrofit initiative needs to establish and maintain trust with homeowners by offering streamlined

processes, including finance and retrofit options, that meet the needs of homeowners. Retrofit options should deliver best case outcomes regarding comfort and return on investment, with benefits clearly understood by each homeowner.

4 *Engaging at trigger points.* The home retrofit initiative needs to engage with homeowners at specific trigger points, which include the point of sale/purchase of a home, the point of advertisement for lease, and the point of renovation.
5 *Strategic communication and stakeholder engagement.* The messaging should use a values-based approach to understand customer motivations and overcome intervention barriers. Furthermore, there is a need for further research on optimised home comfort benefits, which should be communicated to the target market and government.
6 *Providing whole-home assessments and quality control.* An accredited assessment process is necessary to provide homeowners with independent technical guidance, while a scalable but rigorous quality control process ensures consistent quality in the scheme's delivery.

6.4 Conclusions

A home retrofit initiative should focus on creating future-ready homes in two dimensions. First, future-ready homes should be highly energy efficient with a clear path towards electrification, meaning no new gas appliances should be installed, and existing ones should be replaced with efficient electric systems. Second, these homes should also aim to enhance the well-being of their occupants by offering comfort without excessive heating or cooling. These two facets can create a broad vision for future-ready homes.

The three main takeaways from this chapter can be summarised as follows:

1 Home thermal and energy efficiency require a holistic approach beyond installing solar panels. This chapter proposes a guide on the most relevant retrofitting upgrades, explored hierarchically in order of their importance and benefits to the households.
2 These four hierarchical upgrades for retrofitting Australian homes improve thermal efficiency and comfort, including insulation, ceiling fans, efficient lighting and water use, and complete conversion to electric homes.
3 The proposed retrofitting options can improve the quality of existing Australian homes, but they can also be costly, and the costs vary depending on the supply chain and industry demand.

Finally, the chapter discusses the importance of home thermal and energy efficiency in modern homes and the need for retrofitting existing homes to achieve modern performance standards. It highlights the critical role of retrofitting in improving the quality and decreasing the running costs of houses. The chapter specifically focuses on the Australian context, where many homes underperform regarding build quality and overall thermal comfort. The authors propose a fit-for-purpose retrofit solution based on four hierarchical upgrades that aim to transform the Australian home environment. The proposed retrofitting scheme emphasises the need for a holistic intervention beyond quick solutions such as installing solar panels to ensure sustainable operations. Finally, the chapter provides new insights into the importance of retrofitting in achieving sustainable building operations and being more economical in the long run, with remarkable yet hard-to-quantify improvements in health and comfort.

Reference List

Abdalazeem, M. E., Hassan, H., Asawa, T., & Mahmoud, H. (2022). Review on integrated photovoltaic-green roof solutions on urban and energy-efficient buildings in hot climates. *Sustainable Cities and Society*, *82*, 103919. https://doi.org/10.1016/j.scs.2022.103919.

Ade, R., & Rehm, M. (2020). Cold comfort: A post-completion evaluation of internal temperatures and thermal comfort in 6-Homestar dwellings. *Building and Environment*, *167*, 106466. https://doi.org/10.1016/j.buildenv.2019.106466.

Ahmadi, M., Gluesenkamp, K. R., & Bigham, S. (2021). Energy-efficient sorption-based gas clothes dryer systems. *Energy Conversion and Management*, *230*, 113763. https://doi.org/10.1016/j.enconman.2020.113763.

Australian Building Codes Board. (2022a). *Australian Building Codes Board – The National Construction Code*. https://www.abcb.gov.au/

Australian Building Codes Board. (2022b). *National Construction Code, Building Classification*. https://www.abcb.gov.au/sites/default/files/resources/2022/UTNCC-Building-classifications.PDF

Australian Government – Your Home. (2021). *Heating and Cooling*. Retrieved 18/09/2022 from https://www.yourhome.gov.au/energy/heating-and-cooling

Australian Government Energy Rating Calculator. (2023). *Energy Rating Calculator*. https://calculator.energyrating.gov.au/

Bansal, P., Vineyard, E., & Abdelaziz, O. (2011). Advances in household appliances – A review. *Applied Thermal Engineering*, *31*(17), 3748–3760. https://doi.org/10.1016/j.applthermaleng.2011.07.023

Beng Ooi, K., Noguchi, M., & Wah Chau, H. (2017). Sustainable heating or cooling and ventilation of affordable zero-energy housing. *Procedia Engineering*, *205*, 1294–1301. https://doi.org/10.1016/j.proeng.2017.10.383.

Bond, S. (2011). Barriers and drivers to green buildings in Australia and New Zealand. *Journal of Property Investment & Finance*, *29*(4/5), 494–509. https://doi.org/10.1108/14635781111150367

Bulut, M. B., Wilkinson, S., Khan, A., Jin, X.-H., & Lee, C. L. (2022). Thermal performance of retrofitted secondary glazed windows in residential buildings – Two cases from Australia. *Smart and Sustainable Built Environment*, *11*(4), 1182–1192. https://doi.org/10.1108/SASBE-03-2021-0050

Chalmers, M. (2022). Australian homes are so cold that some fall below the WHO's recommended 'safe' temperature. *ABC News*. Retrieved 01/09/2022 from https://www.abc.net.au/news/2022-07-15/why-are-australian-homes-so-cold/101227308

Chew, I., Karunatilaka, D., Tan, C. P., & Kalavally, V. (2017). Smart lighting: The way forward? Reviewing the past to shape the future. *Energy and Buildings*, *149*, 180–191. https://doi.org/10.1016/j.enbuild.2017.04.083.

Chow, T. T., Pei, G., Fong, K. F., Lin, Z., Chan, A. L. S., & He, M. (2010). Modeling and application of direct-expansion solar-assisted heat pump for water heating in subtropical Hong Kong. *Applied Energy*, *87*(2), 643–649. https://doi.org/10.1016/j.apenergy.2009.05.036

Commonwealth Scientific and Industrial Research Organisation. (2016). *Australian Housing Data*. Retrieved 05/09/2022 from https://ahd.csiro.au/

Dalmédico Ioris, I., & Ghisi, E. (2020). Financial feasibility of implementing an automated system to save the cold potable water that would be wasted at the beginning of a hot shower. *Urban Water Journal*, 19(1), 22–31. https://doi.org/10.1080/1573062X.2021.1949479

Daniel, L., Baker, E., & Williamson, T. (2019). Cold housing in mild-climate countries: A study of indoor environmental quality and comfort preferences in homes, Adelaide, Australia. *Building and Environment*, *151*, 207–218. https://doi.org/10.1016/j.buildenv.2019.01.037.

Department of Climate Change, Energy, the Environment and Water. (2018). *Trajectory for Low Energy Buildings*. https://web.archive.org.au/awa/20210603110748mp_/https://energyministers.gov.au/sites/prod.energycouncil/files/publications/documents/Trajectory%20for%20Low%20Energy%20Buildings.pdf

Department of Climate Change, Energy, the Environment and Water. (2022a). *Heating and Cooling*. Retrieved 18/09/2022 from https://www.energy.gov.au/households/heating-and-cooling#toc-anchor-reverse-cycle-air-conditioning

Department of Climate Change, Energy, the Environment and Water. (2022b). *Hot Water Systems*. Retrieved 18/09/2022 from https://www.energy.gov.au/households/hot-water-systems

Department of Climate Change, Energy, the Environment and Water. (2022c). *Quick Wins*. Retrieved 18/09/2022 from https://www.energy.gov.au/households/quick-wins

Dowling, R., Mguirk, P., & Bulkeley, H. (2014). Retrofitting cities: Local governance in Sydney, Australia. *Cities*, *38*, 18–24. https://doi.org/10.1016/j.cities.2013.12.004.

European Commission Joint Research Centre. (2011). *European LED Quality Charter*. Retrieved 18/09/2022 from https://e3p.jrc.ec.europa.eu/sites/default/files/files/documents/Lighting/eu_led_quality_charter.pdf

Fernández-Agüera, J., Domínguez-Amarillo, S., & Campano, M. Á. (2019). Characterising draught in Mediterranean multifamily housing. *Sustainability*, *11*(8), 2433. https://www.mdpi.com/2071-1050/11/8/2433

Fisk, W. J., Singer, B. C., & Chan, W. R. (2020). Association of residential energy efficiency retrofits with indoor environmental quality, comfort, and health: A review of empirical data. *Building and Environment*, *180*, 107067. https://doi.org/10.1016/j.buildenv.2020.107067.

Florides, G. A., Tassou, S. A., Kalogirou, S. A., & Wrobel, L. C. (2002). Measures used to lower building energy consumption and their cost effectiveness. *Applied Energy*, *73*(3), 299–328. https://doi.org/10.1016/S0306-2619(02)00119-8

Fox-Reynolds, K., Vines, K., Minunno, R., Wilmot, K. (2021). *Pathways to Scale: Retrofitting One Million+ Homes* (Research Theme H2: Enhancing home thermal inertia, Issue. https://issuu.com/racefor2030/docs/one_million_homes_final_report_9.12.21

Greenhouse Energy Minimum Standards Regulator. (2022). *Energy Rating*. Retrieved 18/09/2022 from https://www.energyrating.gov.au/

Green Finance Institute. (2022). *Financing energy efficient buildings: The path to retrofit at scale*. Retrieved 18/09/2024 from https://www.greenfinanceinstitute.com/insights/financing-energy-efficient-buildings-the-path-to-retrofit-at-scale/

Grimaldi, S., Englund, A., Partonen, T., Haukka, J., Pirkola, S., Reunanen, A., Aromaa, A., & Lönnqvist, J. (2009). Experienced poor lighting contributes to the seasonal fluctuations in weight and appetite that relate to the metabolic syndrome. *Journal of Environmental and Public Health*, *2009*, 165013. https://doi.org/10.1155/2009/165013.

Gui, X., & Gou, Z. (2022). Household energy technologies in New South Wales, Australia: Regional differences and renewables adoption rates. *Renewable and Sustainable Energy Reviews*, *160*, 112290. https://doi.org/10.1016/j.rser.2022.112290.

Hadengue, B., Morgenroth, E., & Larsen, T. A. (2022). Screening innovative technologies for energy-efficient domestic hot water systems. *Journal of Environmental Management*, *320*, 115713. https://doi.org/10.1016/j.jenvman.2022.115713.

Hardy, A., Glew, D., Gorse, C., & Fletcher, M. (2018). Validating solid wall insulation retrofits with in-use data. *Energy and Buildings*, *165*, 200–205.

Hasama, T., Mihara, K., Sekhar, C., Cheong, K. W. D., & Tham, K. W. (2022). Assessment of airflow and heat transfer around a thermal manikin in a premise served by DOAS and ceiling fans. *Building and Environment*, *214*, 108902. https://doi.org/10.1016/j.buildenv.2022.108902.

Hong, T., Chen, Y., Luo, X., Luo, N., & Lee, S. H. (2020). Ten questions on urban building energy modeling. *Building and Environment*, *168*, 106508. https://doi.org/10.1016/j.buildenv.2019.106508.

Horne, R., & Hayles, C. (2008). Towards global benchmarking for sustainable homes: An international comparison of the energy performance of housing. *Journal of Housing and the Built Environment*, *23*(2), 119–130. https://doi.org/10.1007/s10901-008-9105-1

Jason Windows. (2022). *What Is Double Glazing?* Retrieved 18/09/2022 from https://jasonwindows.com.au/glass/what-is-double-glazing/

Jelle, B. P., Hynd, A., Gustavsen, A., Arasteh, D., Goudey, H., & Hart, R. (2012). Fenestration of today and tomorrow: A state-of-the-art review and future research opportunities. *Solar Energy Materials and Solar Cells*, *96*, 1–28. https://doi.org/10.1016/j.solmat.2011.08.010.

Kumar, D., Alam, M., & Sanjayan, J. G. (2021). Retrofitting building envelope using phase change materials and aerogel render for adaptation to extreme heatwave: A multi-objective analysis considering heat stress, energy, environment, and cost. *Sustainability*, *13*(19), 10716. https://www.mdpi.com/2071-1050/13/19/10716

Lawania, K. K., & Biswas, W. K. (2016). Achieving environmentally friendly building envelope for Western Australia's housing sector: A life cycle assessment approach. *International Journal of Sustainable Built Environment*, *5*(2), 210–224. https://doi.org/10.1016/j.ijsbe.2016.04.005

Marini, D., Buswell, R. A., & Hopfe, C. J. (2021). Development of a dynamic analytical model for estimating waste heat from domestic hot water systems. *Energy and Buildings*, *247*, 111119. https://doi.org/10.1016/j.enbuild.2021.111119.

Mills, B., & Schleich, J. (2014). Household transitions to energy efficient lighting. *Energy Economics*, *46*, 151–160. https://doi.org/10.1016/j.eneco.2014.08.022.

Muddu, R. D., Gowda, D. M., Robinson, A. J., & Byrne, A. (2021). Optimisation of retrofit wall insulation: An Irish case study. *Energy and Buildings*, *235*, 110720. https://doi.org/10.1016/j.enbuild.2021.110720.

Oldfield, P. (2022). *Freezing Indoors? That's Because Australian Homes Are Closer to Tents Than Insulated Eco-Buildings*. Retrieved 01/09/2022 from https://www.theguardian.com/commentisfree/2022/jun/13/freezing-indoors-thats-because-australian-homes-are-closer-to-tents-than-insulated-eco-buildings

Omer, A. M. (2008). Renewable building energy systems and passive human comfort solutions. *Renewable and Sustainable Energy Reviews*, *12*(6), 1562–1587. https://doi.org/10.1016/j.rser.2006.07.010

Pasichnyi, O., Levihn, F., Shahrokni, H., Wallin, J., & Kordas, O. (2019). Data-driven strategic planning of building energy retrofitting: The case of Stockholm. *Journal of Cleaner Production*, *233*, 546–560. https://doi.org/10.1016/j.jclepro.2019.05.373.

Phadnis, N., Yang, R. J., Wijeratne, P. U., Zhao, H., & Liu, C. (2019). *The impact of solar PV design tilt and orientation on project values*. In P. Kaparaju, R. Howlett, J. Littlewood, C. Ekanyake, & L. Vlacic (Eds.), *Sustainability in energy and buildings 2018. KES-SEB 2018. Smart innovation, systems and technologies* (Vol. 131., pp. 301–310). Springer.

Reynolds, K. A., Verhougstraete, M. P., Mena, K. D., Sattar, S. A., Scott, E. A., & Gerba, C. P. (2022). Quantifying pathogen infection risks from household laundry practices. *Journal of Applied Microbiology*, *132*(2), 1435–1448. https://doi.org/10.1111/jam.15273

Romanach, L., Hall, N., & Meikle, S. (2017). Energy consumption in an ageing population: Exploring energy use and behaviour of low-income older Australians. *Energy Procedia*, *121*, 246–253. https://doi.org/10.1016/j.egypro.2017.08.024.

Ruostetsaari, I. (2020). From consumers to energy citizens: Finns' readiness for demand response and prosumerism in energy policy making. *International Journal of Energy Sector Management*, *14*(6), 1157–1175. https://doi.org/10.1108/IJESM-11-2019-0001
Safe Style Windows. (2022). *How Much Energy Are You Losing Through Old Windows?* Retrieved 18/09/2022 from https://www.safestyle-windows.co.uk/ask-an-expert/energy-saving/how-much-energy-is-lost-through-windows/#:~:text=Around%2018%25%20of%20the%20total,is%20lost%20through%20the%20windows.
Schiavon, S., & Melikov, A. K. (2008). Energy saving and improved comfort by increased air movement. *Energy and Buildings*, *40*(10), 1954–1960. https://doi.org/10.1016/j.enbuild.2008.05.001.
Seghier, T. E., Lim, Y.-W., Harun, M. F., Ahmad, M. H., Samah, A. A., & Majid, H. A. (2022). BIM-based retrofit method (RBIM) for building envelope thermal performance optimization. *Energy and Buildings*, *256*, 111693. https://doi.org/10.1016/j.enbuild.2021.111693.
Shen, P., Braham, W., & Yi, Y. (2019). The feasibility and importance of considering climate change impacts in building retrofit analysis. *Applied Energy*, *233-234*, 254–270. https://doi.org/10.1016/j.apenergy.2018.10.041.
Shriram, S., & Ramamurthy, K. (2019). Assessment of CO_2-based demand controlled ventilation requirement for a flexible work environment with ductless split air conditioners. *Science and Technology for the Built Environment*, *25*(7), 805–818. https://doi.org/10.1080/23744731.2019.1569458
Simko, T., & Moore, T. (2021). Optimal window designs for Australian houses. *Energy and Buildings*, *250*, 111300. https://doi.org/10.1016/j.enbuild.2021.111300.
Solar Calculator. (2022). *Solar Panel Efficiency*. https://solarcalculator.com.au/solar-panel-efficiency/
Southwell, D. (2022). Australian homes are now so cold they are falling below WHO safety guidelines – As experts warn houses down under weren't designed for a real winter. *Daily Mail Australia*. Retrieved 01/09/2022 from https://www.dailymail.co.uk/news/article-11028107/Australian-homes-cold-fall-safety-standards.html
Späth, P., & Rohracher, H. (2015). Conflicting strategies towards sustainable heating at an urban junction of heat infrastructure and building standards. *Energy Policy*, *78*, 273–280. https://doi.org/10.1016/j.enpol.2014.12.019.
Stefani, O., Freyburger, M., Veitz, S., Basishvili, T., Meyer, M., Weibel, J., Kobayashi, K., Shirakawa, Y., & Cajochen, C. (2021). Changing color and intensity of LED lighting across the day impacts on circadian melatonin rhythms and sleep in healthy men. *Journal of Pineal Research*, *70*(3), e12714. https://doi.org/10.1111/jpi.12714
Tam, V. W. Y., Le, K. N., & Wang, J. Y. (2018). Cost implication of implementing external facade systems for commercial buildings. *Sustainability*, *10*(6), 1917. https://www.mdpi.com/2071-1050/10/6/1917
Tama, A. (2022). Deployment of electric induction technologies into cooktops plates as a part of energy sustainability. *Smart Grid and Renewable Energy*, *13*(3), 55–74.
Veitch, J. A., & Gifford, R. (1996). Assessing beliefs about lighting effects on health, performance, mood, and social behavior. *Environment and Behavior*, *28*(4), 446–470. https://doi.org/10.1177/0013916596284002
Willis, R. M., Stewart, R. A., Giurco, D. P., Talebpour, M. R., & Mousavinejad, A. (2013). End use water consumption in households: Impact of socio-demographic factors and efficient devices. *Journal of Cleaner Production*, *60*, 107–115. https://doi.org/10.1016/j.jclepro.2011.08.006.
Wills, A. D., Beausoleil-Morrison, I., & Ugursal, V. I. (2021). A modelling approach and a case study to answer the question: What does it take to retrofit a community to net-zero energy? *Journal of Building Engineering*, *40*, 102296. https://doi.org/10.1016/j.jobe.2021.102296.
Wimalasena, N. N., Chang-Richards, A., Wang, K. I. K., & Dirks, K. N. (2022). What makes a healthy home? A study in Auckland, New Zealand. *Building Research & Information*, *50*(7), 738–754. https://doi.org/10.1080/09613218.2022.2043138

World Health Organization. (2018). WHO housing and health guidelines: *Recommendations to promote healthy housing for a sustainable and equitable future*. Retrieved 18/09/2024 from https://www.who.int/publications/i/item/9789241550376

Xu, Y., Raja, S., Ferro, A. R., Jaques, P. A., Hopke, P. K., Gressani, C., & Wetzel, L. E. (2010). Effectiveness of heating, ventilation and air conditioning system with HEPA filter unit on indoor air quality and asthmatic children's health. *Building and Environment*, *45*(2), 330–337. https://doi.org/10.1016/j.buildenv.2009.06.010

Zhang, J., Raufer, R., & Liu, L. (2020). Solar home systems for clean cooking: A cost–health benefit analysis of lower-middle-income countries in Southeast Asia. *Sustainability*, *12*(9), 3909.

7 Energy Efficiency Retrofitting of Existing Building Stock for Net Zero

Sentiment Analysis of Social media Data

Mershack Opoku Tetteh, Amos Darko, Emmanuel B. Boateng, and Albert P.C. Chan

7.1 Introduction

Despite significant advances in building technology, the built environment yet accounts for a disproportionate share of global carbon emissions and energy consumption. The existing building stock is responsible for a significant portion of the associated emissions and energy consumption. According to the World Green Building Council (2021), approximately 75% of the total emissions and energy consumption can be attributed to the operation and maintenance of existing buildings. In contrast, only 25% of these factors are associated with the embodied emissions and energy use of new buildings. Energy efficiency retrofitting of existing buildings (EEREB) is essential and cost-effective to deliver decarbonization. There are also other benefits that come along with this, such as improved air quality, reduced health risks, and cost savings from lower energy consumption and bills. The EEREB initiative has been globally implemented as a priority. For example, in the US, the government has conducted a renovation plan to improve energy performance by 2030 below 2005 levels (The US Department of State, 2021). According to Tan et al. (2018), the German government took measures to improve the energy efficiency of more than 900 older buildings by retrofitting them with energy-efficient heating and cooling systems. Department for Business, Energy and Industrial Strategy (2020) reports that the UK government has set an equally lofty goal of enhancing the thermal and energy performance of the country's present housing portfolio while lowering emissions by 68% by 2030.

To effectively promote widespread adoption of EEREB practices through policy interventions, a complete comprehension of how people make decisions regarding house retrofits is essential. Past research on public opinion on EEREB has not, however, offered a thorough assessment of the topic. This lack of comprehensive evaluation, as highlighted by Galvin (2014a), Tjørring and Gausset (2019), and Huang et al. (2021), hinders our ability to fully grasp the public's sentiments. Consequently, it becomes challenging to develop policies that are both effective and actionable. One common concern in the literature is that factors influencing the general public's retrofit decision-making are not adequately addressed by current policy design. So, while techno-economic models and explanations have largely shaped policy actions, previous research has continuously argued for the

DOI: 10.1201/9781003317890-9

incorporation of social, political, and practice-based factors (He et al., 2019; Kerr and Winskel, 2020). Understanding public opinions and perceptions does not ensure success, but ignorance of it almost certainly will (Boudet, 2019). It is essential for stakeholders, especially retrofit professionals, energy managers and auditors, and legislators, to have a thorough grasp of the public's perception of EEREB to facilitate effective communication. This understanding can serve multiple purposes, such as providing valuable insights into public reactions to new retrofit technologies, facilitating educational initiatives, and ultimately enhancing overall communication strategies. Studies that have evaluated public opinions and perceptions toward EEREB primarily relied on subjective methods such as asking homeowners and residents to report their own needs and perceptions of energy efficiency improvements in existing buildings via questionnaires and surveys (Gram-Hanssen, 2014; He et al., 2019; Kerr et al., 2018). Limited understanding of public requirements, difficulties in collecting real-time opinions and emotions, and the chance of respondent manipulation are all problems with the conventional method (Wu et al., 2021). On top of that, it ignores the concerns of the general public. Further, most of the current research only takes into account the opinions of specialists or subsets of the population that have undergone building retrofits in a particular nation or city. Thus, the policy design link is compromised, and public retrofit decision-behavior is not supported. Tjørring and Gausset (2019), for example, surveyed 12 Danish homeowners who had previously been advised on energy retrofits in order to uncover their impressions of these projects. Only 90 Singaporean residents who had participated in a green retrofitting pilot program were interviewed and analyzed by Hwang et al. (2017). While these studies have improved retrofit policy design in some ways, a more comprehensive and nuanced understanding of public opinion could be even more useful. An inclusive perspective, or a combined set of diverse city-scale population attitudes, is essential for comprehending the public's needs and views toward retrofit, as pointed out by Liu et al. (2022). When it comes to optimizing policies to advance the EEREB practice, a focus in this area is helpful since it allows for the encoding and harnessing of different motivations and needs, including social concerns and cultural values. In a nutshell, the *full* analysis of responses from the general public has not yet been investigated.

The purpose of this chapter is to assess the overall needs, concerns, and perceptions of the general public regarding EEREB. This will be achieved by analyzing comments from the social media platform YouTube. The advancements in big data analytics and the increasing popularity of social media have supported the evaluation of the concerns, attitudes, and motivations of the general public toward a specific service or product in a more objective way than ever before. The vast audience size of YouTube makes it an ideal medium for spreading important messages and promoting eco-friendly lifestyle choices. This chapter provides a rich and varied picture of the public's needs, concerns, and perceptions in relation to EEREB, which is an important addition to the existing body of knowledge. Having a better understanding of this concept can greatly benefit policymakers and authorities in improving the promotion of EEREB in terms of publicity and policy. In other words, policymakers can tailor existing policies to the preferences of the public, resulting in greater energy efficiency benefits for households.

7.2 Retrofit, Public Acceptance, and Policy Design Research

The term "retrofit" refers to an improvement made to old, underperforming structures in order to bring them up to date with modern norms or standards or to get them ready for the future. It also refers to household operations that go beyond standard servicing and repairs and is a metaphor for refurbishing, remodeling, rehabilitation, and restoration (Karvonen, 2013). When referring to energy retrofit programs, the term "public acceptance" refers to the degree to which the public pays attention to them and actively participates in them. According to Devine-Wright (2007), it is affected by how they make decisions and what motivates them. According to Judson and Maller (2014), the best way to decrease energy usage, especially in dwellings, is to encourage people to retrofit their homes. A lot of study has focused on public acceptance and energy retrofitting in existing buildings (He et al., 2019; Tjørring and Gausset, 2019). The expanded theory of planned behavior was applied in the research that was carried out by He et al. (2019) in order to analyze the factors that determine public intention, particularly among residents, with regard to the implementation of green retrofitting in residential structures. Through a questionnaire survey with 507 residents, it was identified that policy, behavior, cognition, perceived behavioral control, and subjective norms are the most important factors influencing the public's intention toward green retrofitting. The understanding of how the public perceives energy retrofits is still limited. In contributing to filling this gap, Tjørring and Gausset (2019) used an extensive anthropological fieldwork study to investigate retrofits from homeowners' day-to-day experiences. It came to light that public social demands, rather than techno-economic needs, are the deciding factors in retrofit decisions. When people have social needs, such as having homes that are part of daily life and support personal social events, relations, status, and values, they may be more inclined toward retrofitting their homes to meet these needs while also improving energy efficiency. In addition, how energy retrofit provides "an answer to public health costs of fuel poverty" can be found in Avanzini et al. (2022). According to Avanzini et al. (2022), individuals experiencing fuel poverty not only face inadequate levels of indoor air quality (IAQ) but also endure thermal discomfort. There is a significant likelihood that individuals may experience various health conditions, including physical ailments such as cardiovascular and respiratory issues, as well as mental health challenges such as depression and stress. This, in turn, can lead to increased hospital admissions and overall negative health outcomes. Retrofit improvements are the only solution to address these challenges. According to Bouzarovski and Petrova (2015), fuel poverty occurs when households have financial constraints, high energy expenditures, and inefficient heating, cooling, and electrical systems that make it difficult, if not impossible, for them to meet their energy needs, including thermal comfort. People may choose to upgrade their buildings in order to enhance their thermal comfort and reduce their energy usage when they feel the need to do so. Addressing the public's need for information on the comfort-related impacts of energy retrofit has been demonstrated to have an impact on people's acceptance of energy retrofit (Ossokina et al., 2021). Having the needs is one thing and the perception of the public toward whether the retrofit can help to meet those needs is another thing. In the case where such perception is positive, retrofit may be viewed positively and supported. This indicates that both public needs and perceptions can shape existing building energy retrofit.

Numerous variables affect the level of public support for energy retrofits, according to prior research. According to Liu et al. (2015), there are several factors that play a role in this process. These include demographics, which include factors such as age, education level, and household type; family dynamics, which include things such as life stage, responsibilities, and preferences of children; beliefs, which are typically seen as drivers, along with past retrofit experiences; home characteristics; authorities and policies, information availability; and social influence. Homeowners who are younger and better educated are more likely to be ecologically concerned and to undertake energy retrofit projects, according to some research (Azizi et al., 2019; Mortensen et al., 2016). But contradictory results have been reported by other studies (Trotta, 2018; Urban and Ščasný, 2012). Furthermore, numerous macro policies have been put in place to encourage the public to accept energy retrofits, including legislation, information and knowledge dissemination, incentives, and training. But, as pointed out by Kerr et al. (2018), these policies' effectiveness differs among countries.

Due to the liberal approach to financial administration, policy options in many nations that concentrate on homeowners' activities have consistently supported market forces and information dissemination instead of legislative norms that mandate homeowners' behavior modification (Kerr and Winskel, 2020). In recent times, several nations have experienced a significant increase in energy retrofit due to the implementation of market-based regulations. However, it is crucial to emphasize that these regulations have primarily been focused on prioritizing cost-effectiveness (European Commission, 2010). The design of macro policies to promote energy retrofit is largely influenced by theories that focus on behavior and decision-making. These theories have been extensively researched by Wilson and Dowlatabadi (2007) and Maller and Horne (2011). According to Jackson (2005), these models depict humans as being primarily concerned with their local interests and driven by a strong desire to pursue cost-effective opportunities. Therefore, policies that stem from this perspective tend to focus excessively on financial aspects and, as a result, do not adequately consider the non-monetary factors that influence behavior (Ackerman, 2008). Behavioral economics and social practice are two theories that run counter to the expectations of rational choice. People are generally considered to be reluctant to make substantial decisions in behavioral economics studies (Gigerenzer and Todd, 1999). Social practice theory, on the other hand, views actors more as a "crossing point for practice." This shifts the emphasis away from people as the primary concern of investigators (Karvonen, 2013). According to the latter perspective, human choices and the amount of energy consumed are influenced more by the socio-technical environment than individual initiative.

Recent studies have emphasized the importance of conducting investigations that are closely related to the dynamics of household life (Kerr et al., 2018; Tjørring and Gausset, 2019). This is because previous explanatory variables have been found to have a direct impact only on actions (Wilson et al., 2013).

While the current studies provide valuable insights into the factors influencing public acceptance of energy retrofit, it is important to note that they have limitations. These studies tend to have a narrow focus and do not fully capture the global public sentiment on this matter. Besides, the approaches used in these studies are traditional, subjective in nature, and largely criticized for the failure to capture the opinions and emotions of the general public in a timely manner (Gasco et al., 2020; Wang et al., 2022). Last, majority of these research are centered on the sentiments of the public who are already interested in retrofit, thereby lacking an *objective* and *holistic* interpretation of public sentiment. This chapter addresses these knowledge gaps by using social media data related to EEREB.

7.3 Methods

7.3.1 Data Extraction

Guided by a set of criteria, the string "energy efficiency retrofitting of existing building" was searched on YouTube on 18 April 2022. According to Balakrishnan and Griffiths (2017), YouTube is the most popular website for presenting videos, which is why it was selected above other social media platforms. As a result, it is suitable for promoting lifestyles that are more sustainable and respectful of the environment. In addition, YouTube has more than 2 billion users who are active on the platform on a monthly basis, and about 43% of all internet users in the world access YouTube on regular basis (Dean, 2021). Among the many popular building and architecture-related channels on YouTube, two stand out: "The BIM" (2.83 million followers) and "Modern Builds" (1.74 million followers). Videos were chosen for further analysis based on three criteria: having over 1000 views, receiving five or more comments, and focusing on the topic of EEREB. This set of criteria was important to ensure a desirable level of interest, engagement, and content appropriacy of the subject matter. The video identifications (IDs) of 22 selected films were compiled from the search results; these videos center on the retrofitting of existing structures to reduce energy consumption. In order to access the YouTube Application Programming Interface (API), the vosonSML package was used in conjunction with the video IDs. The access to the YouTube API was authenticated using an associated Google developer API key. Tools for building analytical networks and accessing data from social media are part of the vosonSML package (Gertzel et al., 2022). No limit was placed on the number of comments that could be collected. A total of 3306 comments (including threads and replies) were collected after the extraction process, amassing over 2 million views on the videos. Overall, the average number of comments was 150,278, minimum was 8, and a maximum of 584. The average number of views was 112,640,455, minimum was 1229, and a maximum of 480,979. Each video looked at retrofitting of existing buildings: to an efficient passive house, to upgrade energy efficiency and indoor air quality; for low energy and serious comfort, for continuity of insulation and ventilation, to reduce carbon footprint, to achieve net zero emissions home, and/or to achieve a deep retrofit.

An Intel i7-9750H processor running at 2.60 GHz, 16GB of memory, Windows 11 Pro, and an NVIDIA GeForce RTX 2060 graphics card with 6GB of RAM were used for all calculations and experiments. With RStudio v2022.07.1, we were able to retrieve data, pre-process it, analyze sentiment, and model topics.

7.3.2 Sentiment Analysis

Investigating how people feel about something—both positively and negatively—and how they rate it is what sentiment analysis is all about. Sentiment analysis is a hotspot for research in data mining and natural language processing (NLP) (Liu, 2012). The exponential growth of opinionated data accessible on social media outlets is congruent with the rising interest in this area of study. For studies including social media, sentiment analysis is becoming central. It is essential to have automated algorithms that can extract subjectivity and polarity from the massive volumes of social media unstructured data. Sentiment analysis is useful as the average human reader encounters difficulty in identifying, extracting, and summarizing comments in lots of written text on social media platforms (Liu, 2012).

To analyze the public opinions on the EEREB, the Syuzhet package was utilized. Sandford University's NLP group created this program. The NRC sentiment dictionary, a vocabulary of emotions used to identify words that convey particular feelings in various settings, is part of this package (Mohammad and Turney, 2010). The NRC sentiment dictionary was utilized in order to ascertain the frequency of both negative and positive emotions, as well as the polarity that is linked with each of these emotions, on the encoded data (Jockers, 2017). This process generated sentiment scores on each comment. The next step was to determine a percentage of the overall sentiment scores for a certain category of sentiment, which was then used to compute a grand sentiment score for each emotion.

7.3.3 Topic Modeling

Topic modeling, a text mining technique, helps to uncover the underlying thematic structure within a collection of documents. This is achieved through the application of a variational expectation-maximization method, as described by Blei (2012) and Blei et al. (2003). Latent Dirichlet Allocation (LDA) is the simplest and most applied topic model (Jiang et al., 2016). The R package "topicmodels" provides an LDA implementation (Hornik and Grün, 2011). In order to discover hidden themes within the data and to organize the terms in the matrix in accordance with each theme, this study applied the topic models approach on the DocumentTermMatrix. Using beta values as a basis, the approach proposed nine subjects and 211 related terms. A higher beta value indicates that the term is more likely to appear in the document. The top ten beta values for each topic are displayed in Table 7.1, along

Table 7.1 Top nine topics for EEREB using beta-top terms and associated probabilities (adapted from Tetteh et al. 2023)

Topic	*Term*	*Beta*	*Topic*	*Term*	*Beta*	*Topic*	*Term*	*Beta*
1	air	0.018	2	house	0.013	3	air	0.012
1	years	0.011	2	insulation	0.012	3	hvac	0.012
1	now	0.011	2	heat	0.007	3	achieve	0.012
1	roof	0.009	2	energy	0.007	3	indoor	0.007
1	environment	0.007	2	years	0.006	3	audit	0.007
1	fresh	0.006	2	windows	0.006	3	culture	0.006
1	refurbishment	0.006	2	efficiency	0.006	3	enviroment	0.005
1	residential	0.006	2	erv	0.006	3	quaility	0.005
1	nzeb	0.005	2	hrv	0.006	3	ghg	0.005
1	renewables	0.005	2	envelope	0.005	3	cleaner	0.004
4	windows	0.009	5	insulation	0.019	6	new	0.010
4	insulation	0.009	5	use	0.013	6	good	0.008
4	water	0.008	5	money	0.008	6	energy	0.008
4	great	0.008	5	cost	0.008	6	move	0.007
4	use	0.006	5	energy	0.007	6	financial	0.006
4	energy	0.006	5	passive	0.006	6	environment	0.006
4	comfort	0.006	5	payback	0.006	6	community	0.006
4	occupants	0.006	5	consumption	0.006	6	organisation	0.006
4	air-conditioner	0.006	5	doubt	0.005	6	engaging	0.006
4	behaviours	0.005	5	affordable	0.005	6	green energy future	0.005
7	work	0.011	8	insulation	0.011	9	like	0.012
7	houses	0.007	8	heat	0.010	9	use	0.008
7	energy	0.007	8	retrofitting	0.010	9	home	0.008
7	years	0.006	8	difficult	0.008	9	much	0.008
7	tool	0.006	8	manage	0.007	9	passive	0.008
7	retrofits	0.006	8	audit	0.007	9	old	0.007
7	digital	0.006	8	expert	0.006	9	alternative	0.007
7	performance	0.005	8	competent	0.006	9	conservation	0.007
7	technology	0.005	8	promote	0.006	9	heritage	0.007
7	software	0.005	8	buildings	0.006	9	deep-retrofit	0.007

with the terms that correspond to them. The next step was for each researcher to provide labels or subjects by performing their own thematic analysis on the given topics and terms. Using a Delphi approach, the process continued till there was consensus on the name of each topic.

7.4 Results

By applying the opinion mining technique, a Latent Dirichlet Allocation model, beta-top terms obtained from EEREB comments were clustered into the following nine topic domains: (1) *Ventilation*, (2) *Energy efficiency*, (3) *Indoor environment quality*, (4) *Comfort and occupant behavior*, (5) *Cost considerations*, (6) *Community engagement*, (7) *Technology use*, (8) *Implementation knowledge*, and (9) *Social impact*. The topics are listed chronologically as shown in Table 7.1 (i.e., 1–9).

Overall, the majority of comments associated with these clusters were positive (63.9%), and the remaining (36.1%) were negative.

7.5 Discussions

To ensure clarity, examples from the data have been included without any edits, even if the commenters' comments unintentionally contain typographical errors. Each comment is assigned a number for reference, such as C#1, C#2, C#3, and so on.

7.5.1 Ventilation

This cluster's terms represent the systems and procedures for controlling ventilation in the occupied areas of buildings. In general, people felt positively about this subject. As an example, C#170 stated: "… a tight home with excellent mechanical ventilation would have no drafts or cold spots. It would be extremely comfortable to live in. And quiet! The home also brings in filtered fresh air rather than air leaking all through the structure, so the indoor environment is also going to be much healthier to live in." Roofing replacements garnered the most attention from the public. For example, C#1066 and C#292, respectively, stated: "Better a roof then the whole house." and "I've been thinking of doing a metal roof since my shingles are past dead, …." The efficiency of a building's envelope has a major impact on its energy usage. To conserve energy, two techniques have been employed: minimizing air leakage through the building envelope and increasing insulation in the roof space (El-Darwish and Gomaa, 2017). The participants also had mixed reactions to the energy retrofit strategy as a whole and the results they anticipated. As an example, C#45 and C#1278 both said: "Why do they need that ventilation system? If they have so many windows, they could just open them." and "… I am a bit worried about living in an airtight space: what if Coronaviruses in the future dictate excessive ventilation?" The terms "nZEB" (nearly zero energy building) and "renewables" were not extensively discussed in relation to this topic. However, they are promising strategies that aim to reduce our reliance on fossil fuels, which are known to contribute to greenhouse gas emissions (Omrany et al., 2022; Tetteh et al., 2022). While some commenters appear more concerned with home ventilation, it seems that most people are under the impression that there is not enough information available on how to upgrade ventilation systems for energy efficiency. A significant obstacle to increasing awareness about energy efficiency-related issues has been the limited understanding and information among homeowners regarding retrofits (Jia et al., 2021). To protect tenants' health and well-being, effective airflow is essential for preventing susceptibility to excessive ambient pollution sources, both during and after the COVID-19 pandemic era.

7.5.2 Energy Efficiency

This cluster explores additional methods for saving energy and reducing energy consumption in buildings. The data indicates that the public mostly prefers to retrofit their homes at their convenience rather than regulatory push. For example,

C#1304 stated: "Whether it is required by law or not, I am taking steps to greatly improve the energy efficiency in my house." This remark implies that not even energy efficiency investments are sufficiently incentivized by strict energy regulations. For instance, a study that was carried out by Kontokosta et al. (2020) on the impact of mandatory audit policy on energy consumption in commercial and multifamily residential buildings in the US demonstrates that homeowners are dissuaded from actively investing in energy efficiency retrofitting when facing policy programs that are mandatory in nature. One of the main concerns is that mandatory policies, such as energy audits, can be expensive and time-consuming for homeowners, particularly those with smaller buildings (Hsu, 2014). Another possible reason could be that homeowners vary in their level of interest in energy efficiency. According to Nair et al. (2010) and Kontokosta (2013), individuals who are highly driven are more likely to consider energy-saving alternatives on their own initiative, even if they are not required to perform a home retrofit. Furthermore, there is a noticeable difference in the level of energy-consciousness among the public. Some individuals have a clear understanding of retrofitting and the energy savings associated with using less energy-consuming systems. Conversely, there are many who are more curious about the non-energy advantages of these efficiency methods. For instance, as stated in C#461 and C#1345, respectively: "Want to help your energy efficiency? Attic fan will help tremendously." and "… it would be extremely comfortable to live in and no chronic health conditions." Shifting the focus toward non-energy benefits is essential. It is important to consider more than only the financial and ecological advantages of energy efficiency upgrades; as pointed out by Weiss et al. (2012), this narrow perspective can slow down their adoption.

7.5.3 Indoor Environmental Quality

Heating, ventilation, and air conditioning—HVAC—system installation was the primary topic of this construct. Having these systems in place is crucial to creating an outstanding indoor environment. For example, C#238 and C#340, respectively, stated: "I do HVAC and have been recommended a lot of blower wheel cleanings lately." and "… I recently got a new HVAC unit …." The majority of individuals also thought of a hybrid approach, which combines the use of an HVAC system with opening windows, to enhance the indoor climate. For instance, C#262 said: "… I have a floating floor basement, which has a lot of crumple zones in the duct work around system. BUT I try to open my windows as much as possible year-round. I only run my air when it's over 88°, or the humidity gets over 20%. And the heat when under a sunny 50° or cloudy 60°." This indicates that the general public not only has an understanding of the potential repercussions of IEQ retrofitting but also has the intention of considerably lowering their energy bills or cutting costs. The public also raised concerns about the IEQ as a result of the HVAC retrofit and how it could potentially impact the health of occupants. For example, C#337 stated: "What is your take on UV lights in the HVAC thinking about SARS, MERS and Virus …." Additionally, many of the comments provided recommendations for enhancing indoor air quality. For instance, C#299 said: "For indoor air quality, in addition to using HVAC systems you may want to address why indoor air quality is poor to begin with. How about spending ... to upgrade building materials that use fewer toxic chemicals that won't off gas into your home. Think about all the glues, formaldehyde, etc. in typical materials." Ortiz et al. (2020) suggest that the quality of the indoor environment is influenced by multiple factors. This requires the use of multidisciplinary approaches that involve examining buildings, HVAC systems, health, and IEQ as interconnected components using an integrated methodology.

7.5.4 Comfort and Occupant Behavior

This section emphasizes how EEREB affects users and occupants. In most cases, EEREB measures lower air exchange, which might be harmful to occupants' health. In this regard, C#170 stated: "The home also brings in filtered fresh air rather than air leaking all through the structure, so the indoor environment is also going to be much healthier to live in. ... to get a better indoor environment is far less expensive than paying for chronic health conditions." Occupants' comfort and satisfaction may increase through improved occupant behavior or by implementing better control systems for temperature and ventilation. For example, C#611 and C#177, respectively, added: "Comfort is always a thing." [sic] and "... better controlled indoor air quality may also reduce the risk of some health conditions. And it also provides a much more comfortable home to live in." This finding aligns with the study conducted by Kerr et al. (2018), which revealed that residents prioritize improving the comfort of their homes over the aesthetic aspects of energy retrofitting. Further, by increasing the utilization of natural lighting, EEREB has the potential to decrease occupants' fatigue and energy consumption, resulting in lower utility bills. This, in turn, enhances the convenience and comfort of homes. According to Azizi et al. (2019), consumers are motivated to adopt energy-efficient measures by factors beyond just the potential for energy savings. In general, occupants want to make their indoor climate better. This might mean a lot of different things, but some common ones include making sure there is enough insulation, better air quality (with or without AC), and more access to natural light.

7.5.5 Cost Considerations

The deployment of EEREB is driven, in large part, by cost considerations (Alabid et al., 2022). Cost being a major concern, it is not unexpected that the public voiced its criticism. For example, C#1553 stated: "The insane cost of retrofits is a problem. What the world needs are a much cheaper way to retrofit old housing without tearing them down to the frame." Overall, most of the comments were centered around the payback period and benefits of energy retrofit. For instance, in cases C#819 and C#1145, it was stated: "There must be a cost/benefit analysis step against retrofitting an old house versus demolition and starting from scratch ...," and "Great enthusiasm, but the payback could be over 50 years. I wonder how much a retrofitted home will costs. ... do you think it makes sense for energy efficiency retrofit?" Although the majority of the comments appear to originate from individuals who are enthusiastic about doing energy retrofits, there have been instances where responses from individuals already considering retrofits have stated that the cost was warranted. For example, C#801 stated: "… for me this value should be seen in house price with comfort part of equation. I would certainly pay far more for a home that was this comfortable." This implies that the potential for energy efficiency retrofitting will be limited to a small number of individuals, complicating efforts to support and promote energy retrofit projects. As mentioned earlier, initiatives that seek to encourage continued EEREB engagement could focus more on non-energy-related factors.

7.5.6 Community Engagement

Previous studies have shown that energy efficiency retrofit will continue to be held back by the government unless there is a strong public effort to accelerate commitments (Karvonen, 2013; Kerr et al., 2018). Although there has been limited discussion on this topic, it is evident that there is a critical need for home energy retrofitting in order to make substantial improvements to the housing stock. This need is apparent based on the few responses received so far. For example, C#3165 and C#1769, respectively, stated: "It will be interesting to see what whole community solutions will come, as that is still a problem for the time being …." and "Lobbying will ensure that massive energy schemes will still serve the majority over intelligently designed community schemes." While a majority of the public still holds the belief that the government should take the lead in driving the movement, as stated in C#3023: "The politicians should be individuals who have actively participated in the community," there are also many who hold the opposite view. This indicates the necessity of creating a more comprehensive strategy for implementing EEREB that concurrently tackles the service system and user routines, attitudes, and motives.

7.5.7 Technology Usage

Technologies such as HVAC system replacement, thermal insulation systems, water heating system replacement roof renovation, window and door replacement are among the important retrofit technologies that are discussed. However, this list is not exhaustive. Views on the usage of and technology were divided among the general population. For example, C#110 and C#246, respectively, stated: "I like the rainwater catchment, cisterns, greywater reuse, and the triple-insulated windows." and "I'm replacing my breaker panel, replacing my furnace and adding an a/c, swapped out leaking water valves, cleaning and sealing ducting, and replacing old, rotten single pane windows that are in my basement." Beyond that, they were enthusiastic about embracing new technology and worried about devices that used too much energy. For example, C#377 stated: "Tools like water misters over a/c units, heat pumps for hot water, air exchange (are well known). … I also hunger a bit for some truly new material and technologies." The public also shared a similar opinion to earlier studies (Li et al., 2017; Tan et al., 2021) that the applicability of retrofit technologies depends on the building characteristics and local climate features. For example, C#693 stated: "… hyper heat is cool. … if you're in a warm climate, you don't even need it." There were many who maintained that current technologies could be implemented anyplace, regardless of the weather or the kind of structure that was present. Nevertheless, as was stated previously, cost is the primary factor that plays a role in this. For example, C#1756 and C#55, respectively, stated: "If you build with modern technology, you can absolutely retrofit any house in any place …." and "… until the cost of this technology comes down it will be hard for the average homeowner to do this." The results show that homeowners who are thinking about retrofitting primarily want a retrofit that significantly reduces their energy use. Enhanced energy efficiency in retrofitting projects is a pressing issue, and there is a need for more contemporary retrofit technologies, particularly renewable technology like solar-energy systems. Public opinion in Portugal on renewable energy technology was investigated in a study that was carried out by Ribeiro et al. (2014). Attitudes toward new renewable energy projects were found to be generally positive, with a special emphasis placed on the utilization of solar power, which is considered to be the most desired technology in terms of both economic and environmental considerations.

7.5.8 Implementation Knowledge

Involving experts in EEREB implementation is crucial (Alam et al., 2019). Getting experts to do retrofit work is a big source of anxiety for most people, according to public feedback. As an example, C#1234 stated: "Although there's a standard, the idea finding a competent and knowledgeable construction company that speaks the standard and then the cost of it, due to the limited numbers of qualified trades, fills me with dread. … but even finding an average builder is a stressful procedure." Many individuals express positive sentiments about the involvement of government-supplied energy auditors in sharing information. For example, C#311 stated: "… you can get an energy audit free from the Mass Saves program. They will also give you lots of freebies like led light bulbs, and they will subsidize most of the cost of insulation or give you a no-interest loan for a new heating system." Another surprising concern raised was the untrustworthy nature of experts who carry out retrofit activities. For example, C#492 stated: "… can I trust the energy audits that are done by the state retrofit programs? And can I trust the work done by the contractors they hire to do the retrofit on the house?" Ultimately, this mistrust creates hesitation among homeowners when deciding whether to invest in energy retrofitting. The need to engage trustworthy external or independent energy auditors for retrofit evaluations and building was another major concern. Alam et al. (2019) emphasized the importance of developing a list of prequalified contractors to enable homeowners to easily locate professionals who possess the necessary technical expertise to perform retrofitting projects.

7.5.9 Social Impact

Retrofitting historic buildings is debatable, as doing so could affect or cover up historic features, changing the building's personality. The majority of the population, especially those who live in culturally sensitive locations, stated that retrofitting procedures such as changing the windows and insulating building facades could alter the structure's individuality and surroundings. For example, C#1768 stated: "You can't just extend ancient rooves, or cover up ancient facades, without destroying the building itself. You can't insulate timber frame buildings externally. The very notion is ridiculous, and the lack of thought about our heritage is seriously disappointing." This raises concerns about the acceptability and compatibility of various retrofit methods with the character of historic buildings (Ascione et al., 2015). Finding a middle ground between preserving cultural artifacts and lowering greenhouse gas emissions has been raised by members of the public on multiple occasions. For example, C#1018 stated: "... and ultimately the climate emergency is at least as important as keeping old buildings looking old, so it should be a consideration, not a veto." In light of these results, it is clear that frameworks facilitating dialogue and compromise between conservation perception and disputes over energy efficiency retrofits are crucial.

7.6 Conclusions and Recommendations

This chapter utilizes social media data to assess the overall public's sentiments of EEREB. A total of 3,306 comments in all were gathered from YouTube. These comments were then pre-processed and analyzed. The comments were summarized by using beta-top terms and then grouped into nine topic domains. These domains include ventilation, energy efficiency, indoor environment quality, comfort and occupant behavior, cost considerations, community engagement, technology use, implementation knowledge, and social impact. Retrofits and energy savings typically center on new or improved systems and equipment rather than enhancing an existing building's energy performance. Personal decisions have a far larger impact on the desire to retrofit than legislation does, although there are some local elements that can significantly affect public perceptions. The results of the study indicated that individuals place a higher importance on comfort, health, and culture compared to energy savings and cost reduction. The public as a whole had negative feelings about EEREB because of the high price tag and the difficulty of doing retrofits. People who are already considering a retrofit have commented on multiple occasions that the cost was reasonable. To further accelerate the general public's commitments and improve their sentiments toward EEREB, the following recommendations are provided:

- Governments should step up publicity efforts to improve the level of public awareness. Providing informative and educational materials that highlight the various non-energy-related advantages of EEREB will greatly enhance public awareness. Public organizations should not only collaborate on awareness training programs funded by the government but also create informative websites, feature finished retrofit projects, and disseminate relevant data and research.
- Homeowners and supply-side players, such as professionals in energy retrofitting (auditors, consultants, contractors, etc.), should be heavily involved in policymaking. This can be done through various communication channels and trusted messengers, in order to increase engagement and attention throughout the entire process, from assessment to implementation.

By offering a more thorough and varied grasp of the public's viewpoints on EEREB, this chapter makes a substantial contribution to the body of information. Therefore, it offers a solid basis for modeling energy efficiency decisions in buildings that already exist. Researchers, relevant stakeholders, and policymakers will find this chapter's findings to be of the utmost relevance. This chapter serves as a valuable reference for researchers studying public sentiments toward new products or services. Energy auditors and contractors will have a better understanding of the needs and decision-making behavior of the general public regarding EEREB. This will enable them to provide more suitable services to meet those needs. Additionally, this could assist policymakers in refining or developing more suitable interventions to promote EEREB. The chapter's reliance on data obtained and evaluated from YouTube is one of its major shortcomings. To better understand public sentiment toward EEREB, future research could broaden the database to incorporate additional social media platforms such as X (used to be called Twitter), Facebook, and Reddit. Also, future studies will seek to examine the correlations among the topics, and whether there are any significant relationships. Furthermore, caution should be exercised before drawing firm conclusions or making wide generalizations regarding the results in regard to a specific geographical region, even though this study has enhanced our knowledge of how individual and sociocultural conditions can influence attitudes toward EEREB.

Acknowledgments

The research detailed in this chapter is a component of a larger study called "Developing an intelligent decision support system for green retrofitting of aged existing buildings in Hong Kong." The study is entirely supported by the Postdoc Matching Fund Scheme of The Hong Kong Polytechnic University (Project ID: P0038800). There may be several publications that share the same background but have distinct aims, scope, and results. The authors (we) would like to acknowledge that the full version of this chapter has been published in the journal *Energy and Buildings* (Tetteh, M. O., Boateng, E. B., Darko, A., and Chan, A. P. (2023). What are the general public's needs, concerns and views about energy efficiency retrofitting of existing building stock? A sentiment analysis of social media data. *Energy and Buildings*, *301*, 113721). The valuable insights obtained through the peer-review process and the constructive feedback received have significantly contributed to the refinement and improvement of the content presented in this work. We are grateful for the opportunity to share this extended version with a broader readership via this book.

References

Ackerman, F. (2008). Critique of cost-benefit analysis, and alternative approaches to decision-making.

Alabid, J., Bennadji, A., and Seddiki, M. (2022). A review on the energy retrofit policies and improvements of the UK existing buildings, challenges and benefits. *Renewable and sustainable energy reviews*, 159, 112161.

Alam, M., Zou, P. X., Stewart, R. A., Bertone, E., Sahin, O., Buntine, C., and Marshall, C. (2019). Government championed strategies to overcome the barriers to public building energy efficiency retrofit projects. *Sustainable Cities and Society*, 44, 56–69.

Ascione, F., Bianco, N., De Masi, R. F., de'Rossi, F., and Vanoli, G. P. (2015). Energy retrofit of an educational building in the ancient center of Benevento. Feasibility study of energy savings and respect of the historical value. *Energy and Buildings*, 95, 172–183.

Avanzini, M., Pinheiro, M. D., Gomes, R., and Rolim, C. (2022). Energy retrofit as an answer to public health costs of fuel poverty in Lisbon social housing. *Energy Policy*, 160, 112658.

Azizi, S., Nair, G., and Olofsson, T. (2019). Analysing the house-owners' perceptions on benefits and barriers of energy renovation in Swedish single-family houses. *Energy and Buildings*, 198, 187–196.

Balakrishnan, J., and Griffiths, M. D. (2017). Social media addiction: What is the role of content in YouTube? *Journal of Behavioral Addictions*, 6(3), 364–377.

Blei, D. M. (2012). Probabilistic topic models. *Communications of the ACM*, 55(4), 77–84.

Blei, D. M., Ng, A. Y., and Jordan, M. I. (2003). Latent Dirichlet allocation. *Journal of Machine Learning Research*, 3(Jan), 993–1022.

Boudet, H. S. (2019). Public perceptions of and responses to new energy technologies. *Nature Energy*, 4(6), 446–455.

Bouzarovski, S., and Petrova, S. (2015). A global perspective on domestic energy deprivation: Overcoming the energy poverty–fuel poverty binary. *Energy Research & Social Science*, 10, 31–40.

Dean, B. (2021). How many people use YouTube in 2022? [New Data]. Available at https://backlinko.com/youtube-users.

Department for Business Energy and Industrial Strategy (2020). "UK sets ambitious new climate target ahead of UN Summit. Climate change and Energy," London – UK. https://www.gov.uk/government/news/uk-sets-ambitious-new-climate-target-ahead-of-un-summit. (September 2022).

Devine-Wright, P. (2007). Reconsidering public attitudes and public acceptance of renewable energy technologies: A critical review. *Beyond Nimbyism: a multidisciplinary investigation of public engagement with renewable energy technologies*, 15.

El-Darwish, I., and Gomaa, M. (2017). Retrofitting strategy for building envelopes to achieve energy efficiency. *Alexandria Engineering Journal*, 56(4), 579–589.

European Commission. (2010). Directive 2010/31/EU of the European Parliament and of the Council on the energy performance of buildings. 2010. Brussels.

Galvin, R. (2014a). Why German homeowners are reluctant to retrofit. *Building Research and Information*, 42(4), 398–408.

Gasco, L., Schifanella, R., Aiello, L. M., Quercia, D., Asensio, C., and de Arcas, G. (2020). Social media and open data to quantify the effects of noise on health. *Frontiers in Sustainable Cities*, 2, 41.

Gertzel, B., Borquez, F., and Ackland, R. (2022). vosonSML: Collecting Social Media Data and Generating Networks for Analysis. VOSON Lab, Virtual Observatory for the Study of Online Networks. https://doi.org/https://github.com/vosonlab/vosonSML

Gigerenzer, G., and Todd, P. M. (1999). *Simple heuristics that make us smart*. Oxford University Press, USA.

Gram-Hanssen, K. (2014). Retrofitting owner-occupied housing: remember the people. *Building Research and Information*, 42(4), 393–397.

He, Q., Zhao, H., Shen, L., Dong, L., Cheng, Y., and Xu, K. (2019). Factors Influencing residents' intention toward green retrofitting of existing residential buildings. *Sustainability*, 11(15), 4246.

Hornik, K., and Grün, B. (2011). Topicmodels: An R package for fitting topic models. *Journal of Statistical Software*, 40(13), 1–30.

Hsu, D. (2014). How much information disclosure of building energy performance is necessary? *Energy Policy*, 64, 263–272.

Huang, C., Ma, J., and Song, K. (2021). Homeowners' willingness to make investment in energy efficiency retrofit of residential buildings in China and its influencing factors. *Energies*, 14(5), 1260.

Hwang, B. G., Shan, M., Xie, S., & Chi, S. (2017). Investigating residents' perceptions of green retrofit program in mature residential estates: The case of Singapore. *Habitat International*, 63, 103–112

Jackson, T. (2005). Motivating sustainable consumption: A review of evidence on consumer behaviour and behavioural change. *Sustainable Development Research Network*, 29(1), 30–40.

Jiang, H., Qiang, M., and Lin, P. (2016). Assessment of online public opinions on large infrastructure projects: A case study of the Three Gorges Project in China. *Environmental Impact Assessment Review*, 61, 38–51.

Jia, L., Qian, Q. K., Meijer, F., and Visscher, H. (2021). Exploring key risks of energy retrofit of residential buildings in China with transaction cost considerations. *Journal of Cleaner Production*, 293, 126099.

Jockers, M. (2017). Package 'syuzhet'. URL: https://cran.r-project.org/web/packages/syuzhet.

Judson, E. P., and Maller, C. (2014). Housing renovations and energy efficiency: insights from homeowners' practices. *Building Research and Information*, 42(4), 501–511.

Karvonen, A. (2013). Towards systemic domestic retrofit: a social practices approach. *Building Research and Information*, 41(5), 563–574.

Kerr, N., Gouldson, A., and Barrett, J. (2018). Holistic narratives of the renovation experience: Using Q-methodology to improve understanding of domestic energy retrofits in the United Kingdom. *Energy Research and Social Science*, 42, 90–99.

Kerr, N., and Winskel, M. (2020). Household investment in home energy retrofit: A review of the evidence on effective public policy design for privately owned homes. *Renewable and Sustainable Energy Reviews*, 123, 109778.

Kontokosta, C. E. (2013). Energy disclosure, market behavior, and the building data ecosystem. *Annals of the New York Academy of Sciences*, 1295(1), 34–43.

Kontokosta, C. E., Spiegel-Feld, D., and Papadopoulos, S. (2020). The impact of mandatory energy audits on building energy use. *Nature Energy*, 5(4), 309–316.

Liu, B. (2012). Sentiment analysis and opinion mining. *Synthesis Lectures on Human Language Technologies*, 5(1), 1–167.

Li, J., Ng, S. T., and Skitmore, M. (2017). Review of low-carbon refurbishment solutions for residential buildings with particular reference to multi-story buildings in Hong Kong. *Renewable and Sustainable Energy Reviews*, 73, 393–407

Liu, G., Ye, K., Tan, Y., Huang, Z., & Li, X. (2022). Factors influencing homeowners' housing renovation decision-making: Towards a holistic understanding. *Energy and Buildings*, 254, 111568.

Liu, W., Zhang, J., Bluemling, B., Mol, A. P., and Wang, C. (2015). Public participation in energy saving retrofitting of residential buildings in China. *Applied Energy*, 147, 287–296.

Maller, C. J., and Horne, R. E. (2011). Living lightly: How does climate change feature in residential home improvements and what are the implications for policy? *Urban Policy and Research*, 29(1), 59–72.

Mohammad, S., and Turney, P. (2010). Emotions evoked by common words and phrases: Using mechanical turk to create an emotion lexicon. Proceedings of the NAACL HLT 2010 workshop on computational approaches to analysis and generation of emotion in text.

Mortensen, A., Heiselberg, P., and Knudstrup, M. (2016). Identification of key parameters determining Danish homeowners' willingness and motivation for energy renovations. *International Journal of Sustainable Built Environment*, 5(2), 246–268.

Nair, G., Gustavsson, L., and Mahapatra, K. (2010). Factors influencing energy efficiency investments in existing Swedish residential buildings. *Energy Policy*, 38(6), 2956–2963.

Omrany, H., Chang, R., Soebarto, V., Zhang, Y., Ghaffarianhoseini, A., and Zuo, J. (2022). A bibliometric review of net zero energy building research 1995–2022. *Energy and Buildings*, 262, 111996.

Ortiz, M., Itard, L., and Bluyssen, P. M. (2020). Indoor environmental quality related risk factors with energy-efficient retrofitting of housing: A literature review. *Energy and Buildings*, 221, 110102.

Ossokina, I. V., Kerperien, S., and Arentze, T. A. (2021). Does information encourage or discourage tenants to accept energy retrofitting of homes? *Energy Economics*, 103, 105534.

Ribeiro, F., Ferreira, P., Araújo, M., and Braga, A. C. (2014). Public opinion on renewable energy technologies in Portugal. *Energy*, 69, 39–50.

Tan, Y., Liu, G., Zhang, Y., Shuai, C., and Shen, G. Q. (2018). Green retrofit of aged residential buildings in Hong Kong: A preliminary study. *Building and Environment*, 143, 89–98.

Tan, Y., Luo, T., Xue, X., Shen, G. Q., Zhang, G., and Hou, L. (2021). An empirical study of green retrofit technologies and policies for aged residential buildings in Hong Kong. *Journal of Building Engineering*, 39, 102271.

Tetteh, M. O., Boateng, E. B., Darko, A., and Chan, A. P. (2023). What are the general public's needs, concerns and views about energy efficiency retrofitting of existing building stock? A sentiment analysis of social media data. *Energy and Buildings*, 301, 113721.

Tetteh, M. O., Darko, A., Chan, A. P., Jafari, A., Brilakis, I., Chen, W., ... and Yevu, S. K. (2022). Scientometric mapping of global research on green retrofitting of existing buildings (GREB): Pathway towards a holistic GREB framework. *Energy and Buildings*, 277, 112532.

The US Department of State. (2021). "The Long-Term Strategy of the United States: Pathways to Net-Zero Greenhouse Gas Emissions by 2050." https://www.whitehouse.gov/wp-content/uploads/2021/10/US-Long-Term-Strategy.pdf. (September 2022).

Tjørring, L., and Gausset, Q. (2019). Drivers for retrofit: A sociocultural approach to houses and inhabitants. *Building Research and Information*, 47(4), 394–403.

Trotta, G. (2018). Factors affecting energy-saving behaviours and energy efficiency investments in British households. *Energy Policy*, 114, 529–539.

Urban, J., and Ščasný, M. (2012). Exploring domestic energy-saving: The role of environmental concern and background variables. *Energy Policy*, 47, 69–80.

Wang, Y., Wang, G., Li, H., Gong, L., and Wu, Z. (2022). Mapping and analyzing the construction noise pollution in China using social media platforms. *Environmental Impact Assessment Review*, 97, 106863.

Weiss, J., Dunkelberg, E., and Vogelpohl, T. (2012). Improving policy instruments to better tap into homeowner refurbishment potential: Lessons learned from a case study in Germany. *Energy Policy*, 44, 406–415.

Wilson, C., Chrysochoidis, G., and Pettifor, H. (2013). Understanding homeowners' renovation decisions: Findings of the VERD project.

Wilson, C., and Dowlatabadi, H. (2007). Models of decision making and residential energy use. *Annual Review of Environment and Resources*, 32, 169–203.

World Green Building Council. (2021). WorldGBC Net Zero Carbon Buildings Commitment. Available at: https://worldgbc.org/wp-content/uploads/2023/03/WorldGBC-NZCBCmt_Intro_2021_hs-1.pdf. (Vol. September 2021) (Accessed: July 2022).

Wu, Z., Zhang, Y., Chen, Q., and Wang, H. (2021). Attitude of Chinese public towards municipal solid waste sorting policy: A text mining study. *Science of the Total Environment*, 756, 142674.

Part III

Sustainable Building Assessment and Certification Systems

8 Sustainable Buildings within Low-Carbon Energy Systems

David Jenkins, Peter McCallum, and Kumar Biswajit Debnath

8.1 Introduction to Building Modelling for Sustainable Buildings

The way building simulation and modelling are used in different countries varies with the national energy policy in that country, type of building stock, prevalence of building technologies (e.g. heating- vs. cooling-dominated countries), and climate. The specific energy assessment framework adopted in a country, the assessors (and training programmes), and the role of assessment in defining sustainability are therefore not uniform and some of these differences can make us reflect on what a sustainable building actually is. This section will note some applications of modelling from specific policy examples attempting to drive a market transformation of the building stock.

Furthermore, individual building energy assessment is arguably playing a greater role in other sectors. Sustainable buildings must make sense, and function, within wider energy systems in order to meet carbon targets. The interaction of different sectors of energy modelling is therefore noted.

This chapter also applies building energy analysis to a very specific case study occurring during the period of COVID lockdown. As well as demonstrating the importance of external factors to our assessment of building energy use, this brings in the value of empirical energy consumption data, particularly for cases where the understanding of user behaviour is paramount.

Finally, an indication is given of where building energy modelling goes next; as software improves, computational power increases, and the targets to be met become more challenging, how might building modelling respond?

8.1.1 What Is Building Modelling For?

Building modelling, in a general sense, gives the promise of providing energy consumption estimates for buildings that do not yet exist, or existing buildings in future scenarios that have yet to be experienced. However, this promise comes with the caveat that, for traditional, theoretical building modelling, the outputs are not real. This difference between modelled and measured energy consumption is often referred to as the Performance Gap and, rather than being a failure of physical models, is simply a consequence of the difficulty in accurately accounting for every action relating to every building technology over a duration of time.

DOI: 10.1201/9781003317890-11

This chapter will attempt to clarify that "building modelling" covers a range of very different tools and methods. It is important to, firstly, distinguish between these methods and, then, identify for what applications they are suitable. Defining a model as wrong or inaccurate is meaningless without a strong understanding of what that model is being used for.

8.1.2 Energy Compliance of Buildings

One of the most frequent, and visible, areas of building modelling is with energy compliance standards of buildings. In Europe, through the Energy Performance in Buildings Directive (EPBD) (European Commission, 2002), this is seen through the established use of Energy Performance Certificates (EPC), though other countries have successful energy rating schemes under different guises. Generally, EPCs rely on steady-state modelling of relatively simple building archetypes. Some modelling of non-domestic buildings can include dynamic simulation (such as in the United Kingdom), or requiring more complex descriptions of the building itself (e.g. HVAC equipment definitions).

The EPBD actually allows for real energy consumption to be used for generating an EPC, stating "energy performance of a building shall be determined on the basis of the calculated or actual energy that is consumed to meet the different needs associated with its typical use"; however, it is more common to see countries avoid the use of real energy consumption due to the difficulty in standardising this across different buildings (and assessors of those buildings). For example, although rating schemes in Australia (NABERS, 2021) and a Display Energy Certificate (UK Government, 2015) in some UK buildings (albeit outside of EPCs) use real energy consumption data, a review of EPCs (Sayfikar and Jenkins, 2022) shows most schemes taking steady-state model routes.

As well as identifying market transformation (in relation to energy efficiency of the building stock), EPCs (and equivalent energy rating schemes elsewhere) are commonly used to support and guide policy, particularly with regard to financial instruments used to assist low-carbon refurbishment, examples of which are noted elsewhere (Jenkins *et al.*, 2021). For such modelling, standardisation is a prime objective; that is the desire to have a consistent and harmonised approach to how building modelling, and the wider assessment, is conducted. This generally lends itself to simpler modelling, operating at quite broad temporal resolutions. Therefore, performance metrics are often based on the annual energy consumption (or carbon emissions) of a building. Compromises have to be made on the gathering of inputs also. As an example, for UK residential buildings, such compromises result in the Standard Assessment Procedure (SAP) (BRE, 2019), a Reduced Data version (RdSAP) of which (RdSAP [BRE, 2012]) is used for existing buildings. This relies on a series of "look-up" tables for parameters such as U-values (Semple and Jenkins, 2020), so an assessor may make an assumption on such parameters without having detailed construction data of that property.

Many critiques of, in particular, EPCs have been published, whether relating to the accuracy of outputs (Bordass, 2013; de Wilde, 2014) or the consistency of the assessment itself (UK Government, 2014). However, it is difficult to argue against the historical success of the EPBD in raising the visibility of building energy efficiency to users of buildings. One, perhaps, legitimate concern is the reliance it places on steady-state modelling of buildings – essentially, models that do not understand energy use as a function of time. Coupled with the ubiquity of the use of such models, there is a danger that we miss out on other opportunities when attempting to understand other aspects of buildings. Later in this chapter, the value of dynamic simulation (which attempts to understand the transient nature of building performance) and statistical modelling will be noted, with the key message that our choice of building modelling must be tailored to the expected application. As we review some of the challenges of designing low-carbon energy systems, where the timing of demand (compared to the timing of supply) is so important, there should be an understanding that we need a larger toolkit of options to assess and analyse our buildings.

8.1.3 Detailed Design Tools

Dynamic simulation (Clarke, 2001), as already noted, does have a place within standard energy ratings such as EPCs. Using the United Kingdom again as a case study, commercial models (such as Integrated Energy Systems-Virtual Environment (IES-VE) and Thermal Analysis Software (TAS)) are acceptable options for generating EPCs for some non-domestic buildings, albeit within a National Calculation Methodology that attempts to standardise how that simulation is carried out. However, dynamic simulation has the potential to better understand the complex interaction of systems within a building, even if it is still prone to the previously noted Performance Gap. A building services engineer may therefore use simulation to optioneer different design solutions for a client that goes beyond the requirements of a standard energy compliance scheme. Or, to put this another way, it is not the job of an EPC to design a detailed passive ventilation scheme or a control system of a heating/cooling system, but an engineering team must do this during the design process. This distinction between modelling for compliance and modelling for design is an important one but dynamic simulation can have a role in both.

One advantage of dynamic simulation, and other forms of modelling, is the ability to understand, and output, energy use at a higher temporal resolution. Whilst not required for energy ratings or energy bill estimates, this is a requirement for any assessment of how buildings use energy within an energy network (e.g. electricity grid).

8.2 Role of Buildings Within Wider Energy Systems

The growing understanding of what constitutes a "smart" energy system has placed a greater onus on the symbiotic relationship between energy supply and demand. The needs of buildings, transport, and industry have, in most industrialised countries, been met by the design and operation of supply-side solutions – a combination of, ideally, a diversified selection of energy supply technologies, interconnection, and large-scale storage (e.g. pumped hydro). However, both research and practice now point to the opportunities for changing and controlling demand to meet supply-side constraints. In turn, this requires a more comprehensive understanding of activities occurring in buildings, and the buildings themselves.

8.2.1 Sharing Information between Buildings and Energy Systems

An energy modeller working on larger energy systems is unlikely to be using the same tools as a traditional building energy modeller. Furthermore, it is not necessarily the case that their respective models can, or should, be formally integrated in a hard-wired way; different models operate at different resolutions, different software platforms, and with very different input requirements (Jenkins, 2018). However, by sharing assumptions and (for example) wider energy scenarios, such models can be used together to provide consistent advice and support to different actors within the wider energy sector.

Therefore, two very different models can reference the same future climate scenario or electricity carbon intensity. The use of quite standard future climate scenarios (IPCC, 2022), and analysis to format these scenarios into useable weather and/or climate information (Met Office, 2019), can aid this task considerably.

In the case of the carbon intensity of energy provision, projected changes in the decarbonisation of supply can be reflected in the modelling of building carbon emissions; though accounting for this within formal building energy compliance can be difficult when models used with (for example) EPCs are not updated frequently enough – and so carbon accounting of buildings start to lag the true carbon cost of energy (as has been seen in the United Kingdom, where the aforementioned SAP has not traditionally had carbon intensity factors updated regularly [Building Research Establishment, 2022]).

Future scenarios also, clearly, go beyond just climate and carbon intensity. More detailed scenarios (National Grid, 2022) can bring discussion on the range, and combination, of different demand- and supply-side technologies that may impact both the built environment and wider energy systems. Having such clear, and referable, scenarios means that modellers within different parts of the energy system can use similar technology pathways; for example, preparing the supply side for an assumed rise in electrified (heat pump) heating within the built environment across a stated timescale.

From these scenarios, ultimately, we also need the ability to estimate demand characteristics that are useful for those designing and updating energy systems. This is touched on in Sections 8.2.3 and 8.2.4 but requires an understanding of scale, resolution, and model limitation.

8.2.2 Defining Smart

A common term used to describe more integration between buildings and energy systems is "smart" (e.g. smart meters). One example of an attempt to produce useful building modelling output for energy system applications is the Smart Readiness Indicator (SRI) (European Commission, 2020). This index, designed with EPC-type output in mind (see Section 8.1), addresses building smartness in terms of the (i) ability to adapt energy performance, (ii) ability to provide a healthy indoor environment through smart control, and (iii) ability to respond to external signals for purposes of demand flexibility. This latter aspect is particularly important to the operation of future, low-carbon, energy systems. Rather than attempting to explicitly model the demand profiles of a smart building (which, in any case, would be beyond the EPC-based models being used by most countries), this indicator reflects the types of metering and controls in use within a building – to some extent indicating the smartness potential of that building. Whilst this is unlikely to comprehensively indicate how that building is being used, it could be used to target certain buildings (or groups of buildings) that are more likely to be suitable for smart control options that could benefit local energy system management. As a very recent, and voluntary, indicator from EPBD discussions, the impact of the SRI is yet to be seen and testing is currently underway in a number of European countries (European Commission, 2022).

8.2.3 Building Tools for Energy System Integration

There exist a multitude of genuine energy system models, such as the MARKAL/TIMES family of models (Taylor *et al.*, 2014), that attempt to incorporate buildings in a relatively top-down approach, whilst offering the ability to advise a much wider array of energy policy support in relation to technology selection across sectors. However, the ability of traditional building models to inform energy system level can be limited, due in part to some of the previously noted issues around scale and resolution.

Section 8.4 refers to some innovations that allow modellers to take a different approach to modelling built environment energy demand for energy systems. As with any attempt to gauge the success of a model, in any sector, the suitability of its application depends on how the generated outputs will be used. So, a steady-state model producing annual energy metrics is not suitable for quantifying peak energy demand. A dynamic model can produce higher resolution energy demand outputs and this can, in theory, be used to generate estimates of peak demand for individual and aggregated buildings. However, that has to be understood in the context of the assumptions used within those dynamic models; if a modeller wishes to understand the impact of behavioural or cultural choices on peak energy demand (or other demand characteristics), this is difficult without a form of empirical energy data. The use of such data, distinct from the analysis of dynamically simulated energy demand, is provided in a case study in Section 8.3.

8.3 A UK Case Study of Energy Demand during the COVID Pandemic

Extreme disruptive events such as COVID-19 pandemic-induced lockdowns profoundly affected the public and their behaviours. When these disruptions persist, they often influence substantial shifts in behaviour, fostering the adoption of alternative practices. This could manifest in prolonged periods of remote work, learning, or living for a considerable part of the population. Such alterations were anticipated to influence electricity consumption patterns significantly. For modellers interested in the causation of energy demand, such events were therefore of great interest, where one might expect apparent, measurable changes to energy demand characteristics. For building modellers specifically, the role of buildings (and their users) is crucial to understanding the broader energy picture.

The initial COVID-19 lockdown in the United Kingdom commenced on March 23, 2020, with a gradual relaxation of restrictions starting mid-May. After a surge in cases, the second and third lockdowns unfolded in November 2020 and January 2021, respectively. Despite ongoing studies, substantial uncertainties remain regarding the enduring effects of COVID-19 on the United Kingdom's energy demand (Hook *et al.*, 2020). The National Grid Electricity System Operator (ESO) reported a decline in total energy consumption during the lockdown period compared to the pre-COVID demand (National Grid, 2020). This reduction was primarily attributed to widespread remote work, children being home due to school closures, and decreased industrial activities. However, more conclusive evidence would be needed detailing how the lockdown influenced residential electricity consumption patterns.

Therefore, the following study by the authors focused on scrutinising the initial phase of the COVID-19 lockdown and its immediate aftermath spanning from April to August 2020 in the United Kingdom. This specific timeframe enabled a comprehensive investigation into the initial fluctuations in electricity consumption during the lockdown, the gradual reversal of these changes upon easing restrictions, and the subsequent return to more conventional practices as the United Kingdom emerged from lockdown. The objective was to discern whether specific shifts observed during this period could have a lasting impact on the long-term energy consumption patterns within buildings.

8.3.1 Methodology

This study delved into residential data from the Cornwall Local Energy Market (LEM) project by Centrica PLC. The initiative garnered support through partial funding from the European Regional Development Fund, constituting a component of the European Structural and Investment Funds Programme for 2014–2020 (Nicholls and Kane, 2021). The dataset utilised in this study encompassed detailed 1-minute records of electricity demand and generation sourced from 100 residences in the southwestern region of the United Kingdom from April 2018 to August 2020. The overarching objective of the project was to formulate and establish a LEM trading platform tailored for households equipped with Solar Photovoltaic (PV) systems and behind-the-meter Battery Energy Storage Systems (BESS) (Figure 8.1). Thus, the database served as a valuable resource for exploring the dynamics and intricacies of energy consumption and generation within this context.

While the primary dataset was originally compiled for LEM trading experiments, this study predominantly focuses on energy consumption data. Deliberately excluding the PV and BESS performance metrics, the aim was to shift emphasis towards understanding the broader effects of COVID-19 on the general population. Moreover, the main time-series dataset was complemented by additional site-by-site metadata, providing details on dwelling characteristics and household composition.

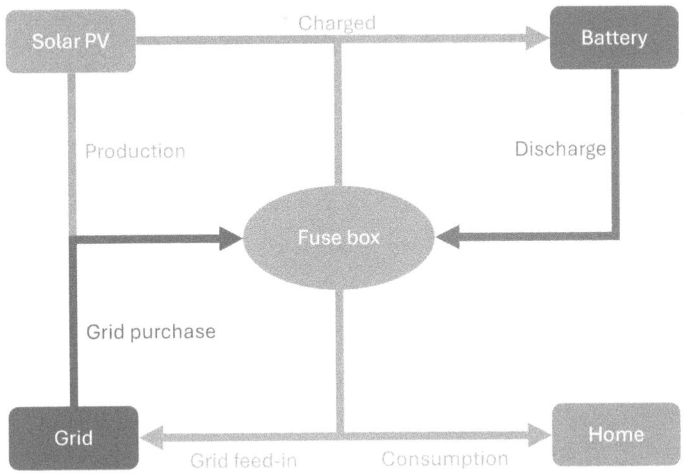

Figure 8.1 Energy systems and flow in the monitored houses

We analysed energy consumption data, accessed through the MySonnenBatterie (MSB) portal, at dual tiers to examine how residential electricity usage evolves in the face of highly disruptive scenarios, specifically during the COVID-19 lockdown. This dual-level analysis aimed to delve into the intricacies of household energy consumption patterns amidst unprecedented and challenging circumstances. We used Python language in the Jupyter compiler to analyse with different library packages: Numpy, Pandas, Scipy, Statistics, Math, Time, Matplotlib, and Seaborn. We used Python for big data analysis because it showed potential for large dataset analysis (Siebert *et al.*, 2021; Sivapriyan *et al.*, 2021; Velosa *et al.*, 2023). Initially, an examination was conducted on the consolidated 1-minute data of electricity consumption derived from a sample of 100 households. The aim was to scrutinise alterations in overall residential demand during periods of disruption, specifically for April–August 2020, and a comparative assessment with the corresponding timeframe in 2019. Additionally, an exploration was undertaken into the combined consumption patterns on both weekdays and weekends within the examined months for both years.

Moreover, we organised household consumption data based on the primary heating fuel used – gas, oil, and electricity. This categorisation aimed to explore the collective impact of lockdowns on the aggregated electricity consumption within these specific fuel clusters. Independent t-tests were subsequently employed to scrutinise the statistical significance of differences observed between periods (weekdays and weekends) and differences in heating types.

As a complementary facet of our analysis, we examined the minutely electricity consumption data from three randomly selected houses to gain deeper insights into specific household activities, such as remote work and schooling, during lockdown periods. This high-resolution analysis allowed us to characterise and discern the nuanced impact of lockdown measures on individual homes. Independent t-tests at this granular level were applied to ascertain statistical significance, enabling the inference of occupant activities. This analysis contributed to a richer understanding of the intricate relationship between a disruptive event like the COVID-19 lockdown and household responses and facilitated more generalised conclusions. The aggregation of such consumption was what we observed when studying, for example, national grid data, but individual demand profiles were required to understand the correlation.

8.3.2 *Impact of COVID-19 Lockdown on the Electricity Consumption*

Figure 8.2A shows separate electricity consumption patterns during the weekdays and weekends in April–August 2019 and 2020. It showed weekends and weekdays dominating the shift towards lower and upper consumption bins, respectively, in April and May 2020 compared to 2019. Furthermore, the average electricity consumption change towards lower bins was higher than weekday consumption during April and May 2020. However, the weekdays and weekends significantly shifted towards lower consumption bins for June–August. In June, the unimodal distribution from 2019 changed into a bimodal one in 2020, which denoted two distinct types of consumption patterns emerging within 100 houses. The average electricity

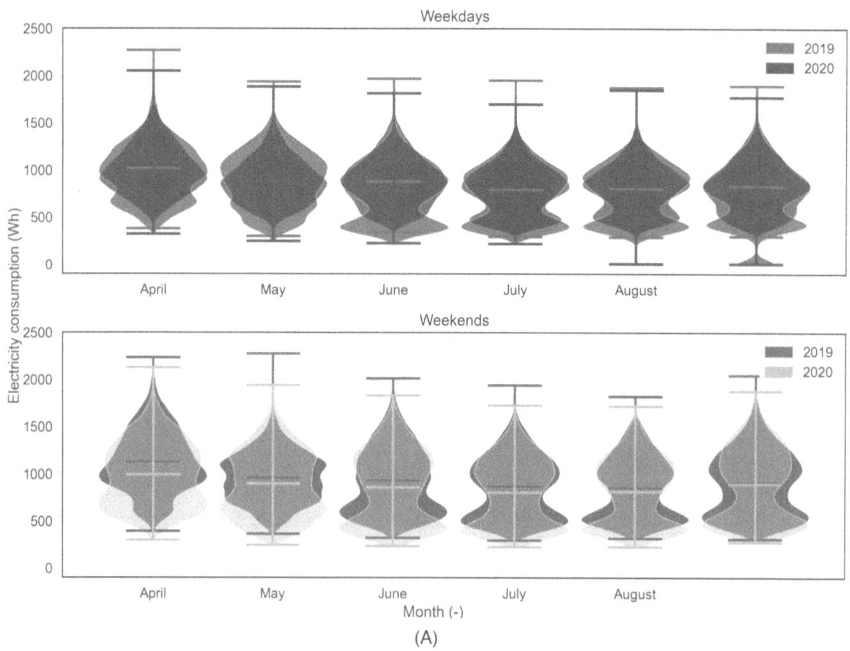

Figure 8.2 (A) Total minutely electricity consumption distribution of 100 houses for weekdays and weekends in April–August of 2019 and 2020; (B) Total minutely electricity consumption distribution for houses with gas, oil, and electricity for heating in April–August of 2019 and 2020. (*Continued*)

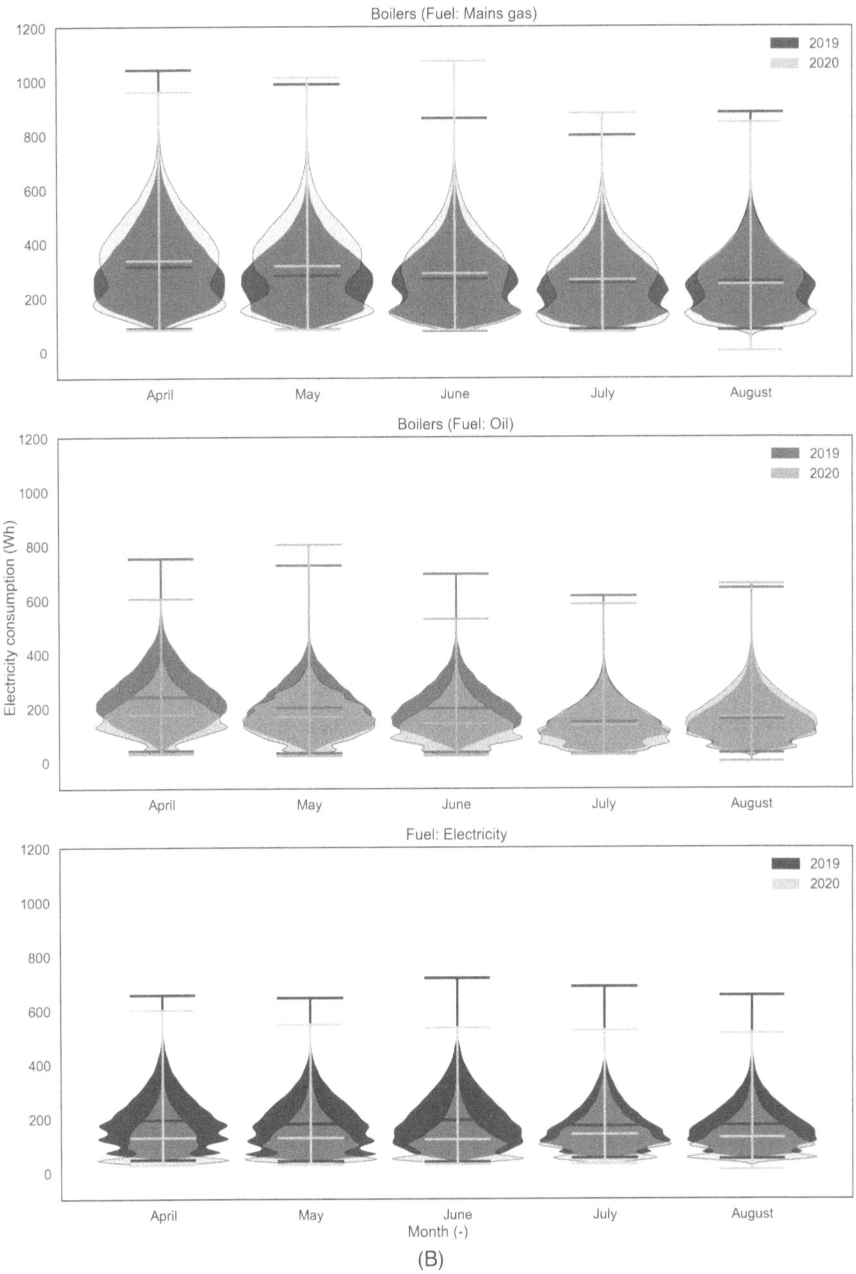

Figure 8.2 (Continued)

consumption on the weekdays and weekends of April–August 2020 was reduced by 3.9–11.6% and 1.9–12.3% compared to 2019, except for May 2020 weekdays when the average electricity consumption was 0.45% higher than in May 2019. The slightly elevated demand might be due to the staged easing of lockdown restrictions from mid-May, which might require further investigation.

For clustering the houses, we used the heating fuel, producing categories of gas (32 houses), oil (18 houses), and electricity (19 houses). Notably, houses utilising gas boilers exhibited the most pronounced alterations in 2020 when contrasted with those reliant on oil and electricity (Figure 8.2B). The houses with gas boilers displayed a bimodal distribution in April and May 2020, diverging from the pattern observed in 2019. This distribution exhibited shifts towards both higher and lower consumption bins. However, the predominant shift was towards higher consumption, resulting in an average electricity consumption increase ranging from 4.1% to 12.4% from April to July 2020 compared to the corresponding period in 2019. Notably, in August 2020, the average electricity consumption was 4.3% lower than in 2019. The distribution of gas boiler houses retained its bimodal nature from April to August 2020. Conversely, residences relying on oil and electric heating predominantly shifted towards a lower consumption distribution in 2020 than in 2019. The average electricity consumption in houses with oil heating fuel was 7.9–28.7% lower in April–July 2020 than in 2019, except in August 2020, when the consumption increased by 4.6% more than in 2019. In the case of houses with electric boilers, the average electricity demand reduced by 18.6–37.2% in April–August 2020 compared to 2019. Due to a lack of detailed demographic and activity data, although it was not clear from the data, the potential causes of the bimodal distribution of energy consumption in 2020 for houses with gas boilers might be the age of the houses and occupancy age groups (their activity requiring electricity and thermal comfort preference) who had to stay home for a more extended period due to pandemic-induced lockdown; however, it would require further research.

Out of the 100 residences, a random selection was made to thoroughly analyse the consumption patterns of three houses – specifically, House No. 2, 32, and 85. These houses, featuring floor areas of 105, 239, and 93 m^2, respectively, were equipped with double-glazed windows and walls filled with cavity insulation. Houses 2 and 85 were semi-detached, while House 32 was a detached unit. Demographically, House 2 accommodated four individuals: two adults aged 16 and above and two adults aged 65 and above. Similarly, House 85 housed four occupants: two adults aged 16 and above and two children. Unfortunately, no information was available regarding the age or number of occupants in House 32.

House 85 exhibited a lesser variance in the distribution of electricity consumption compared to Houses 2 and 32. Noteworthy alterations in the distribution of electricity consumption during weekdays were observed in the morning, afternoon, and evening from April to August 2020 when contrasted with the corresponding timeframe in 2019 (Figure 8.5A). Specifically, there was a discernible shift towards higher weekday consumption in April, May, and July 2020. Conversely, the distribution of electricity demand indicated lower consumption during weekend afternoons in April 2020. However, a notable shift occurred towards higher consumption in the afternoons of May, June, and July 2020 (Figure 8.5B).

House 2 exhibited a noteworthy alteration in electricity consumption patterns on weekdays and weekends for all months in 2020, except for August, compared to the corresponding periods in 2019 (Figure 8.3). Specifically, there was an upward shift in the distribution of electricity consumption during the afternoon, evening, and early night hours for both weekdays and weekends in 2020. On weekend mornings, the distribution trended towards higher consumption in April and May but shifted towards lower consumption in June–August 2020 compared to the corresponding months in 2019.

In contrast, House 32 displayed a substantially different electricity consumption distribution than House 2. The most prominent shift in distribution occurred during the late-night/early-morning period (12–5 a.m.) on weekends, with a distinct trend towards lower consumption in April, May, and August but an inclination towards higher consumption in June and July for the year 2020 (Figure 8.4). Notably, there was a relatively modest shift during weekdays in 2020. However, the consumption distribution experienced a significant change in the early evenings during weekdays in May and July and throughout April–August 2020, excluding July.

We employed independent t-tests to assess weekdays and weekends in Houses 2, 32, and 85. The null hypothesis posited that there was no statistically significant difference in mean consumption between 2019 and 2020. A p-value below 0.05 indicated rejecting the null hypothesis at a 95% confidence interval for Houses 2 and 85 on weekdays and weekends. Notably, a statistically significant disparity existed in mean consumption on weekdays and weekends across all analysed months in 2019 and 2020. For House 32, excluding June weekdays, the p-value remained below 0.05, aligning with the conclusions drawn for House 2. However, in the case of House 32 during June weekdays, the p-value exceeded 0.05, indicating no significant difference between 2019 and 2020 (Figure 8.4).

Our examination of the minutely detailed electricity consumption data from 100 residences in the South West United Kingdom revealed a substantial upswing in electricity usage during the lockdown period in 2020 compared to a corresponding timeframe in 2019. While there was a decline in electricity demand post-lockdown, a lasting transformation in the residential electricity consumption pattern became evident, attributable to the persistent trend of remote work. Our analysis further indicated a shift in electricity demand from unimodal distribution in the summer to bimodal, particularly noticeable on weekdays and weekends in April and May. The lockdown-induced alteration suggested two distinct demand clusters, possibly influenced by the widespread adoption of remote work and schooling from home. Additionally, there was a discernible shift towards a lower energy consumption distribution between April and August 2020, prompting inquiries into the nature of home activities during the lockdown and their implications for energy demand. Further investigation into house clusters based on heating fuel types unveiled a more pronounced distribution change in

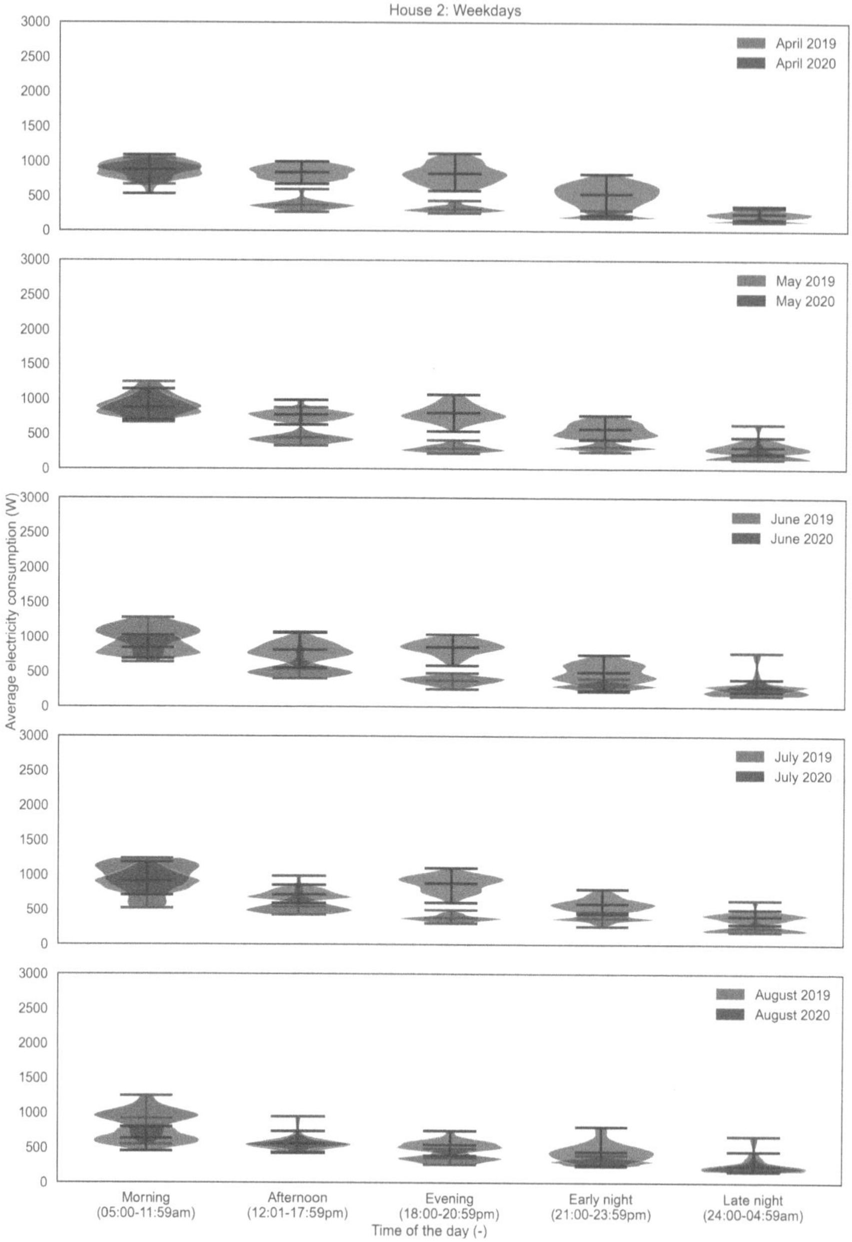

(A)

Figure 8.3 Average 1-minutely electricity consumption during different parts of the day on (A) weekdays and (B) weekends in House 2 (*Continued*)

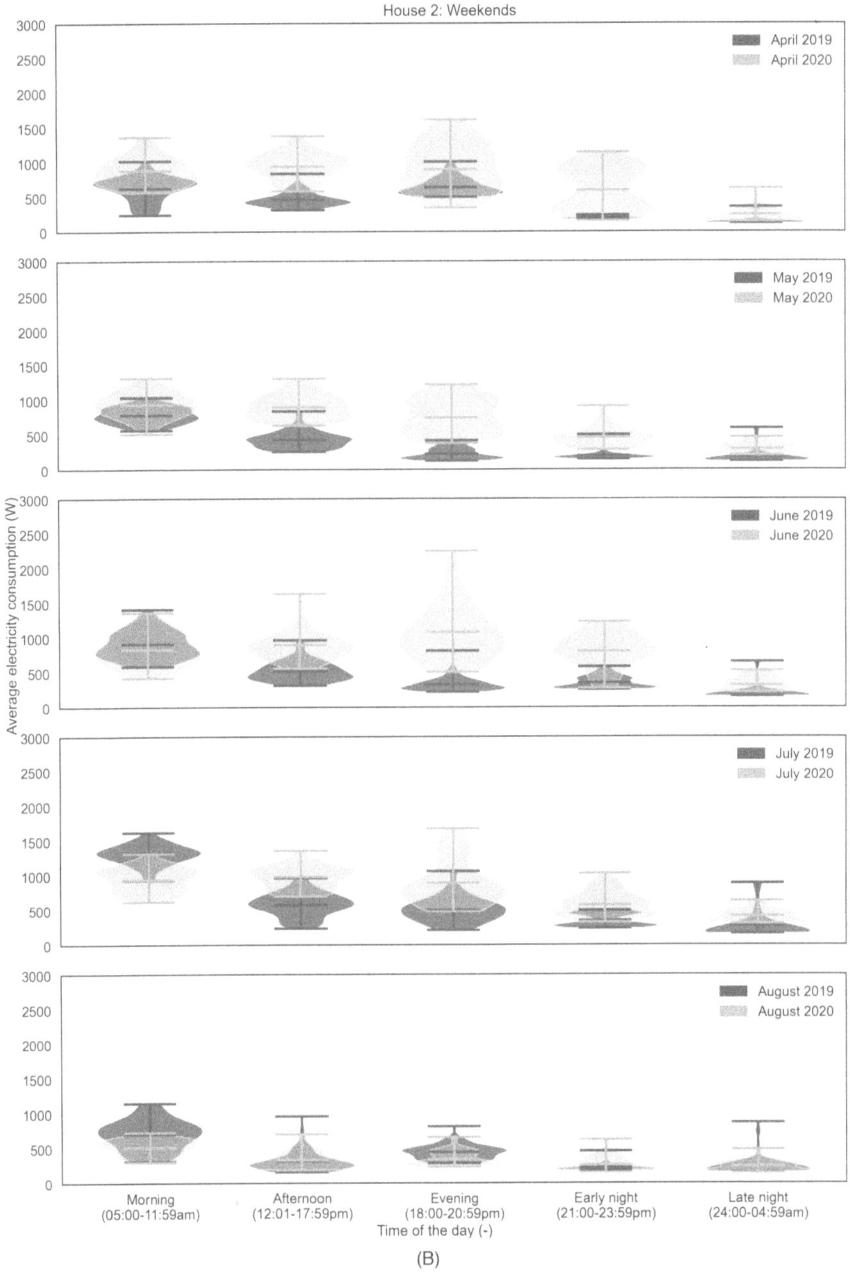

Figure 8.3 (Continued)

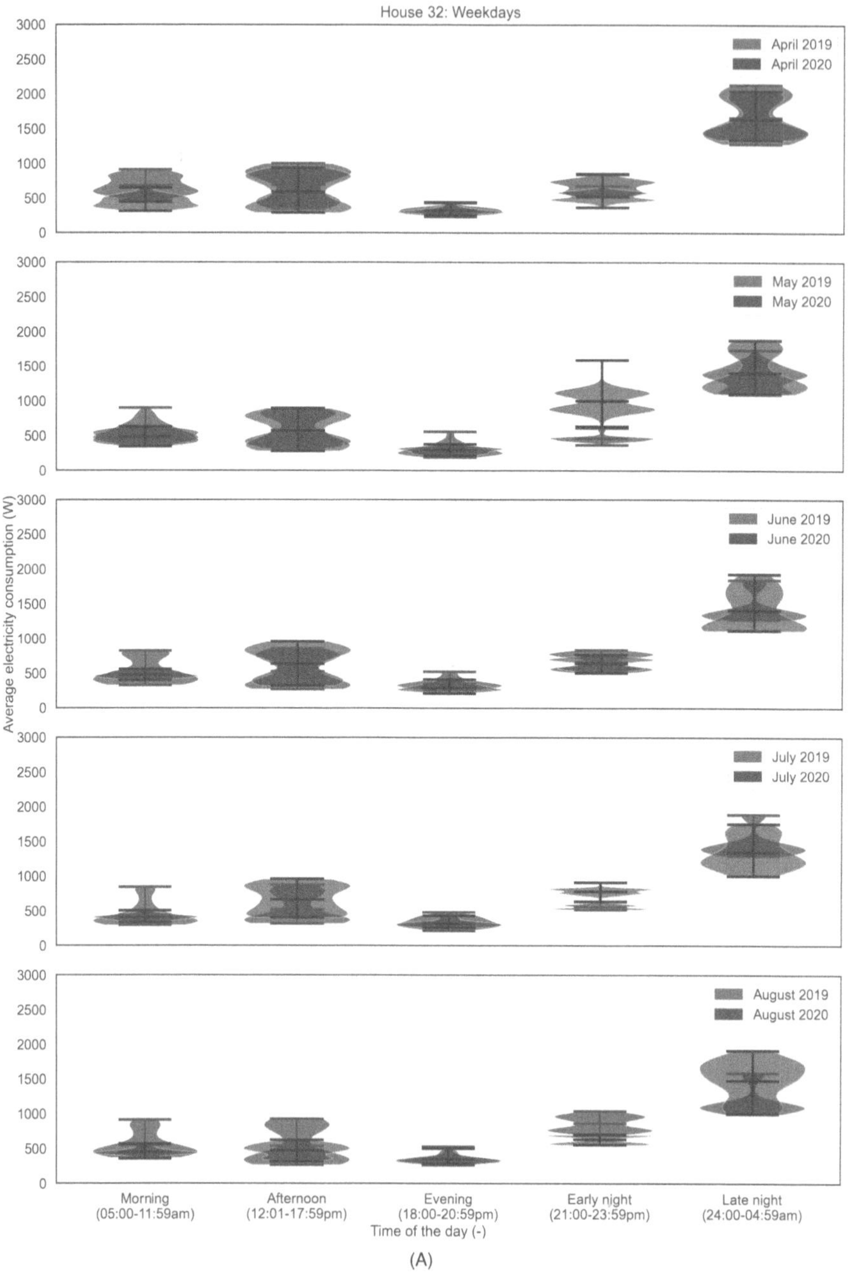

Figure 8.4 Average one-minutely electricity consumption during different parts of the day on (A) weekdays and (B) weekends in House 32 (Continued)

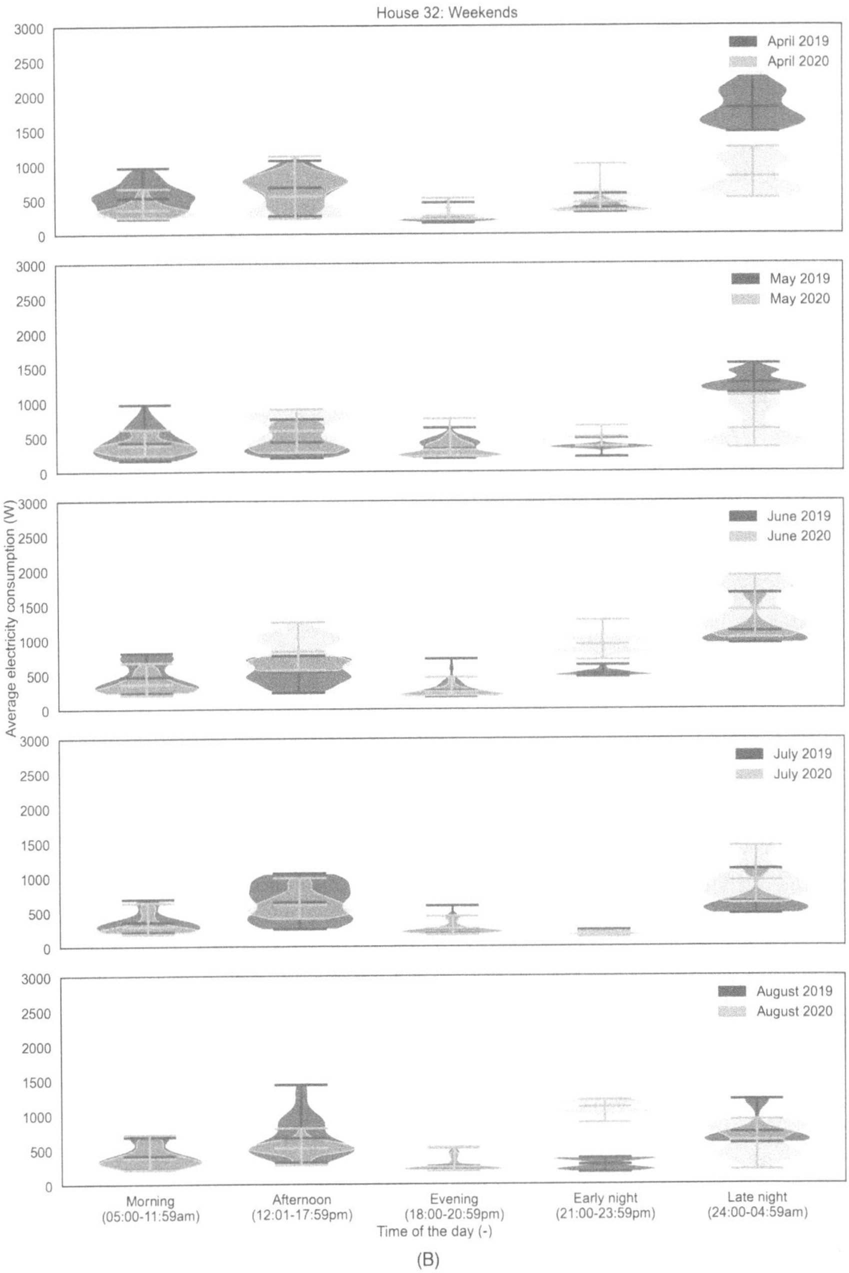

Figure 8.4 (Continued)

houses equipped with gas boilers than those relying on oil- and electricity-powered heating systems. Despite the analysis being conducted during the summer and autumn months, minimising the use of heating for thermal comfort, the observed distribution disparity raised critical questions regarding household energy consumption practices.

Examination of three specific residences unveiled an additional dimension of the lockdown's influence on electricity consumption patterns. Houses 2 and 85 exhibited statistically significant disparities in their electricity consumption data between 2019 and 2020, specifically from April to August. Notably, the energy usage profiles of these two houses differed substantially. In the case of House 2, on weekdays from April to July 2020, there was a discernible shift in the distribution of electricity demand towards higher consumption in the afternoon to early night hours, a pattern observed during the afternoon for House 85 (April, May, and July 2020). Conversely, House 32 witnessed a notable shift in electricity demand during the late-night period. The observed transition of electricity demand towards the afternoon in Houses 2 and 85 on weekdays mirrored the trends outlined in the National Grid data (National Grid, 2020), highlighting alterations in morning routines due to the lockdown and the prevalence of remote work. However, it's crucial to note that not every household experienced similar impacts regarding changes in daily routines induced by the lockdown. This analysis prompted a more intriguing question about variations in household practices during lockdown, warranting a deeper investigation into the nuanced dynamics at play.

As well as illustrating the complex relationships between disruptive events and energy behaviour, the study also shows the different conclusions that were gained when studying aggregated (i.e. multi-building) data (Figure 8.2) and individual dwelling data (Figure 8.3–8.5). For example, the aggregated average electricity consumption on the weekdays and weekends of April–August 2020 was reduced by 3.9–11.6% and 1.9–12.3% compared to 2019, except for May 2020. However, in the selected individual dwellings, the average electricity consumption changes were diverse in different parts of the day. House 2's average electricity consumption in the afternoon, evening, early night, and late night on weekdays was 1.86–171.5% higher in April–August 2020 than in 2019. The average electricity consumption increased by 1.1% in April 2020 mornings but reduced by 4.7–31.1% in May–August 2020 mornings compared to 2019. Only in August 2020 did the average electricity consumption decrease by 2.3% compared to 2019.

It was an essential point for both studies of empirical energy data and theoretical building performance modelling. Aggregation was often the key to wider scale generalisations of the work. Still, building-specific studies allowed for a much better appreciation of causation and the complexities and pitfalls of quantifying this. Both were important, but due to the aggregation process, the high (temporal)-resolution diversity of energy demand characteristics at individual building level would not be seen with aggregated data (as seen in the concept of "After Diversity

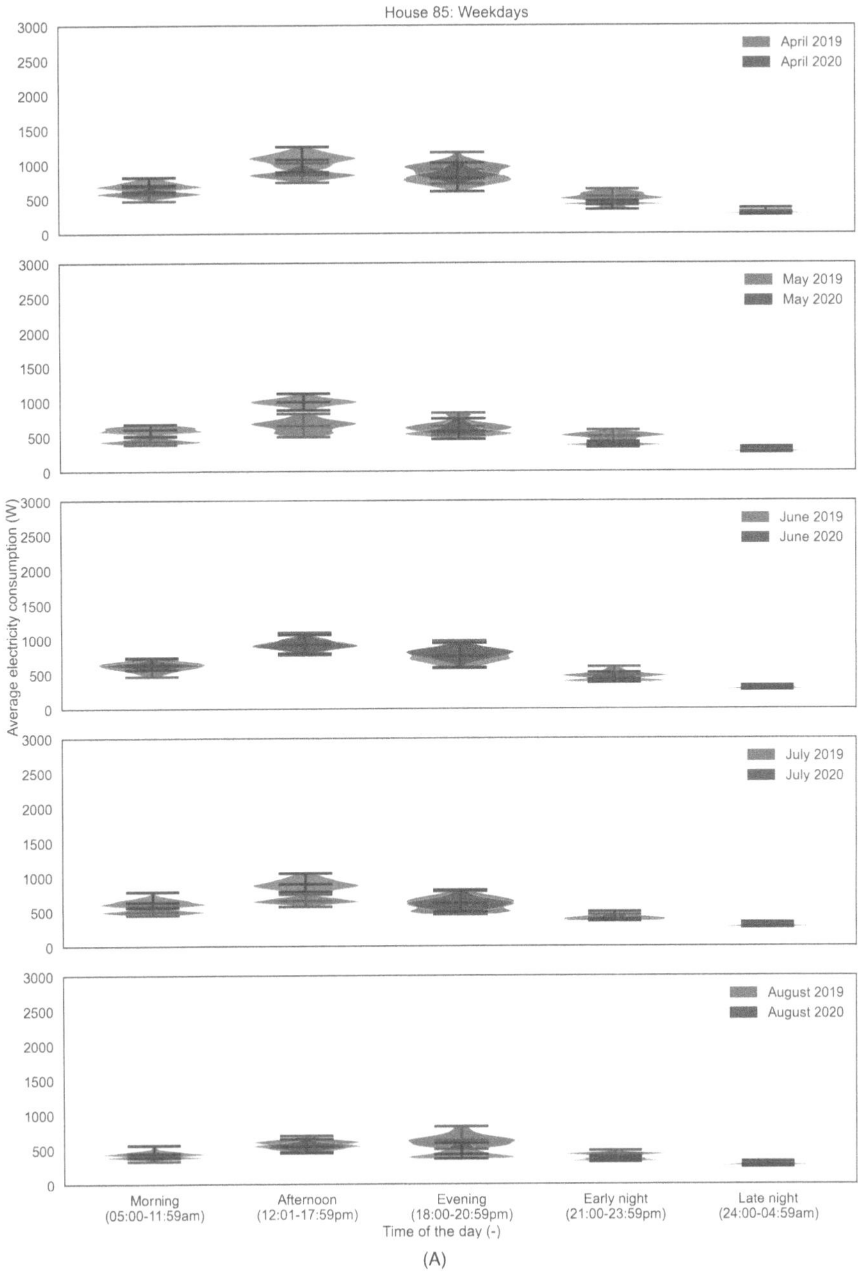

Figure 8.5 Average 1-minutely electricity consumption during different parts of the day on (A) weekdays and (B) weekends in House 85 (*Continued*)

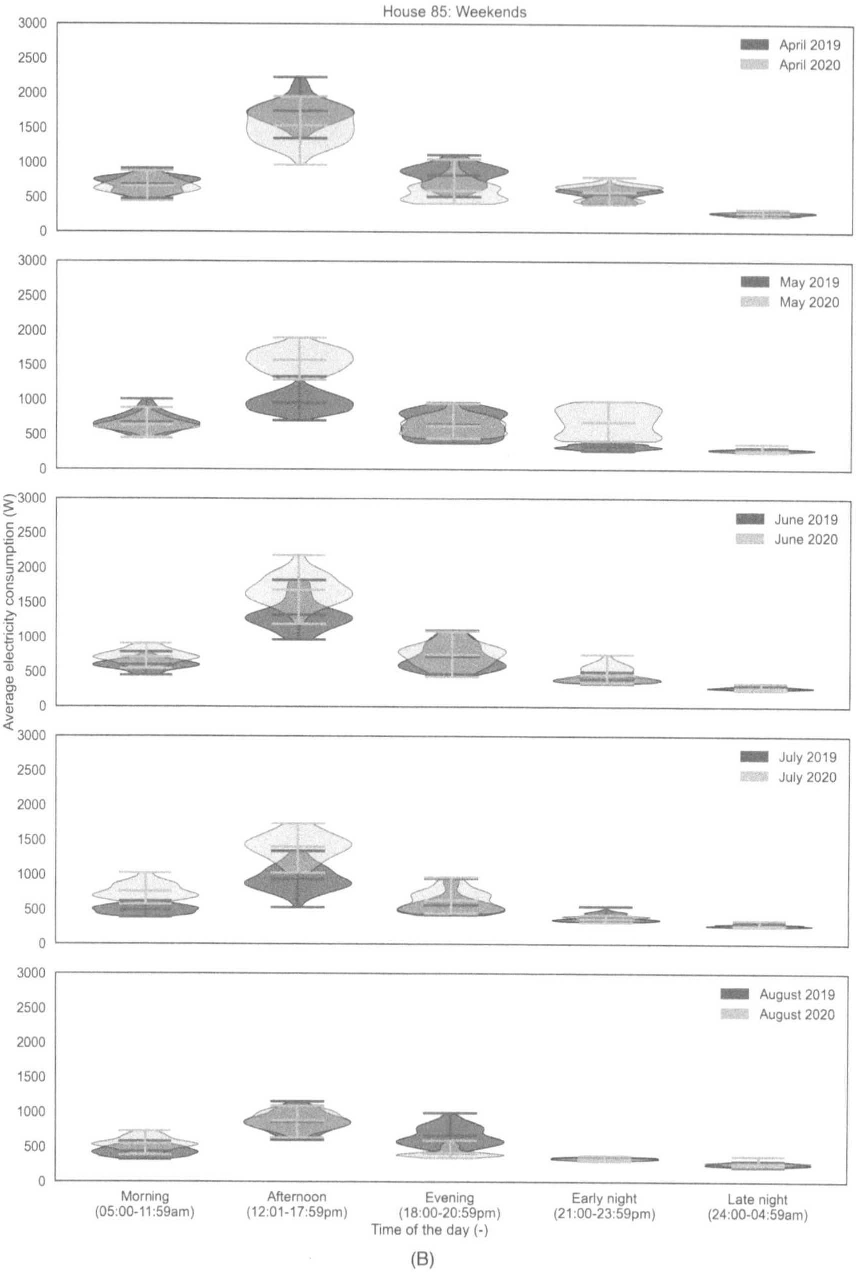

Figure 8.5 (Continued)

Maximum Demand", used within the electricity industry to quantify peak demand after accounting for that aggregated diversity).

8.4 Future Applications of Building Modelling

Dynamic building simulation has already been highlighted as a critical area of current and future practice in sustainable building design. This type of modelling entails a physics-based approach to studying heat pathways in buildings, accounting for conduction, convection, and radiation and relating these to the complete building fabric representation that defines the building in question. This model's "dynamic" nature relates to the time-stepped calculations, which incrementally update room temperatures and humidities, surface temperatures, thermal fluxes, and heat capacitance (thermal mass), based on the evolving internal and external environments. The simulations typically cover one year and record a wide range of outputs at 30- or 60-minute time intervals.

There has been a long and relatively rich history of application of these models in the academic sector, which continues at present(de Wilde, 2023; Harish and Kumar, 2016; Jebaraj and Iniyan, 2006). Professional bodies and national research institutions have also taken a long-standing interest in these models for setting practitioner guidance(Lee and Burnett, 2008; Roderick *et al.*, 2009; Schwartz and Raslan, 2013). In reality, however, real projects are influenced by dynamic simulation in one of three ways:

1 for engineering design – designing to extremes (coldest day, hottest day, busiest day, etc.)
2 for engineering design – to optioneer designs that meet the client's needs (comparing notional capital cost, operational cost, and energy implications for natural ventilation, mechanical ventilation, and full air conditioning, for example)
3 for formally assessing the *relative* energy performance, against similar buildings.

In the United Kingdom, point 3 relates to the local interpretation of the EPBD and provides notional running costs and carbon emissions (Delorme and Higley, 2020). Every building owner should be aware, however, that these are based on synthesised time-sensitive inputs and should be interpreted with caution.

The remainder of this section describes the current scope of these models in more detail before describing how, and for what purpose, the scope of dynamic simulation might extend in the future. Some of these future use-cases have already been partially developed, exploring areas such as power system network model integration and wider energy system planning (McCallum *et al.*, 2019). It is always worth bearing in mind, however, that dynamic simulation is both expensive and imperfect, whether used for traditional or experimental applications. A good reason is required for a client to pay or a regulatory body to enforce a dynamic simulation. In any case, the specific context of any simulation outputs must always be considered carefully.

8.4.1 The Current Scope and Application of Dynamic Building Simulation

The original application of dynamic simulation was for engineering design. From the 1970s onwards (Clarke, 2001), research models from academia gave rise to simulation packages that allowed engineers to design buildings for their clients. The goal of these activities was purely to engineer effective solutions that minimise capital cost whilst also minimising the space required for the plant. Operational cost and energy performance studies were not uncommon (Castaldo and Pisello, 2018), but only as far as clients would engage and pay for what was essentially work beyond the minimum requirement to deliver a building (Hensen and Lamberts, 2019). In Europe, EPBD pushed a new agenda on energy efficiency, and dynamic simulation began to be used for this purpose, particularly in circumstances where a model already existed for traditional plant sizing evaluations and other optioneering assessments.

This tangent of application highlights a subtle but key feature in how dynamic simulation is used – designing a building to respond under extremes (the coldest winter, hottest summer, most severe occupancy, etc.) creates a very different type of model output to one that is used in estimating dynamic energy consumption. The latter involves establishing a reasonable set of occupancy patterns, incorporating dynamic schedules of thermal gains, such as small power and lighting, potentially adding more complex sensible and latent gains from various building activities, and linking this to weather and carbon intensities (UK Government, 2022) that are typical by some statistical measure. In the UK context, much of the templating of these transient inputs is standardised by the National Calculation Methodology(UK NCM, 2023) and CIBSE weather libraries (DesignBuilder, 2023). And whilst the practice of building and running this kind of model can be quite similar for these two distinct purposes, the outputs should be treated very differently. Furthermore, a very complex picture of uncertainties can arise as a result of the time-sensitive inputs described above (Tian *et al.*, 2018).

The energy assessment performance gap and related criticism can, in part, be traced back to the abstract simulation inputs used for describing time-sensitive building activity. In one respect, this is the best we can expect for building systems that may not yet have been built, or existing buildings which have very limited data (on occupancy schedules, for example). In another respect, this (NCM) framework is ideal and serves the precise purpose for which it was designed – to benchmark and compare the *relative* performance of similar buildings. In general, when building simulation tools are adopted for new purposes, the inputs determine how the outputs should be used.

It is worth noting that within the current scope of dynamic simulation, it is possible to find what can be described as "campus modelling" (IES-VE, 2021). This involves using what are essentially the same tools as for the dynamic simulation of a building but with an expanded scale to cover large campus areas. This could be for modelling large university sites, hospitals, or entire airport terminals. Whilst this is a complex and challenging task, it generally involves the same principles and modelling activities as traditional dynamic building simulation.

Some of these tools may consider metered energy demands from separate buildings and may consider some of the connectivity of site utilities. These types of studies are generally funded directly by clients who manage large non-residential sites and have sole responsibility for energy and heat provision, potentially planning moderate-scale energy generation with on-site energy centres or large renewable energy installations.

8.4.2 Expanding Dynamic Modelling to Building Stocks

Sensitive adoption of new model inputs is necessary for broadening the scope of dynamic simulation to large building stocks. Given that regulators mandate energy efficiency assessments on a building-basis, not a stock-basis, using NCM templates in a stock-level model offers very little meaningful value. More generally, a relevant question is – "Why even use dynamic simulation to model a building stock?"

Two areas of active development for stock-level dynamic simulation are:

1 for feeding simulated demand into a power system network model
2 for feeding simulated demand into a "whole-system" energy model.

For both applications, some of the intricacies and considerations are explored in more detail in Section 8.4.3. In general, there are many non-trivial issues that need to be addressed and the application is ultimately very specific to a small number of organisations (such as Distribution Network Operators [DNOs], utility-scale generators, and authorities). In this sense, the application environment for these types of models is very different to that of traditional dynamic simulation, which sees widespread use across many organisations and projects. With the lack of a conventional market for both desktop and cloud-based applications, software development in the above areas is often undertaken in-house by the end-user organisation. To elicit anything of value from a dynamic stock model, there are two major areas where traditional dynamic simulation requires significant adaptation(McCallum *et al.*, 2019):

1 scalable, time-sensitive inputs (occupancy, gains, and plant schedules) need specific tailoring to the real-world case being studied
2 the chosen dynamic simulation tool must be integrated within an archetyping framework, and within some data harmonisation scheme that can ingest Geographic Information System (GIS) data and bulk building stock data, automatically linking the two.

There are major challenges that overlap both above themes. Some of these relate to the availability and coverage of data and extend to issues around data licencing and privacy (BEIS, 2020). There are other issues that are systemic modelling difficulties - how to discern non-residential and residential in mixed-use buildings; or, how to model aspects that cannot be calibrated against a ground truth (a building being under-heated because the occupier cannot afford to heat it, for example). On the two areas introduced above, the following provides some expanded detail on the necessary adaptations for stock-level modelling.

8.4.2.1 Time-Sensitive Inputs

To provide a stock simulation with any contextual relevance to the area of study, time-sensitive model inputs should be linked to real data from that locale. This onerous task requires both data and data methods to extract useful features to calibrate the stock model. Some data sources, such as gas records, lend themselves well to behavioural pattern recognition (McCallum *et al.*, 2020b) that can be fed directly into simulations. Understanding comfort temperature setpoints, on the other hand, would require much more advanced data methods to discern useful model inputs (Calama-González *et al.*, 2022).

Processing electricity consumption records poses different challenges. The temporal resolution of meter records makes detailed disaggregation infeasible. Certain loads, specifically relating to electric heating, may more easily suit disaggregation due to their duration and shape (McCallum *et al.*, 2020b). There is also a more tangible connection between these kinds of loads to the thermal focus of dynamic building simulation. In practice, any robust form of dynamic stock model is likely to feature a blend of data-linked thermo-physical modelling and an ensemble of relevant electricity data.

At the scales necessary for stock modelling, the composition and implementation of time-sensitive inputs and electricity data should be automated. Furthermore, a statistical representation of time-sensitive input is favoured, to compose statistical inputs for many thousands of buildings using data from a small subset of those buildings. Whilst considerable effort is required to develop adequate temporal inputs, this type of model can leverage aggregation quite effectively to provide useful results.

8.4.2.2 Archetyping and Data Harmonisation

One of the ways that stock-level building simulation has been demonstrated (McCallum, Jenkins and Vatougiou, 2020a; Perwez *et al.*, 2022) is by using GIS as the data basis for archetyping. Other stock-level characterisations include the use of LiDAR (Issermann, Chang and Kow, 2021). In contrast to campus modelling, much larger numbers of buildings are involved (typically dwellings) and a consistent method is required for automating model creation around the relevant building forms. This is an inherently approximate process, but again this leverages aggregation to ultimately provide energy patterns for moderate to large building stocks that reflect some meaningful set of scenarios, offering additional value to traditional "energy profiles".

The principles of this model design are to trace all individual buildings within the region of interest and to link (geo-reference) these to coordinates of known properties. As part of this process, formal methods such as ray tracing can be used to automate the paring of shapes and properties. This association is necessary to ensure that intended buildings are introduced to the stock and that these buildings can be related to their material and system attributes gathered from other data sources (in the United Kingdom, there is no persistent link between GIS data and property databases). In the case of residences, many properties may exist within a single footprint; routines are required within GIS interpretation modules to establish individual property floor plans and adjacencies, both lateral and vertical.

Beyond tracing all relevant floor plans, dwelling archetyping can take the form of plan parameterisation. The objective of this type of model is to describe every floor plan in terms of a minimum set of parameters – width, depth, number of storeys, apex and eaves heights, and orientation. All combinations of horizontal plans that can be described as T-shaped, L-shaped, and C-shaped can also be recorded unambiguously using just three extra dimensions (McCallum *et al.*, 2020a). Once this type of archetyping description is augmented with building fabric and system features for any given dwelling, the final description will not only describe that dwelling but also all other dwellings that are effectively identical.

The computations required for these data-integration processes are straightforward and fast, and the entire stock can be processed in this way to leave a long list of dwellings, filled with inherent replication. This is the basis for archetyping in this type of model. Where survey-based data, such as EPCs, are used to provide fabric and system attributes, this data can be extrapolated and sanitised to provide reliable descriptions of the stock. For conducting scenario-based evaluations, attributes such as heating systems can be interchanged on some probabilistic basis (i.e. some percentage of properties moving from gas boiler to heat pump) before simulations are carried out using the dynamic thermal simulation. Carrying out these thermal simulations is still computationally expensive, but this archetyping process allows for widespread recycling of simulation results and the automation removes all manual overhead in generating and dispatching these model runs.

8.4.3 Linking to Other Models

In Section 8.4.2, two areas were identified for linking dynamic simulation to energy system models – for providing demand scenarios to operational network models and whole-system planning models. As mentioned, there are clear benefits in this potential offering, to provide energy demand scenarios that have a theoretical link to physical processes, caused by climate change, gradual adoption of new building technology, and widespread changes in how buildings are used in the future. This contrasts traditional demand profiles, which are inherently constrained by the historical data used to derive them. The big question is – how to integrate models, and to what extent is this even possible?

In the context of archetype modelling, proposing to link dynamic building simulation to models from other engineering disciplines requires the development of appropriate Application Programming Interfaces (APIs). Some commercial simulation packages offer integrated suites of in-house modules that can be used in the co-design of buildings and energy systems at campus scale, typically for projects with a single client (IES-VE, 2021). Any activity, however, which considers the concerns of District Network Operators (DNOs), Transmission System Operators (TSOs), or policymakers, requires integration with software tools that are far beyond the responsibility of one software vendor. Irrespective of whether future building stock simulations are delivered by commercial packages or sets of open-source community-built tools, the API is the essential communication channel between two pieces of software which are developed and maintained by different parties.

8.4.3.1 Sustainable Multi-Sectoral Model Integration

Some commentaries on whole-system model integration discuss the idea of hard-wiring models from different disciplines. Whilst this practice has an ideological basis, the reality of attempting to design and orchestrate a monolithic whole-system tool would be both impractical and damaging to the overall objectives of energy system model integration. The language of hard-wiring or soft-wiring is itself ambiguous and unhelpful in terms of code development and can cause confusion with the very inflexible practice of hard-coding in computer science. In spite of this, hard-wiring is generally considered to be the uniting of two or more models so that they work together as an advanced integrated tool. Soft-wiring, by contrast, is the manual injection of outputs from one model to the next, as inputs, to achieve some more basic form of integration.

This definition of hard-wiring makes no suggestion of how the models are designed: as intertwined models that are designed to always work together in a specific software environment; or, tools that are principally stand-alone, that talk a universal language which other models speak. The former often appears as the foremost idea when hard-wiring is mentioned in the context of energy system model integration. The reality of this idea, however, would entail the curation and maintenance of some universal software library by a *single* organisation, bypassing any rational basis for the ongoing expansion and development of models by individual developers across the whole energy sector, in industrial, academic, and policy settings. To allow archetype-based dynamic simulation tools to talk to established energy system models like the UK-TIMES model, and the diverse range of emerging tools for energy network planning and operation, API-based design is a necessity (Morrison, 2018). An important area of development in this context includes the Functional Mock-up Interface standard (FMI, 2023). In this modular environment, maintenance and ongoing development are self-sustaining, where independent developers are free to use their chosen combination of tools, with complete freedom to propose new alternatives to incumbents. The most recent developments from academia and private enterprise can also result in a direct impact on large-scale industrial and policy development.

8.4.3.2 API Design

Designing the APIs between building stock simulations and energy systems tools requires some careful consideration. Model developers from different disciplines must coordinate the detailed format of inputs and outputs that are of most value between models, with key decisions including temporal resolutions, the frequency and volume of data exchanges, and the efficient use of statistical representation.

Whilst dynamic simulations take manually configured inputs from various data sources, many of these inputs could be replaced in the future by live data or external models from established online APIs for weather (Pang *et al.*, 2016) and grid carbon intensity (National Grid ESO, 2023), for example. In the present discussion, dynamic simulations are seen as "upstream models"; after ingesting information via API relating to weather and carbon intensity, they principally send data outwards to support "downstream models" evaluating the power network and/or the energy mix.

Examples of outbound API content may include:

- energy demand time-series, aggregated on various scales as required, with temporal resolution tailored to the application (potentially down to 15 minutes)
- homogenised and extrapolated (partially synthesised) stock data
- time-series of distributed, behind-the-meter generation
- battery state-of-charge evaluations, linked to domestic generation and self-consumption
- behavioural and occupancy patterns, for relating non-residential and residential demand (working patterns), with potential links to agent-based transport models
- statistical information on internal room temperatures, to study under-heating.

The above outputs would be tailored to any chosen scenario by the "downstream modeller". In the processing of such an API, the downstream modeller would make a bespoke request ("backwards") and the upstream model would send the relevant outputs ("forwards").

Whilst the API request itself from a downstream energy system model to an upstream demand model is in some sense information being sent in reverse, the request is always in the form of a question. It is conceivable, however, that modelled results could flow in reverse to allow the demand model to benefit from insight derived from a network model.

Inbound API content could potentially include:

- demand response signals, to react to the management of the local grid
- time-of-use tariffs, derived from LEM models
- grid-scale battery storage state-of-charge
- consumer reaction to local carbon intensities.

The potential of these model integration mechanisms is significant, but there are vast challenges to overcome before these future ideas become reality. On the basis that this overall API-based architecture is universally accepted and funded, model developers need to produce software that can run in the cloud, or at least on a locally distributed network (most likely within universities). Significant model developments often emerge from engineering departments in industry and academia that don't traditionally target formal computer science skillsets and instead employ scientific computing as an incidental activity, targeting knowledge limited to stand-alone Python or MATLAB-based programming. This needs to change in order to achieve integrated modelling (Morrison, 2018). The complex integration ideas that are imagined by model developers require robust computer science fundamentals to provide the integration infrastructure, including server-based API development and relational database management. Once this happens, domain specialists can then provide a rich array of engineering models that provide new insight.

8.4.4 Insights and Outlook

From the descriptions provided on the modelling landscape earlier in this chapter, a number of key points are summarised here, capturing insights on the present state of modelling, along with the future outlook.

- From their origins as tools for assessing single buildings, physics-based dynamic simulation models have been used increasingly to explore energy demand predictions for large-scale building stocks. This transition, however, is not straightforward; despite the long legacy of these methods in academia and industry, it is the characterisation of simulation inputs for diverse stocks of residential and non-residential buildings that continues to restrict their use in applied contexts.
- Entirely new methods of stock-level building simulation are being explored as alternatives to traditional physics-based approaches. These include grey/black-box methods which lean heavily on modern data science disciplines and lend themselves well to the rapidly evolving data landscape associated with energy systems and buildings (de Wilde, 2023).
- The link between the nature of a specific model and the applied role that it plays can be subtle. Model outputs that are used to design a building to extremes differ considerably from outputs used to assess energy performance, despite the apparent similarities in the modelling process. The spatial scope of a model (i.e. from one to many thousands of buildings) implies certain non-trivial aspects relating to the purpose and objective of modelling at the outset. There is (presently) no mandate for regulating energy performance on a stock-basis where multiple building owners are involved. Modelling large-scale building stock to determine energy demand patterns is not, therefore, carried out by building owners or designers but by large organisations that manage infrastructure and planning.
- The spatial scope, end-user base, and use-case of models impact the software landscape. Industry-standard building modelling tools can have many

thousands of users and are built around user interfaces with a relatively low barrier to entry. Bespoke stock models using experimental methods are often developed as in-house codes or open-source libraries; groups that develop and operate these models are typically resourced with programming expertise and the wide remit and fluid nature of these models means that standardised user interfaces are rare.

- As integrated whole-system modelling is not an activity undertaken by an individual or a siloed specialist group, the architecture of a fully integrated model needs to reflect the distributed nature of the required knowledge and expertise. This includes detailed knowledge of energy demand, energy generation, supply and distribution, power networks, energy markets, and the related policy and social science fields. This entails distinct models that need to run (potentially simultaneously) in different computational environments by different specialist groups. For distinct computers to exchange the necessary information, specialist infrastructure is required to support this, either within a local area network or, more likely, over the internet. Open-source standards such as The Functional Mock-up Interface (FMI, 2023) offer the greatest promise in achieving this, to help build robust APIs to stitch together arrays of distinct models to attempt to model the whole energy system.

8.5 Summary

This chapter has summarised some traditional applications of building modelling, and compared that to the need for building modelling in areas relating to energy systems. As we see energy demand become increasingly important in our assessment of low-carbon energy systems, the opportunities presented by state-of-the-art modelling of buildings should be exploited. Furthermore, if we are to produce sustainable buildings at a scale not seen before (both in terms of raising new building standards and retrofitting our existing buildings) then building modelling has a role to play to characterise these buildings; energy demand signatures of the building stock should, if the low-carbon transition is successful, be significantly different to what we have accepted as default in years past. Thus building modelling will provide a vital toolkit for new designs, new policies, and new operational approaches for buildings and the energy systems that support them.

References List

BEIS (2020) *Non-domestic National Energy Efficiency Data-Framework (ND-NEED), 2020 - GOV.UK*. Available at: https://www.gov.uk/government/statistics/non-domestic-national-energy-efficiency-data-framework-nd-need-2020 (Accessed: 6 December 2021).

Bordass, B. (2013) 'Does anyone understand the performance gap ?', *Building.co.uk*, pp. 1–2.

BRE (2012) *Appendix S: Reduced Data SAP for existing dwellings*.

BRE (2019) *Standard Assessment Procedure (SAP 10)*. Available at: https://www.bregroup.com/sap/sap10/.

Building Research Establishment (2022) *SAP10 - Standard Assessment Procedure - BRE Group*. Available at: https://bregroup.com/sap/sap10/ (Accessed: 12 September 2022).

Calama-González, C.M., Suárez, R. and León-Rodríguez, ÁL. (2022) 'Thermal comfort prediction of the existing housing stock in southern Spain through calibrated and validated parameterized simulation models', *Energy and Buildings*, 254, p. 111562. Available at: https://doi.org/10.1016/j.enbuild.2021.111562.

Castaldo, V.L. and Pisello, A.L. (2018) 'Uses of dynamic simulation to predict thermal-energy performance of buildings and districts: A review', *Wiley Interdisciplinary Reviews: Energy and Environment*, 7(1), p. e269. Available at: https://doi.org/10.1002/wene.269.

Clarke, J.A. (2001) *Energy Simulations in Building Design*. Oxford: Butterworth-Heinemann.

de Wilde, P. (2014) 'The gap between predicted and measured energy performance of buildings: A framework for investigation', *Automation in Construction*, 41, pp. 40–49. Available at: https://doi.org/10.1016/j.autcon.2014.02.009.

de Wilde, P. (2023) 'Building performance simulation in the brave new world of artificial intelligence and digital twins: A systematic review', *Energy and Buildings*, 292, p. 113171. Available at: https://doi.org/10.1016/j.enbuild.2023.113171.

Delorme, L. and Higley, K. (2020) *Implementation of the EPBD United Kingdom – England*.

DesignBuilder (2023) *CIBSE Weather Data Sets*. Available at: https://designbuilder.co.uk/software/cibse-weather-data-sets.

European Commision (2002) 'Directive 2002/91/EC of the European Parliament and of the Council of 16 December 2002 on the energy performance of buildings', *Official Journal Of The European Union* [Preprint]. Available at: https://doi.org/10.1039/ap9842100196.

European Commission (2020) *Final report on the technical support to the development of a smart readiness indicator for buildings : final report*. Publications Office. Available at: https://doi.org/doi/10.2833/41100.

European Commission (2022) *Smart Readiness Indicator (SRI)*. Available at: https://ec.europa.eu/newsroom/ener/newsletter-archives/37893 (Accessed: 12 September 2022).

FMI (2023) *Functional Mock up Interface Standard*. Available at: https://fmi-standard.org.

Harish, V.S.K.V. and Kumar, A. (2016) 'A review on modeling and simulation of building energy systems', *Renewable and Sustainable Energy Reviews*, 56, pp. 1272–1292. Available at: https://doi.org/10.1016/j.rser.2015.12.040.

Hensen, J.L.M. and Lamberts, R. (2019) 'Building performance simulation – challenges and opportunities', in *Building Performance Simulation for Design and Operation*. Second edition. | Abingdon, Oxon ; New York, NY: Routledge, 2019.: Routledge, pp. 1–10. Available at: https://doi.org/10.1201/9780429402296-1.

Hook, A. *et al.* (2020) 'A systematic review of the energy and climate impacts of teleworking', *Environmental Research Letters*, 15(9), p. 093003. Available at: https://doi.org/10.1088/1748-9326/ab8a84.

IES-VE (2021) *Case-study: Heriot-Watt*.

IPCC (2022) *Global Climate Projections*. Available at: https://www.ipcc.ch/report/ar4/wg1/global-climate-projections/ (Accessed: 6 September 2022).

Issermann, M., Chang, F.-J. and Kow, P.-Y. (2021) 'Interactive urban building energy modelling with functional mockup interface of a local residential building stock', *Journal of Cleaner Production*, 289, p. 125683. Available at: https://doi.org/10.1016/j.jclepro.2020.125683.

Jebaraj, S. and Iniyan, S. (2006) 'A review of energy models', *Renewable and Sustainable Energy Reviews*, 10(4), p. 281. Available at: https://doi.org/10.1016/j.rser.2004.09.004. Available at: https://doi.org/10.1016/j.rser.2004.09.004.

Jenkins, D. (2018) 'Integrating building modelling with future energy systems', *Building Services Engineering Research and Technology*, 39(2), pp. 135–146.

Jenkins, D.P. *et al.* (2021) 'Changing the approach to energy compliance in Residential buildings – Re-imagining EPCs', *Energy and Buildings*, 249, p. 111239. Available at: https://doi.org/10.1016/j.enbuild.2021.111239.

Lee, W.L. and Burnett, J. (2008) 'Benchmarking energy use assessment of HK-BEAM, BREEAM and LEED', *Building and Environment*, 43(11), p. 1882. Available at:

https://doi.org/10.1016/j.buildenv.2007.11.007. Available at: https://doi.org/10.1016/j.buildenv.2007.11.007.

McCallum, P. *et al.* (2019) 'A multi-sectoral approach to modelling community energy demand of the built environment', *Energy Policy*, 132, pp. 865–875. Available at: https://doi.org/10.1016/j.enpol.2019.06.041.

McCallum, P., Jenkins, D.P. and Vatougiou, P. (2020a) 'An urban-scale residential stock model for grid-constrained regions', in *Building Simulation and Optimisation (BSO) Conference*. Online.

McCallum, P., Jenkins, D.P. and Vatougiou, P. (2020b) 'Control Event Matrices for Stochastic Heating Use in Urban-Scale Energy Simulations', in *IBPSA Conference on Urban Simulation – uSIM2020*.

Met Office (2019) *UKCP18 Guidance: Data availability, access and formats*.

Morrison, R. (2018) 'Energy system modeling: Public transparency, scientific reproducibility, and open development', *Energy Strategy Reviews*, 20, pp. 49–63. Available at: https://doi.org/10.1016/j.esr.2017.12.010.

NABERS (2021) *NABERS homepage*. Available at: https://www.nabers.gov.au/ (Accessed: 11 January 2022).

National Grid (2020) *4 ways lockdown life affected UK electricity use*. Available at: https://www.nationalgrid.com/uk/stories/grid-at-work-stories/4-ways-lockdown-life-affected-uk-electricity-use (Accessed: 14 September 2022).

National Grid (2022) 'Future Energy Scenarios'.

National Grid ESO (2023) *UK Carbon Intensity API*. Available at: https://carbonintensity.org.uk.

Nicholls, D. and Kane, D. (2021) *Cornwall LEM Residential Electricity Dataset With Solar Production and Battery Storage, 2018-2020*. UK Data Service. Available at: https://doi.org/10.5255/UKDA-SN-854578.

Pang, X. *et al.* (2016) 'Building energy simulation in real time through an open standard interface', *Energy and Buildings*, 117, pp. 282–289. Available at: https://doi.org/10.1016/j.enbuild.2015.10.025.

Perwez, U. *et al.* (2022) 'Multi-scale GIS-synthetic hybrid approach for the development of commercial building stock energy model', *Applied Energy*, 323, p. 119536. Available at: https://doi.org/10.1016/j.apenergy.2022.119536.

Roderick, Y. *et al.* (2009) 'Comparison of energy performance assessment between LEED, BREEAM, and Green Star', in *11th International IBPSA Conference, Glasgow, 27–30 July*.

Sayfikar, M. and Jenkins, D. (2022) 'Review of approaches to EPC assessment across chosen member states'. Available at: https://www.crosscert.eu/fileadmin/user_upload/crossCert_D3.1_Review_of_approaches_to_EPC_assessment_v4.4.pdf (Accessed: 15 November 2023).

Schwartz, Y. and Raslan, R. (2013) 'Variations in results of building energy simulation tools, and their impact on BREEAM and LEED ratings: A case study', *Energy and Buildings*, 62, pp. 350–359. Available at: https://doi.org/10.1016/j.enbuild.2013.03.022.

Semple, S. and Jenkins, D. (2020) 'Variation of energy performance certificate assessments in the European union', *Energy Policy*, 137, p. 111127. Available at: https://doi.org/10.1016/j.enpol.2019.111127.

Siebert, J., Groß, J. and Schroth, C. (2021) 'A Systematic Review of Packages for Time Series Analysis', in *The 7th International conference on Time Series and Forecasting*. Basel Switzerland: MDPI, p. 22. Available at: https://doi.org/10.3390/engproc2021005022.

Sivapriyan, R., Elangovan, D. and Lekhana, K.S.N. (2021) 'Review of Python for Solar Photovoltaic Systems', in, pp. 103–112. Available at: https://doi.org/10.1007/978-981-15-5258-8_12.

Taylor, P.G. *et al.* (2014) 'Energy model, boundary object and societal lens: 35 years of the MARKAL model in the UK', *Energy Research & Social Science*, 4, pp. 32–41. Available at: https://doi.org/10.1016/j.erss.2014.08.007.

Tian, W. *et al.* (2018) 'A review of uncertainty analysis in building energy assessment', *Renewable and Sustainable Energy Reviews*, 93, pp. 285–301. Available at: https://doi.org/10.1016/j.rser.2018.05.029.

UK Government (2014) *Green Deal Assessment Mystery Shopping Research - GOV.UK*. Available at: https://www.gov.uk/government/publications/green-deal-assessment-mystery-shopping-research (Accessed: 30 June 2020).

UK Government (2015) *Display Energy Certificates and advisory reports for public buildings*. Available at: https://www.gov.uk/government/publications/display-energy-certificates-and-advisory-reports-for-public-buildings (Accessed: 15 November 2023).

UK Government (2022) *Government conersion factors for company reporting of greenhouse gas emissions*.

UK NCM (2023) *UK's National Calculation Method for Non-Domestic Buildings*. Available at: https://www.uk-ncm.org.uk.

Velosa, N. *et al.* (2023) 'PROCSIM: An open-source simulator to generate energy community power demand and generation scenarios', *Energies*, 16(4), p. 1611. Available at: https://doi.org/10.3390/en16041611.

9 Life Cycle Assessment in the Building and Construction Sector

Yahong Dong

9.1 Introduction

The building and construction sector has significant environmental impacts and consumes large amounts of energy and materials. Worldwide, construction is responsible for 60% of raw material consumption (Bribián et al., 2011). Buildings consume 36% of global energy use and generate 39% of energy-related carbon dioxide emissions (Abergel et al., 2019). As an inevitable construction material, the annual consumption of concrete increases by 30 billion tons, and the demand for concrete is continually growing (Monteiro et al., 2017). In the use stage, residential and commercial buildings consume 60% of the world's electricity (UNEP, 2022).

The massive environmental degradations and energy consumptions throughout the life cycle of buildings and construction works should be evaluated systematically so that the environmental hotspots can be detected to facilitate a comprehensive interpretation of the environmental impacts. Life cycle assessment (LCA), as a widely recognized environmental assessment instrument, enables the quantitative assessment of the environmental impacts of a product. In ISO 14040 (ISO, 2006), LCA is defined as "*compilation and evaluation of the inputs, outputs and the potential environmental impacts of a product system throughout its life cycle*". The whole LCA of a product is also known as the "cradle-to-grave" assessment, encompassing raw material acquisition, manufacturing, transportation, use, and disposal.

An LCA study is composed of four iterative phases: goal and scope definition, life cycle inventory (LCI), life cycle impact assessment (LCIA), and interpretation (ISO, 2006). In goal and scope definition, several important aspects are described, including the study goal, the system boundary, a functional unit, cut-off criteria, and the target audience. In the second phase – LCI, data collection and inventory calculation are conducted to quantify the inputs and outputs of a product system. In the third phase – LCIA, indicators of impact categories are derived from LCI results through a set of characterization factors that convert the number of substances into indicator values, followed by the three optional procedures, i.e., normalization, grouping, and weighting. In the fourth phase – interpretation, results from LCI and LCIA are considered together and the results should be evaluated to guarantee they are consistent with the requirements of goal and scope definition.

DOI: 10.1201/9781003317890-12

Over the past three decades, there has been a widespread implementation of LCA in the building and construction sector, with an increasing trend of scientific publications in this field (He and Yu, 2020; Yılmaz and Seyis, 2021). The study areas of LCA in the building and construction sector include life cycle energy analysis, carbon emission assessment, LCA of refurbishments, dynamic LCA (DLCA), uncertainty analysis, integration of LCA in building rating systems, integration with life cycle costing (LCC), and integration with building information modeling (BIM) (Nwodo and Anumba, 2019). Although LCA has been recognized as a mature assessment tool, several deficiencies still exist when LCA is applied in the building and construction sector, such as incomplete scope, different data sources, lack of transparency, and inconsistent results from LCIA methods (Vilches et al., 2017). The problem of data scarcity can have an impact on the scope, leading to the omission of certain materials or processes due to the unavailability of data (Khasreen et al., 2009). Many LCA tools are hindered by a lack of transparency (Roberts et al., 2020), and the assumptions of LCA models are usually not clearly described (Price and Kendall, 2012). The enormous inconsistency in the existing methods of LCA calls for further exploration, e.g., a "meta-comparisons" approach based on directional statistics as proposed by Heijungs and Dekker (2022).

The interests in LCA as an assessment tool to evaluate the environmental impacts of buildings and construction are not only arisen in academia but also in the industry and government (Sala et al., 2016). Many countries have launched national and regional policies to mandate LCA in the building and construction sector, including France, Denmark, the United Kingdom, and China (Park and Kim, 2017; Yang et al., 2021; Larsen et al., 2023). For example, in December 2020, the UK government mandated the adoption of whole-life carbon assessments for all public works projects and programs, including building, civil engineering, construction, and equipment projects (HM Government, 2022). Since April 1, 2022, the implementation of the national standard GB/T55015 "General Specification for Building Energy Efficiency and Renewable Energy Utilization" in China has made it obligatory for all new buildings, renovation projects, and expansion works to undergo a life cycle carbon assessment (LCCA) (MOHURD, 2021). Therefore, it is necessary to make adaptations in the building and construction sector to incorporate LCA into regular practice in accordance with government regulations. In addition to scientific research, demands from industry and government turned out to be a new driving force to promote the development of LCA of buildings and construction. Regulatory instruments, managerial support, and organizational measures are crucial in influencing firms within the construction industry to adopt eco-innovative practices (Hazarika and Zhang, 2019).

The increasing awareness of LCA as an assessment technique necessitates benchmark systems of environmental impacts of buildings and construction. Benchmarking is a systematic process to measure the environmental performance of buildings in terms of building types, geographic regions, building structures, and life cycle stages. An LCA benchmark can facilitate relevant stakeholders' evaluation of the position of their environmental impacts in the market and foster the sector to improve environmental performance (Trigaux et al., 2021). Inconsistency of system boundaries, functional units, accuracy of data, LCA approaches, and assumptions make it very difficult to conduct an appropriate comparison of the environmental impacts of buildings (Gervasio et al., 2018).

In recent years, the single-dimensional environmental assessment of LCA has evolved to the multi-dimensional life cycle sustainability assessment (LCSA), which is capable of evaluating impacts of the triple bottom line (TBL), i.e., environment, economy, and society. LCSA is the combination of three life cycle techniques, including environmental LCA, LCC, and social LCA (S-LCA) (Kloepffer, 2008). The applications of LCSA in the building and construction sector are increasing fast. However, the combination of three techniques into an overarching LCSA framework leads to various challenges (Costa et al., 2019), such as deficiencies in the four phases, incompleteness of assessment, lack of standard methods, and lack of transparency, which may decrease the reliability and feasibility of LCSA and hinder its applications in the building and construction sector. Therefore, there is a need to conduct a comprehensive investigation into the statistical and qualitative development of LCA and LCSA in the building and construction sector. This can be accomplished through a combination of bibliometric analysis and critical analysis.

This chapter investigates the state-of-the-art development of LCA in the building and construction sector. First, the research trend is reviewed through bibliometric analysis and a critical analysis. Second, benchmark systems based on LCA are introduced and the benchmarks along life cycle stages are presented. Third, the transition from LCA to LCSA is described and the LCC and S-LCA are elucidated. Fourth, future opportunities of LCA and LCSA in the building and construction sector are discussed, followed by the last section of conclusions.

9.2 Research Trend: A Bibliometric Analysis

The fast growth of LCA in the building and construction sector calls for a timely investigation to reflect the state-of-the-art research trend of the literature (Geng et al., 2017). Bibliometric analysis is regarded as an appropriate method to explore the large volume of publications based on statistics. A bibliometric analysis enables quantitative assessment of the evolution of a specific research area and visualization of a large amount of literature information.

In order to investigate the recent research trend of LCA in the building and construction sector, a bibliometric analysis was conducted. Web of Science was selected as the search engine since it is a comprehensive global citation database that refines search results and avoids duplication of work. The search strings were "life cycle assessment" AND ("building*" OR "construction") and the search scope covered titles of publications. There was no constraint on languages and document types. The time span of the search was until December 31, 2021. The search was carried out on September 17, 2022, which resulted in 10,141 publications from all available databases in the system, including Web of Science Core Collection, MEDLINE, Chinese Science Citation Database, KCI-Korean Journal Database, and SciELO Citation Index.

According to Figure 9.1, an increasing trend of publications was observed from 1990 to 2021. The yearly publications were less than 10 before 1998, and

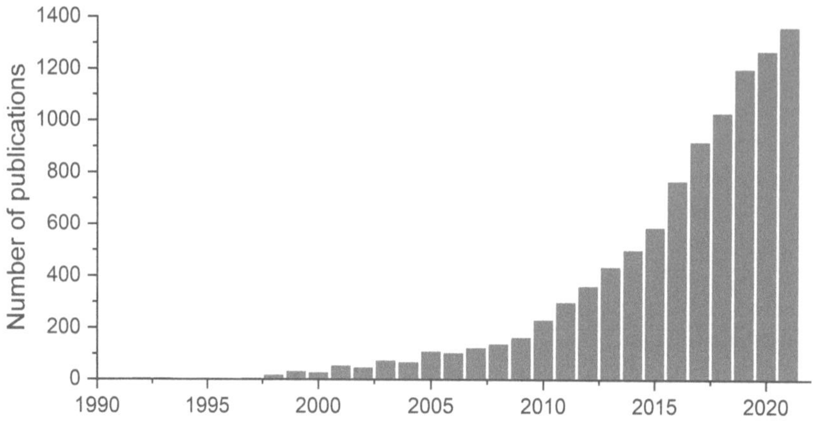

Figure 9.1 The number of publications from 1990 to 2021

the number increased to more than 100 since 2005. From 2018, more than 1,000 publications in the LCA of buildings and construction were found in each year and the number of publications keeps growing steadily.

There were 9,810 publications written in English, accounting for 96.7%, followed by Korean, German, Spanish, French, Chinese, and Portuguese. Among all the countries, the United States contributed 1,742 publications and ranked at the first place, followed by China, Italy, Spain, the United Kingdom, Australia, etc., indicating the research of LCA of buildings and construction was most active in North America, East Asia, and Europe.

With respect to the affiliation, 244 institutions in the world had publication records. There were 11 institutions that contributed more than 100 publications, i.e., ETH Zurich, University of California, Norwegian University of Science and Technology, Universidade De Lisboa, Hong Kong Polytech University, Instituto Superior Technico, Polytechnic University of Milan, Helmholtz Association, Udice French Research Universities, Technical University of Denmark, and State University System of Florida.

The top five research areas of the publications were engineering, environmental science ecology, business economics, construction building technology, and energy fuels since the building and construction sector heavily relies on engineering principles and environmental science plays a crucial role in assessing the environmental impacts and sustainability aspects of construction activities. The top five major concepts were conservation, computational biology, pollution assessment control and management, sanitation, and waste management, as these concepts are essential considerations of the environmental assessment of the sector. The top ten journals were *Journal of Cleaner Production* (1,017), *Sustainability* (538), *International Journal of Life Cycle Assessment* (370), *Energy and Buildings* (347), *Building and Environment* (271), *Resources Conservation and Recycling* (236), *Renewable and Sustainable Energy Reviews* (207), *Construction and Building Materials* (191), *Energies* (169), and *Journal of Building Engineering* (145). These journals are expected to contribute continuously to the research on LCA of buildings and construction in the future.

9.3 Development of LCA in the Building and Construction Sector

Throughout the life cycle of a building or a construction work, various environmental impacts are generated, such as greenhouse gas (GHG) emissions, emissions of particulate matter at a construction site, volatile organic compounds (VOC) from construction materials, noise pollution, water consumption, land use change, etc. (Mehra et al., 2022). LCA facilitates the quantitative analyses of diverse environmental impact categories at different life cycle stages. In addition, LCA is an efficient technique that can provide a quick assessment based on existing databases. Over the past three decades, the methods of LCA have been developed to incorporate comprehensive scopes, diverse data sources, detailed granularity, updated LCIA approaches, sophisticated technologies, etc. to improve the quality and reliability of LCA in the building and construction sector (Anand and Amor, 2017; van der Giesen et al., 2020). These include the integration of LCA with geographic information systems (GIS) (Mastrucci et al., 2015) and artificial intelligence (AI) (D'Amico et al., 2019), as well as the development of LCI databases for the sector (Martínez-Rocamora et al., 2016).

As a result of anthropogenic activities, the global climate change issue has reached an alarming level of severity. Worldwide, over 150 countries pledged to achieve carbon neutrality in the next 30–40 years (Ozawa et al., 2022). LCA is an esteemed technique to account for GHG emissions in the building and construction sector. However, the implementation of LCA in the industry presents new challenges stemming from a limited understanding of LCA within the sector. Consequently, this lack of understanding hinders effective communication among various stakeholders involved, including architects, developers, contractors, suppliers, consultants, engineers, and government entities. Thus, it is imperative to reorganize communications channels to integrate carbon accounting or ecological footprint considerations into the conventional workflow. Nevertheless, the challenges in the LCA research, including inappropriate data input, inconsistent methods, and lack of an agreed benchmarking system, may be amplified in the construction practice.

In this section, the methodological development of LCA in the building and construction sector is reviewed with respect to the four phases of LCA. The state-of-the-art of each phase is critically analyzed and the challenges are discussed. Moreover, the implementation of LCA in the building and construction sector is further explored with a particular focus on the interactions between scientific research and relevant policies.

9.3.1 Methodological Development

9.3.1.1 Goal and Scope Definition

The study goal of an LCA of buildings and construction varied across literature, such as comparison between materials/components and assessment of environmental impacts of a building design (Cabeza et al., 2014). For example, Werner and Richter (2007) investigated the LCA of wooden building products and compared these products to the functionally equivalent products from other construction materials. Aktas and Bilec (2012) studied the lifetime impacts of a residential building using LCA. Moreover, a review article (Soust-Verdaguer et al., 2017) depicted that recent studies attempted to explore the integration of BIM and LCA. For instance, Röck et al. (2018) proposed a BIM-integrated assessment and applied the approach to evaluate the environmental impacts of different types of building construction. In order to incorporate temporal variation, many studies attempted to adopt DLCA (Su et al., 2021). For instance, Negishi et al. (2018) summarized dynamic temporal characteristics of building systems at three levels: temporal evolution at the level of the building, temporal parameters of the building's end-users, and change in the background system of the building. Transforming from conventional LCA to DLCA can promote the utilization of LCA to assess the environmental performance of buildings and other built systems (Collinge et al., 2013).

LCA can be applied to study different objects in the building and construction sector. With respect to the typology of buildings, more than 45% of the literature studied residential buildings, while 27% considered office buildings with concrete structures (Safari and AzariJafari, 2021). Moreover, LCA can also be applied to study campuses (Deda et al., 2023), hospitals (Seifert et al., 2021), governmental buildings (Azzouz et al., 2017), neighborhood (Nematchoua et al., 2021), construction materials (Zhang et al., 2019), components (O'Neill et al., 2021), etc.

System boundary is a critical element that should be clearly stated in the goal and scope definition. According to EN 15978 (EN, 2011), the life cycle of a building includes the product stage, construction process stage, use stage, end of life (EoL) stage, and benefits and loads beyond the system. A full life cycle is also called the "cradle to grave" of buildings and most of the literature adopted this system boundary (Anand and Amor, 2017). In addition to the "cradle to grave" LCA, a few studies conducted partial LCA, e.g., "cradle to gate", "cradle to end of construction", "cradle to use", etc. Nevertheless, the selection of system boundaries is usually subjective, and the elimination of processes is often due to the unavailability of data. Therefore, the completeness of an LCA model is of paramount importance to the quality of an LCA study. The completeness of the LCA model of buildings can be evaluated through an Index of Completeness (IoC) (Dong et al., 2021a). This index quantifies the ratio between the LCA results obtained from the studied building and those derived from a complete model, which is based on the impact assessment results.

Functional unit is the quantitative basis to measure inputs and outputs of a study system. Different functional units were observed in previous studies. The most adopted functional units of buildings are one house or one square meter of construction/floor area, while other functional units are volume and heat delivery, heated floor area, etc. (Anand and Amor, 2017). In terms of construction materials, the functional units can be the volume or weight of the studied material (Turk et al., 2015; Hossain et al., 2016). Nevertheless, the adoption of different functional units in different studies has imposed challenges to the fair comparison of LCA results. For example, de Simone Souza et al. (2021) studied the influence of FU on the LCA of buildings and found that the high-standard dwellings exhibited the highest global warming indicator (GWi) values per dwelling unit and occupant, while the lower standard dwellings displayed the highest GWi values when assessed per square meter.

9.3.1.2 Life Cycle Inventory

LCI is the second phase of an LCA study. According to ISO 14044, procedures in LCI include preparation for data collection, data collection, validation of data, relating data to unit process, relating data to a functional unit, data aggregation, refining the system boundary, and complete inventory. For the LCA studies of buildings and construction, the data sources include site-specific data, estimated data, emission factors from government and literature, and other upstream or downstream data from databases. It is a common practice that an LCA study involves more than one data source since one data source cannot provide all the required information for the LCA model. For instance, databases such as Ecoinvent and US LCI are commonly utilized to complement site-specific data obtained from developers or contractors. However, inconsistent data sources may bring problems, like different levels of data quality, and downgrade the reliability of an LCA study. For example, by replacing the data inputs of the concrete datasets in Ecoinvent and US LCI with local data, the LCA results can be changed up to 20% (Dong et al., 2015).

LCA can be classified into three types: process LCA (P-LCA), input-output LCA (IO-LCA), and hybrid LCA (H-LCA). P-LCA is the conventional LCA method that builds up LCA models based on data for each process unit. The bottom-up approach of P-LCA requires detailed site-specific data along the life cycle stages within the system boundary. Although P-LCA has been mostly adopted, the large truncation error is an unavoidable problem, and this is mainly due to the difficulties in data collection. IO-LCA is a top-down approach based on the national sector-by-sector matrix that quantifies the interactions between sectors, as well as the input and output of sectors. IO-LCA, which adopts the national data of sectors, is more comprehensive by nature (Kyrö et al., 2011). However, an IO-LCA often fails to capture the details of process units, leading to difficulties in conducting further analysis. Bilec et al. (2006) proposed an H-LCA method that integrated both process-based data and input-output data and developed the hybrid construction model. Interestingly, whether H-LCA can outstand P-LCA is still under debate. Yang (2017) argued that due to the restrictive assumptions (e.g., fixed input-output coefficients and unlimited supply of inputs), H-LCA may not provide adequate results for product promotion. Pomponi and Lenzen (2018) questioned Yang's conclusions and claimed that H-LCA can significantly decrease the truncation errors. Further research on how to integrate process-based data and input-output data is desired to improve the accuracy and reliability of LCA. In the building and construction sector, over 60% of studies adopted P-LCA, and about 30% of studies adopted H-LCA, while only 5% of studies used IO-LCA (Bahramian and Yetilmezsoy, 2020).

9.3.1.3 Impact Assessment

LCIA is the third phase of an LCA study. There are two LCIA approaches, namely, midpoint and endpoint approaches. The midpoint approach analyzes environmental impacts prior to the endpoint of a cause-effect chain to quantify the emissions of pollutants or extractions of resources and calculate the indicators of impact categories (Bare et al., 2000). The endpoint approach looks at the end of the cause-effect chain and evaluates the impacts on human health, ecosystems, and resources. The midpoint approach was adopted in most LCA studies of buildings and construction, due to its comprehensive nature, while an endpoint approach tends to offer less detailed information. Similar to LCA in other sectors, energy and emissions are the most popular metrics in the building and construction sector (Anand and Amor, 2017). Other principal midpoint impact categories include eutrophication, acidification, ozone depletion, particulate matter, and human toxicity (Jang et al., 2022).

A well-established LCIA method incorporates either midpoint or endpoint, or both approaches into a systematic paradigm. Adoption of an LCIA method is a common practice in the building and construction sector. For example, Azari (2014) studied the environmental performance of building envelopes using the Tool for Reduction and Assessment of Chemicals and Other Environmental Impacts (TRACI, an LCIA method developed by the US Environmental Protection Agency) and analyzed six midpoint impact categories, i.e., fossil fuel, global warming, acidification, eutrophication, ozone depletion, and smog formation. Pajchrowski et al. (2014) compared four different material structures and analyzed the full life cycle of a building using IMPACT 2002+. Abd Rashid et al. (2017) studied a semi-detached residential house in Malaysia using CML 2001 (developed by the Institute of Environmental Sciences, Leiden University) and analyzed four midpoint impact categories, including acidification, eutrophication, global warming, and ozone depletion. Buyle et al. (2014) studied an apartment building using the ReCiPe Endpoint method to compare the attributional and consequential LCA (an LCA typology related to allocation). Other available LCIA methods that are widely adopted include Eco-indicator, Environmental Product Declaration (EPD), Environmental Design of Industrial Products (EDIP), and the UNEP-SETAC toxicity model (USEtox).

The popularity of these LCIA methods is mainly caused by two reasons. First, the LCIA methods provide ready-to-use model factors, such as characterization factors, normalization factors, and weighting factors, which can enable a fast assessment. Second, these LCIA methods are provided in the widely adopted LCA software, such as SimaPro™ and GaBi™, making the LCIA methods accessible to LCA participators. However, the adoption of LCIA methods may generate non-representative LCA results, in particular for the regional impact categories. The characterization models in LCIA methods are developed based on fate models and regression models in certain regions. If an LCIA method was based in Europe, the regional impact categories, such as eutrophication, acidification, ecotoxicity, and human toxicity, may not be representative in other places.

Large variations were observed in previous studies. Dong et al. (2021b) analyzed 105 buildings in the publications over the past ten years. The production stage (material production and construction) of buildings generated 307–1,487 kg CO_2 eq/m^2. The use stage emitted 1.3–78.8 kg CO_2 eq/m^{2}a. The GHG emission at the end-of-life stage ranged from 0.87 to 49.7 kg CO_2 eq/m^2. Significant distinctions among LCA studies of buildings and construction works were also detected by other studies. For example, Säynäjoki et al. (2017) studied 116 cases and the results of buildings varied from 0.03 to 2.00 tCO_2 eq/m^2. The wide spectrum of LCA results of buildings is partially caused by the real differences between building cases, and partially due to the inconsistency in methods, data, boundaries, etc.

9.3.1.4 Interpretation

As the last phase of LCA, interpretation revisits LCI and LCIA results in accordance with goal and scope definition. In addition, completeness, sensitivity, and consistency of LCA should be evaluated. For the LCA research on buildings and construction, most studies did not separate interpretation from impact assessment and involved the elements of interpretation in the results or discussion sections (e.g., Wen et al., 2015), while few provided a distinct section of interpretation (e.g., Vitale et al., 2017). However, interpretation is an indispensable phase to highlight essential outcomes of an LCA study and support decision-making processes.

Sensitivity analysis, which evaluates the change of results by adjusting input variables, was mostly conducted in the LCA case studies of buildings and construction works. For example, Satola et al. (2020) explored the effect of climate change on the life cycle performance of a container-based housing unit by changing the mean temperature from 16.2°C to 18.4°C and found that climate change will lead to additional energy consumption in 2050. Other variables to be studied in sensitivity analysis include structural materials, recycled content, waste treatment methods, building designs, construction methods, etc. For instance, in a study conducted by Häfliger et al. in 2017, a sensitivity analysis was performed on construction materials utilized in multi-family houses in Switzerland. The researchers discovered that the selection of "Wood and wood products", "Insulation materials", and "Windows and doors" exhibited a significant sensitivity to various modeling choices. However, different sensitivity methods can lead to varied interpretations. Pernetti et al. (2021) compared two sensitivity analysis methods, namely, differential sensitivity analysis (DSA) and elementary effects method (EE), to study 11 nearly zero-energy buildings in Europe. They found that DSA is a straightforward method that relies on a well-established baseline. However, it does not take into account the interactions between different factors. On the other hand, the EE approach has the potential to provide more reliable and comprehensive results by considering a dynamic baseline. However, it requires more effort in terms of implementation and computational resources. Unfortunately, there has been no guidance on how to conduct a sensitivity analysis in LCA of the building and construction sector, and further studies are needed in this aspect.

9.3.2 Applications of LCA

Worldwide, national policies of LCA in the building and construction sector commenced in recent years. For example, the UK government has required whole-life carbon assessments for all public work projects and programs since December 2020. Meanwhile, all UK contracting authorities were requested to provide strategic plans to achieve net-zero carbon of estate and infrastructure. In France, RE2020 came into force in January 2022 to reduce the life cycle carbon footprint of buildings, aiming to achieve the carbon neutrality goal by 2050. In the United States, LCA has been included in a number of standards and certification systems, such as IgCC/Standard 189.1, GBI Green Globes, and Leadership in Energy and Environmental Design (LEED). In February 2021, the US Green Building Advisory Committee made a recommendation for all buildings to adopt a material approach and implement EPDs for at least 75% of their products. Additionally, it was suggested that large projects should employ a whole-building LCA approach with the goal of achieving a 20% reduction in carbon emissions compared to a baseline building specific to the project. In April 2022, China mandated the national standard GB/T51366 – *Standard for Building Carbon Emission Calculation* for all new buildings in the country to conduct the LCCA and submit a carbon accounting report for a proposed building project. The national policies of LCA in the building and construction sector can significantly promote the application of LCA in the industry and create new demands for LCA development in the building and construction sector.

Over the years, the green building schemes have successfully enhanced the application of green technologies and awareness of environmental protection in the industry. As a mature instrument, LCA becomes an indispensable element in the existing green building schemes. The LEED is a world-popular green building assessment scheme developed by US Green Building Council (USGBC). A green building can receive three credits if an LCA is completed and 10% environmental impact is reduced. The Building Research Establishment Environmental Assessment Method (BREEAM) is the world's longest green building assessment scheme established by the Building Research Establishment (BRE) in 1990. In BREEAM, nine points are related to LCA, of which six points can be awarded by conducting the LCA of embodied environmental impacts of materials and three innovation points can be obtained by conducting a full LCA of buildings. The Building Environmental Assessment Method (BEAM Plus) was developed by the Hong Kong Green Building Council (HKGBC) and 1 credit can be obtained when LCA is conducted to evaluate the embodied energy assessment of construction materials.

When LCA is applied in the industrial practice, problems may take place, especially in case the relevant stakeholders have little knowledge of LCA. For instance, the requirement of mandated LCCA in China has forced the entire sector to adapt to the new regulation. At the national level, the government has implemented a range of policies and strategies and has mandated regional governments (provinces or cities) to adhere to these new regulations. However, regional governments may face challenges when attempting to implement the regulations, as stakeholders within the industry may have a limited understanding of what LCA entails and how to integrate it into the existing system. The challenges also lie in the lack of local LCI databases, lack of guidance on the data quality, and lack of standardization in LCA reports for buildings. In China, the requirement to conduct LCCAs for new buildings is mandated, with GB/T51366 serving as the national standard for this purpose. However, the sector faces significant challenges in collecting all the necessary data and creating reports that adhere to GB/T51366 due to the scarcity of information and difficulties in effectively communicating with various stakeholders.

9.4 Benchmark Systems

The intensive applications of LCA in the building and construction sector bring about the need to benchmark the environmental impacts of buildings. Until now, benchmarks based on LCA for buildings were developed in the United States, Europe, New Zealand, Korea, etc. There are several methods to develop a benchmark, such as top-down and bottom-up approaches, absolute or relative values, and comparison of a whole building or building parts (Hollberg et al., 2019). Through a comprehensive review of the existing benchmark systems, Trigaux et al. (2021) concluded that the bottom-up approach was mostly adopted as compared with other benchmark methods. In addition, per unit area per year (m^2a) was selected as the reference unit in most existing benchmark systems. However, challenges have arised in the establishment of LCA benchmarks of buildings since LCA results of buildings from different studies have significant discrepancies (Säynäjoki et al., 2017), such as inconsistent system boundaries, varied data sources, the different requirements on data quality, and adoption of different LCIA methods.

Dong et al. (2021b) analyzed 105 buildings worldwide, published over the past decade. The midpoint results of seven impact categories of the studied buildings were statistically summarized and the derived medians were compared with existing benchmarks in Europe and Korea. Findings from the study revealed interesting insights regarding the climate change impact of the production stage in building LCAs. The median value for this stage was found to be 490 kg CO_2 eq/m^2. In comparison, the Korean benchmark is much higher at 815 kg CO_2 eq/m^2, while the Italian benchmark was significantly lower at 380 kg CO_2 eq/m^2. Although large variations were observed in the reviewed buildings, the medians of the 105 buildings were comparable to the published benchmarks.

As shown in Table 9.1, the benchmarks were provided for three life cycle stages, namely, production, use, and EoL. The production stage included raw material acquisition, manufacturing, material transportation, and construction. The use stage incorporated operation and maintenance processes. Life spans of buildings can vary significantly and the benchmarks for the use stage were given on a yearly basis. EoL dealt with building demolition, waste disposal, waste transportation, and recycling. Given 50 years as the typical life span of a building, the production stage accounted for 68%–90% of eutrophication, ozone depletion, and human toxicity, while this stage represented only 12%–29% for climate change, energy depletion, acidification, and particulate matter. Since the benchmarks were derived from buildings across the world, the values can be referred to by relevant studies to guarantee the quality and reliability of LCA of buildings.

9.5 From LCA to LCSA

The building and construction sector is not only influential to the environment but also has huge impacts on the economy and society. Worldwide, the sector shares 5%–15% of GDP and 5%–10% employment (Robinson et al., 2021). In a fast-developing economy, the building and construction sector contributes much more

Table 9.1 Benchmarks of buildings (Dong et al., 2021b)

Impact categories	*Stage*	*Unit*	*Median*
Climate change	Production	kg CO_2 eq/m^2	490
	Use	kg CO_2 eq/m^{2}a	27.7
	EoL	kg CO_2 eq/m^2	28.9
Energy depletion	Production	MJ/m^2	6,946
	Use	MJ/m^{2}a	341
	EoL	MJ/m^2	360
Acidification	Production	kg SO_2 eq/m^2	1.3
	Use	kg SO_2 eq/m^{2}a	0.182
	EoL	kg SO_2 eq/m^2	0.246
Eutrophication	Production	kg PO_4 eq /m^2	0.215
	Use	kg PO_4 eq /m^{2}a	0.002
	EoL	kg PO_4 eq /m^2	0.002
Ozone depletion	Production	kg CFC-11 eq/m^2	9.10E–06
	Use	kg CFC-11 eq/m^{2}a	1.90E–08
	EoL	kg CFC-11 eq/m^2	4.10E–08
Particulate matter	Production	kg $PM_{2.5}$ eq/m^2	0.14
	Use	kg $PM_{2.5}$ eq/m^{2}a	0.014
	EoL	kg $PM_{2.5}$ eq/m^2	0.03
Human toxicity	Production	kg 1,4-DB eq/m^2	284
	Use	kg 1,4-DB eq/m^{2}a	1.11
	EoL	kg 1,4-DB eq/m^2	0.81

toward the economy. In China, for instance, the building and construction sector accounts for over 25% of GDP and contributed to 29.3 trillion RMB in 2021 (Zhu et al., 2022). In addition, buildings and construction works provide essential space for living, working, transportation, manufacturing, and other human activities. The social impacts are critical to the relevant stakeholders. Moreover, the world is not only suffering the environmental degradation but also encountering economic and social crises, such as unemployment, poverty, inequality, and access to basic services. To meet the emerging needs of sustainable development, LCA has evolved to LCSA that encompasses the TBL, namely, environment, economy, and society. The social and economic aspects are incorporated into the LCSA framework following the ISO 14040/44 framework: goal and scope definition, inventory analysis, impact assessment, and interpretation. While LCSA has the potential to evaluate a broader scope, the challenges due to the integration of the three dimensions remain unsolved, such as data availability and quality, methodological complexity, trade-offs and conflicts, subjectivity, and lack of standardization. These identified challenges of LCSA have been discussed in recent review articles (Costa et al., 2019; Larsen et al., 2022).

LCSA has been increasingly adopted in the building and construction sector, in particular in the past decade, with an average number of publications of over ten articles per year (Bhyan et al., 2022). The complex nature of buildings and construction works, together with the thorny challenges to combine the quantitative and non-quantitative evaluations in LCSA, provoke an emerging need to explore the implementation of LCSA in the building and construction sector. In this regard, in this section, the application and development of LCC, S-LCA, and LCSA are introduced, and the limitations of each technique are discussed in the context of building and construction.

9.5.1 *Life Cycle Costing*

LCC studies the costs associated with the life cycle of a product. According to the coverage of analysis, there are three types of LCC, namely, conventional LCC, environmental LCC, and societal LCC (Hunkeler et al., 2008). A conventional LCC evaluates the costs directly covered by producers or users. An environmental LCC includes those accounted for in conventional LCC, as well as the external costs caused by environmental degradation. A societal LCC encompasses all environmental LCC plus further external costs by anyone in the society.

An increasing trend of LCC publications in the building and construction sector can be observed, especially after 2000 (Goh and Sun, 2016). In recent years, LCC is often applied together with LCA and BIM since the integration of LCC, LCA, and BIM can optimize design, monitor performance, and compare potential schemes (Lu et al., 2021). For example, Santos et al. (2020) proposed an automatic BIM-based LCA/LCC tool and applied the developed method to an office building in the Netherlands and confirmed the combination of BIM and LCA/LCC can improve the understanding of both environmental and economic aspects.

LCC is different from LCA. The results of LCC are more affected by the location of buildings, the regional development level, suppliers, and financial and economic elements (e.g., inflation, interest rate). Therefore, a fair comparison between LCC results from distinct studies may not be feasible. For example, a green industrial building in Sri Lanka cost 209 USD/m^2 in construction, 23.9 USD/m^2a in operation and maintenance, and 0.85 USD/m^2 in EoL (1 LKR = 0.0027 USD) (Weerasinghe et al., 2021). However, in a developed economy like Hong Kong, the average construction cost can be over 1,000 USD/m^2 (Arcadis, 2023).

9.5.2 *Social Life Cycle Assessment*

S-LCA evaluates social impacts to relevant stakeholders through a product's life cycle. According to the *Guidelines for Social Life Cycle Assessment of Products* (Andrews, 2009), S-LCA follows the ISO 14044 and contains four standard phases, namely, goal and scope definition, LCI, impact assessment, and interpretation. Different from LCA, data in S-LCA can be quantitative, semi-quantitative, or qualitative. Given the various types of data in S-LCA, quantification is usually difficult (Larsen et al., 2022), in particular in the normalization of the input and output flows according to a functional unit. In addition, S-LCA not only assesses negative impacts but also evaluates positive social impacts (Benoît Norris et al., 2020), imposing difficulties in inventory analysis and impact assessment of an S-LCA. With respect to impact assessment, six stakeholders are considered, namely, worker, local community, consumer, society, children, and value chain actors (Benoît Norris et al., 2020). Under each stakeholder, there are distinct subcategories that can be estimated. Table 9.2 provides the recommended stakeholder categories and

Table 9.2 Stakeholders and subcategories of S-LCA (Benoît Norris et al., 2020)

Stakeholders	*Subcategories*
Worker	Freedom of association and collective bargaining, Child labor, Fair salary, Working hours, Forced labor, Equal opportunities/ discrimination, Health and safety, Social benefits/social security, Employment relationship, Sexual harassment, Smallholders including farmers
Local community	Access to material resources, Access to immaterial resources, Delocalization and migration, Cultural heritage, Safe and healthy living conditions, Respect of indigenous rights, Community engagement, Local employment, Secure living conditions
Value chain actors	Fair competition, Promoting social responsibility, Supplier relationships, Respect of intellectual property rights, Wealth distribution
Consumer	Health and safety, Feedback mechanism, Consumer privacy, Transparency, End-of-life responsibility
Society	Public commitments to sustainability issues, Contribution to economic development, Prevention and mitigation of armed conflicts, Technology development, Corruption, Ethical treatment of animals, Poverty alleviation
Children	Education provided in the local community, Health issues for children as consumers, Children concerns regarding marketing practices

the corresponding subcategories for each stakeholder in the *Guidelines for Social Life Cycle Assessment of Products and Organizations 2020* (Benoît Norris et al., 2020). The identified stakeholders and subcategories can be applied to establish a social LCIA (sLCIA) method, in which the social indicators of subcategories are calculated, and this step is called characterization. In the next, normalization and weighting are conducted to integrate the results to stakeholders, and the social performance of building and construction projects is evaluated.

S-LCA can be applied to either construction materials or buildings. For example, Hosseinijou et al. (2014) studied the concrete and steel building materials in Iran. The results showed that steel/iron had better performance than concrete/cement since negative social impacts of safe and healthy living conditions were found in cement production. Dong and Ng (2015) developed an S-LCA model for building construction and proposed an sLCIA method that encompassed characterization, normalization, and weighting. The S-LCA model was then applied to a public residential housing project in Hong Kong. They found that precast concrete had negative impacts on fair salary and local employment since the precast factory was located outside Hong Kong. Recently, S-LCA has often adopted in parallel to LCA and LCC, or incorporated into an overarching LCSA framework. For example, Janjua et al. (2020) developed triple bottom indicators for the LCSA of residential buildings in Australia and incorporated social indicators into the LCSA framework. Figueiredo et al. (2021) conducted S-LCA, LCA, and LCC by integrating BIM with LCSA for sustainable material choice for construction projects.

9.5.3 Life Cycle Sustainability Assessment

LCSA has been increasingly applied in the building and construction sector. The number of publications peaked in 2021 and is expected to grow in the future (Dong et al., 2023). Most of the existing studies of LCSA in the building and construction sector focused on buildings, while a few studies worked on construction materials, building components, and infrastructure. For example, Balasbaneh and Sher (2021) studied the different construction methods for a residential building and concluded that prefabrication demonstrated the best performance in environmental and social performance, while *in-situ* construction had the best economic performance. Case studies for LCSA of buildings can be found in Australia, China, Malaysia, Singapore, Canada, the United States, etc. (Leroy-Parmentier et al., 2023). As an example of LCSA of non-building cases in the sector, Arshad et al. (2021) proposed a sustainability assessment tool based on LCSA for major highway projects in Pakistan. By reviewing 27 LCSA case studies in the building and construction sector, Dong et al. (2023) summarized research topics in this field, and they are pertaining to the following aspects:

i To build an LCSA framework;
ii To assess sustainability performance;
iii To determine the indicators;
iv To select among alternative options;
v To benchmark the performance;
vi To develop a new LCSA method.

The research on LCSA thrives over the past ten years and substantial improvement can be observed. Early LCSA studies strived to include more than one dimension of TBL. However, due to the limitations of methods and data, many early LCSA studies only involved two dimensions. For example, Ostermeyer et al. (2013) conducted LCSA through integrating LCA and LCC, while social indicators were not evaluated. With intensive development, recent LCSA studies tend to include all three dimensions of sustainability. The purpose of LCSA studies also transformed, from conducting a sustainability assessment toward integration with advanced technologies, such as BIM, digital twin, AI, internet of things (IoT), and big data (Costa et al., 2019; Larsen et al., 2022). For example, Boje et al. (2023) developed an LCSA framework for buildings by integrating BIM and digital twin technologies to facilitate a more comprehensive LCSA. Mathern et al. (2019) proposed a framework for structural design optimization by integrating machine learning and AI. They found that AI presents a paradigm shift in the optimization and evaluation of structures during the early stages of planning and design, providing novel opportunities for innovation and advancement.

Despite the significant progress of LCSA development, there are several challenges calling for further investigation. Valdivia et al. (2021) summarized the major problems in LCSA, including only two pillars considered, inconsistent with ISO 14040, lack of interconnections among three dimensions, contradictory assumptions, unclear statements, unclear definitions of data and methods, inconsistent level of details, non-transparent weighting, etc. In response to the existing challenges, the UNEP Life Cycle Initiative proposed ten LCSA principles that deal with key challenges of LCSA, including understanding the areas of protection, alignment with ISO 14040, completeness, consideration of stakeholders, product utility, materiality of system boundaries, consistency, transparency, explicit trade-offs' communication, and consideration of positive and negative impacts (Valdivia et al., 2021). Moreover, the building and construction sector encounters specific problems, such as less awareness of non-residential buildings and inconsistent data on BIM and LCSA. These problems can undermine the accuracy and reliability of LCSA results, which ultimately hinders informed decision-making regarding sustainability aspects in the life cycle of buildings. Addressing these issues is crucial for enhancing the effectiveness and credibility of LCSA as a tool for assessing and optimizing the environmental, social, and economic impacts of buildings throughout their life cycle.

9.6 Future Opportunities

With the increasing awareness of climate change and the continuous endorsement of carbon reduction, LCA is expected to be a prevailing environmental assessment approach in the future. There are many opportunities in the LCA of buildings and construction. Although LCA has been recognized as a mature technique, many problems are unsolved with respect to the methods of LCA. For example, the available LCIA methods were developed in Europe, the United States, and Japan, while most of the countries in the world do not have a local LCIA method that can reflect the regional impact. It is critical to improve LCA methods, especially to provide a localization approach for the regions without a specified LCIA method. Data quality is usually difficult to measure, leading to unreliable LCA results. In order to improve the quality of LCA, spatial and temporal information can be incorporated into LCA, giving rise to two new LCA variations: GIS-LCA and DLCA (Pigné et al., 2020; Bernegger et al., 2022). Transparency is required by ISO 14044, but it may not be possible to fully follow this rule due to confidential considerations. The paradox of transparency requirement and the reluctance to disclose information will remain one of the greatest obstacles to LCA communication. The government and associations should encourage the sector to improve the transparency of LCA. Technologies are under fast development. The integration of LCA and digital twins can greatly improve the visualization of LCA results. The integration of LCA and BIM may solve the problem of lack of data. In the building and construction sector, integration of LCA with new technologies will be a prevailing research area in the next decades.

Benchmarking LCA results is indispensable in controlling the environmental impacts of buildings. However, existing benchmarks are mostly developed in Europe, the United States, and Korea, and a benchmark system is still lacking in other countries. It is suggested to develop regional benchmark systems of buildings based on LCA. In addition, a benchmark should consider impact categories in addition to climate change and energy consumption. It is necessary to establish benchmarks according to building types, structural materials, green and non-green buildings, etc.

LCSA is under fast development and a lot of efforts should be endorsed. First, S-LCA is the least studied technique among the three dimensions. It is suggested to conduct S-LCA case studies and explore the methods of S-LCA for buildings and construction. Second, integration of the three dimensions into a harmonized LCSA is a challenging task. The system boundaries, functional units, assumptions, and data types may vary across the three dimensions. Therefore, further effort should be paid to propose a consistent LCSA framework. Third, more case studies of LCSA are demanded in the building and construction sector.

9.7 Concluding Remarks

LCA is a prevailing environmental assessment tool that has been intensively applied in the building and construction sector. This chapter conducted a comprehensive study on the recent development of LCA through bibliometric analysis and critical analysis, reviewed the benchmark systems, described the transition from LCA to LCSA, and discussed the future opportunities. The outcomes of this study revealed that studies on LCA in the building and construction sector are growing intensively. Benchmarks are necessary to improve the understanding of results, facilitating environmental protection. LCSA is under fast development and the challenges in this field will bring about further development of methods, data, and harmonization among environmental, social, and economic aspects.

References List

Abd Rashid, A.F., Idris, J. and Yusoff, S., 2017. Environmental impact analysis on residential building in Malaysia using life cycle assessment. *Sustainability*, 9(3), 329.

Abergel, T., Dulac, J., Hamilton, I., Jordan, M. and Pradeep, A., 2019. *Global status report for buildings and construction—Towards a zero-emissions, efficient and resilient buildings and construction sector*. United Nations Environment Programme.

Aktas, C.B. and Bilec, M.M., 2012. Impact of lifetime on US residential building LCA results. *The International Journal of Life Cycle Assessment*, 17(3), 337–349.

Anand, C.K. and Amor, B., 2017. Recent developments, future challenges and new research directions in LCA of buildings: A critical review. *Renewable and Sustainable Energy Reviews*, 67, 408–416.

Andrews, E.S., 2009. *Guidelines for social life cycle assessment of products: Social and socio-economic LCA guidelines complementing environmental LCA and life cycle costing, contributing to the full assessment of goods and services within the context of sustainable development*. UNEP/Earthprint.

Arcadis, 2023. *Construction cost handbook, China & Hong Kong 2023*. Acradis Hong Kong Limited.

Arshad, H., Thaheem, M.J., Bakhtawar, B. and Shrestha, A., 2021. Evaluation of road infrastructure projects: A life cycle sustainability-based decision-making approach. *Sustainability*, 13(7), 3743.

Azari, R., 2014. Integrated energy and environmental life cycle assessment of office building envelopes. *Energy and Buildings*, 82, 156–162.

Azzouz, A., Borchers, M., Moreira, J. and Mavrogianni, A., 2017. Life cycle assessment of energy conservation measures during early stage office building design: A case study in London, UK. *Energy and Buildings*, 139, 547–568.

Bahramian, M. and Yetilmezsoy, K., 2020. Life cycle assessment of the building industry: An overview of two decades of research (1995–2018). *Energy and Buildings*, 219, 109917.

Balasbaneh, A.T. and Sher, W., 2021. Life cycle sustainability assessment analysis of different concrete construction techniques for residential building in Malaysia. *The International Journal of Life Cycle Assessment*, 26(7), 1301–1318.

Bare, J.C., Hofstetter, P., Pennington, D.W. and De Haes, H.A.U., 2000. Midpoints versus endpoints: The sacrifices and benefits. *The International Journal of Life Cycle Assessment*, 5(6), 319–326.

Benoît Norris, C., Traverzo, M., Neugebauer, S., Ekener, E., Schaubroeck, T. and Russo Garrido, S., 2020. *Guidelines for social life cycle assessment of products and organizations, United Nations Environment Programme (UNEP), 2020.*

Bernegger, H.J., Laube, P., Ochsner, P., Meslec, M., Rahn, H., Junghardt, J., Aurich, I. and Ashworth, S., 2022, December. A new method combining BIM and GIS data to optimise the sustainability of new construction projects in Switzerland. *In IOP Conference Series: Earth and Environmental Science*, 1122(1), 012052.

Bhyan, P., Shrivastava, B. and Kumar, N., 2022. Systematic literature review of life cycle sustainability assessment system for residential buildings: Using bibliometric analysis 2000–2020. *Environment, Development and Sustainability*, 25, 13637–13665.

Bilec, M., Ries, R., Matthews, H.S. and Sharrard, A.L., 2006. Example of a hybrid life-cycle assessment of construction processes. *Journal of Infrastructure Systems*, 12(4), 207–215.

Boje, C., Menacho, ÁJ.H., Marvuglia, A., Benetto, E., Kubicki, S., Schaubroeck, T. and Gutiérrez, T.N., 2023. A framework using BIM and digital twins in facilitating LCSA for buildings. *Journal of Building Engineering*, 76, 107232.

Bribián, I.Z., Capilla, A.V. and Usón, A.A., 2011. Life cycle assessment of building materials: Comparative analysis of energy and environmental impacts and evaluation of the eco-efficiency improvement potential. *Building and Environment*, 46(5), 1133–1140.

Buyle, M., Braet, J. and Audenaert, A., 2014. Life cycle assessment of an apartment building: Comparison of an attributional and consequential approach. *Energy Procedia*, 62, 132–140.

Cabeza, L.F., Rincón, L., Vilariño, V., Pérez, G. and Castell, A., 2014. Life cycle assessment (LCA) and life cycle energy analysis (LCEA) of buildings and the building sector: A review. *Renewable and Sustainable Energy Reviews*, 29, 394–416.

Collinge, W.O., Landis, A.E., Jones, A.K., Schaefer, L.A. and Bilec, M.M., 2013. Dynamic life cycle assessment: Framework and application to an institutional building. *The International Journal of Life Cycle Assessment*, 18(3), 538–552.

Costa, D., Quinteiro, P. and Dias, A.C., 2019. A systematic review of life cycle sustainability assessment: Current state, methodological challenges, and implementation issues. *Science of the Total Environment*, *686*, 774–787.

D'Amico, A., Ciulla, G., Traverso, M., Brano, V.L. and Palumbo, E., 2019. Artificial neural networks to assess energy and environmental performance of buildings: An Italian case study. *Journal of Cleaner Production*, 239, 117993.

de Simone Souza, H.H., de Abreu Evangelista, P.P., Medeiros, D.L., Albertí, J., Fullana-i-Palmer, P., Boncz, M.Á., Kiperstok, A. and Gonçalves, J.P., 2021. Functional unit influence on building life cycle assessment. *The International Journal of Life Cycle Assessment*, 26, 435–454.

Deda, D., Gervásio, H. and Quina, M.J., 2023. Bibliometric analysis and benchmarking of life cycle assessment of higher education Institutions. *Sustainability*, 15(5), 4319.

Dong, Y.H. and Ng, S.T., 2015. A social life cycle assessment model for building construction in Hong Kong. *The International Journal of Life Cycle Assessment*, 20(8), 1166–1180.

Dong, Y.H., Ng, S.T., Kwan, A.H. and Wu, S.K., 2015. Substituting local data for overseas life cycle inventories–a case study of concrete products in Hong Kong. *Journal of Cleaner Production*, 87, 414–422.

Dong, Y., Liu, P., Hossain, M.U., Fang, Y., He, Y. and Li, H., 2021a. An index of completeness (IoC) of life cycle assessment: Implementation in the building sector. *Journal of Cleaner Production*, 283, 124672.

Dong, Y., Ng, S.T. and Liu, P., 2021b. A comprehensive analysis towards benchmarking of life cycle assessment of buildings based on systematic review. *Building and Environment*, 204, 108162.

Dong, Y., Ng, S.T. and Liu, P., 2023. Towards the principles of life cycle sustainability assessment: An integrative review for the construction and building industry. *Sustainable Cities and Society*, 104604.

EN, 2011. *EN 15978:2011. Sustainability of construction works—Assessment of environmental performance of buildings—Calculation method.* CEN.

Figueiredo, K., Pierott, R., Hammad, A.W. and Haddad, A., 2021. Sustainable material choice for construction projects: A life cycle sustainability assessment framework based on BIM and Fuzzy-AHP. *Building and Environment*, 196, 107805.

Geng, S., Wang, Y., Zuo, J., Zhou, Z., Du, H. and Mao, G., 2017. Building life cycle assessment research: A review by bibliometric analysis. *Renewable and Sustainable Energy Reviews*, 76, 176–184.

Gervasio, H., Dimova, S. and Pinto, A., 2018. Benchmarking the life-cycle environmental performance of buildings. *Sustainability*, 10(5), 1454.

Goh, B.H. and Sun, Y., 2016. The development of life-cycle costing for buildings. *Building Research & Information*, 44(3), 319–333.

Häfliger, I.F., John, V., Passer, A., Lasvaux, S., Hoxha, E., Saade, M.R.M. and Habert, G., 2017. Buildings environmental impacts' sensitivity related to LCA modelling choices of construction materials. *Journal of Cleaner Production*, 156, 805–816.

Hazarika, N. and Zhang, X., 2019. Factors that drive and sustain eco-innovation in the construction industry: The case of Hong Kong. *Journal of Cleaner Production*, 238, 117816.

He, X. and Yu, D., 2020. Research trends in life cycle assessment research: A 20-year bibliometric analysis (1999–2018). *Environmental Impact Assessment Review*, 85, 106461.

Heijungs, R. and Dekker, E., 2022. Meta-comparisons: How to compare methods for LCA? *The International Journal of Life Cycle Assessment*, 27(7), 993–1015.

HM Government, 2022. *The construction playbook – Government guidance on sourcing, and contracting public works projects and programmes*, GOV.UK., December 2022.

Hollberg, A., Lützkendorf, T. and Habert, G., 2019. Top-down or bottom-up?–How environmental benchmarks can support the design process. *Building and Environment*, 153, 148–157.

Hossain, M., Poon, C.S., Lo, I. and Cheng, J.C., 2016. Evaluation of environmental friendliness of concrete paving eco-blocks using LCA approach. *The International Journal of Life Cycle Assessment*, 21(1), 70–84.

Hosseinijou, S.A., Mansour, S. and Shirazi, M.A., 2014. Social life cycle assessment for material selection: A case study of building materials. *The International Journal of Life Cycle Assessment*, 19(3), 620–645.

Hunkeler, D., Lichtenvort, K. and Rebitzer, G., 2008. *Environmental life cycle costing.* CRC press.

ISO, 2006. *ISO 14040: International standard. Environmental management – Life cycle assessment – Principles and framework.* International Organization for Standardization (ISO).

Jang, H.J., Ahn, Y.H. and Tae, S.H., 2022. Proposal of major environmental impact categories of construction materials based on life cycle impact assessments. *Materials*, 15(14), 5047.

Janjua, S.Y., Sarker, P.K. and Biswas, W.K., 2020. Development of triple bottom line indicators for life cycle sustainability assessment of residential bulidings. *Journal of Environmental Management*, 264, 110476.

Kaveh Safari, Hessam AzariJafari, 2021. Challenges and opportunities for integrating BIM and LCA: Methodological choices and framework development. *Sustainable Cities and Society*, 67, 102728, https://doi.org/10.1016/j.scs.2021.102728.

Khasreen, M.M., Banfill, P.F. and Menzies, G.F., 2009. Life-cycle assessment and the environmental impact of buildings: A review. *Sustainability*, 1(3), 674–701.

Kloepffer, W., 2008. Life cycle sustainability assessment of products. *The International Journal of Life Cycle Assessment*, 13(2), 89–95.
Kyrö, R., Heinonen, J., Säynäjoki, A. and Junnila, S., 2011. Climate impact of housing companies – A hybrid LCA approach. In *The 2nd international conference on environmental science and technology*, Singapore.
Larsen, V.G., Antoniucci, V., Tollin, N., Sattrup, P.A., Jens, K., Birkved, M., … and Marella, G., 2023. A methodological framework to foster social value creation in architectural practice. *Sustainability*, 15(3), 1849.
Larsen, V.G., Tollin, N., Sattrup, P.A., Birkved, M. and Holmboe, T., 2022. What are the challenges in assessing circular economy for the built environment? A literature review on integrating LCA, LCC and S-LCA in life cycle sustainability assessment, LCSA. *Journal of Building Engineering*, 50, 104203.
Leroy-Parmentier, N., Valdivia, S., Loubet, P. and Sonnemann, G., 2023. Alignment of the life cycle initiative's "principles for the application of life cycle sustainability assessment" with the LCSA practice: A case study review. *The International Journal of Life Cycle Assessment*, 28, 704–740.
Lu, K., Jiang, X., Yu, J., Tam, V.W. and Skitmore, M., 2021. Integration of life cycle assessment and life cycle cost using building information modeling: A critical review. *Journal of Cleaner Production*, 285, 125438.
Martínez-Rocamora, A., Solís-Guzmán, J. and Marrero, M., 2016. LCA databases focused on construction materials: A review. *Renewable and Sustainable Energy Reviews*, 58, 565–573.
Mastrucci, A., Popovici, E., Marvuglia, A., De Sousa, L., Benetto, E. and Leopold, U., 2015, September. GIS-based life cycle assessment of urban building stocks retrofitting – A bottom-up framework applied to Luxembourg. In *EnviroInfo and ICT for sustainability 2015* (pp. 47–56). Atlantis Press.
Mathern, A., Ek, K. and Rempling, R., 2019, September. Sustainability-driven structural design using artificial intelligence. In *Proceedings of the IABSE congress New York City—The evolving metropolis*, New York, NY, USA.
Mehra, S., Singh, M., Sharma, G., Kumar, S., Navishi and Chadha, P., 2022. Impact of construction material on environment. *Ecological and Health Effects of Building Materials*, Springer, Cham. https://doi.org/10.1007/978-3-030-76073-1_22
MOHURD, 2021. *GB/T 55015-2021. General specification for energy efficiency and renewable energy use in buildings*. Ministry of Housing and Urban-Rural Development of the People's Republic of China. (In Chinese)
Monteiro, P.J., Miller, S.A. and Horvath, A., 2017. Towards sustainable concrete. *Nature Materials*, 16(7), 698–699.
Negishi, K., Tiruta-Barna, L., Schiopu, N., Lebert, A. and Chevalier, J., 2018. An operational methodology for applying dynamic life cycle assessment to buildings. *Building and Environment*, 144, 611–621.
Nematchoua, M.K., Sadeghi, M. and Reiter, S., 2021. Strategies and scenarios to reduce energy consumption and CO_2 emission in the urban, rural and sustainable neighbourhoods. *Sustainable Cities and Society*, 72, 103053.
Nwodo, M.N. and Anumba, C.J., 2019. A review of life cycle assessment of buildings using a systematic approach. *Building and Environment*, 162, 106290.
O'Neill, R., Window, A., Kenway, S. and Dargusch, P., 2021. Integrated operational and life-cycle modelling of energy, carbon and cost for building façades. *Journal of Cleaner Production*, 286, 125370.
Ostermeyer, Y., Wallbaum, H. and Reuter, F., 2013. Multidimensional Pareto optimization as an approach for site-specific building refurbishment solutions applicable for life cycle sustainability assessment. *The International Journal of Life Cycle Assessment*, 18, 1762–1779.
Ozawa, A., Tsani, T. and Kudoh, Y., 2022. Japan's pathways to achieve carbon neutrality by 2050 – Scenario analysis using an energy modeling methodology. *Renewable and Sustainable Energy Reviews*, 169, 112943.

Pajchrowski, G., Noskowiak, A., Lewandowska, A. and Strykowski, W., 2014. Wood as a building material in the light of environmental assessment of full life cycle of four buildings. *Construction and Building Materials*, 52, 428–436.
Park, K. and Kim, K., 2017. BIM application and adoption in the UK housing sector. *Integrated Building Information Modelling*, 46–81. Bentham Science Publishers. https://doi.org/10.2174/9781681084572117010005
Pernetti, R., Garzia, F. and Oberegger, U.F., 2021. Sensitivity analysis as support for reliable life cycle cost evaluation applied to eleven nearly zero-energy buildings in Europe. *Sustainable Cities and Society*, 74, 103139.
Pigné, Y., Gutiérrez, T.N., Gibon, T., Schaubroeck, T., Popovici, E., Shimako, A.H., Benetto, E. and Tiruta-Barna, L., 2020. A tool to operationalize dynamic LCA, including time differentiation on the complete background database. *The International Journal of Life Cycle Assessment*, 25, 267–279.
Pomponi, F. and Lenzen, M., 2018. Hybrid life cycle assessment (LCA) will likely yield more accurate results than process-based LCA. *Journal of Cleaner Production*, 176, 210–215.
Price, L. and Kendall, A., 2012. Wind power as a case study: Improving life cycle assessment reporting to better enable meta-analyses. *Journal of Industrial Ecology*, 16, S22–S27.
Roberts, M., Allen, S. and Coley, D., 2020. Life cycle assessment in the building design process – A systematic literature review. *Building and Environment*, 185, 107274.
Robinson, G., Leonard, J. and Whittington, W., 2021. *Future of construction – A global forecast for construction to 2030*. Marsh & GuyCarpenter, Oxford Economics.
Röck, M., Hollberg, A., Habert, G. and Passer, A., 2018. LCA and BIM: Visualization of environmental potentials in building construction at early design stages. *Building and Environment*, 140, 153–161.
Safari, K. and AzariJafari, H., 2021. Challenges and opportunities for integrating BIM and LCA: Methodological choices and framework development. *Sustainable Cities and Society*, 67, 102728.
Sala, S., Reale, F., Cristobal-Garcia, J., Marelli, L. and Pant, R., 2016. Life cycle assessment for the impact assessment of policies. *Life thinking and assessment in the European policies and for evaluating policy options*, EUR 28380 EN, doi:10.2788/318544.
Santos, R., Costa, A.A., Silvestre, J.D., Vandenbergh, T. and Pyl, L., 2020. BIM-based life cycle assessment and life cycle costing of an office building in Western Europe. *Building and Environment*, 169, 106568.
Satola, D., Kristiansen, A.B., Houlihan-Wiberg, A., Gustavsen, A., Ma, T. and Wang, R.Z., 2020. Comparative life cycle assessment of various energy efficiency designs of a container-based housing unit in China: A case study. *Building and Environment*, 186, 107358.
Säynäjoki, A., Heinonen, J., Junnila, S. and Horvath, A., 2017. Can life-cycle assessment produce reliable policy guidelines in the building sector? *Environmental Research Letters*, 12(1), 013001.
Seifert, C., Koep, L., Wolf, P. and Guenther, E., 2021. Life cycle assessment as decision support tool for environmental management in hospitals: A literature review. *Health Care Management Review*, 46(1), 12–24.
Soust-Verdaguer, B., Llatas, C. and García-Martínez, A. 2017. Critical review of BIM-based LCA method to buildings. *Energy and Buildings*, 136, 110–120.
Su, S., Zhang, H., Zuo, J., Li, X. and Yuan, J., 2021. Assessment models and dynamic variables for dynamic life cycle assessment of buildings: A review. *Environmental Science and Pollution Research*, 28, 26199–26214.
Trigaux, D., Allacker, K. and Debacker, W., 2021. Environmental benchmarks for buildings: A critical literature review. *The International Journal of Life Cycle Assessment*, 26(1), 1–21.
Turk, J., Cotič, Z., Mladenovič, A. and Šajna, A., 2015. Environmental evaluation of green concretes versus conventional concrete by means of LCA. *Waste Management*, 45, 194–205.
UNEP, 2022. *Energy efficiency for buildings*. UNEP Division of Technology, Industry and Economics, Paris, France.

Valdivia, S., Backes, J.G., Traverso, M., Sonnemann, G., Cucurachi, S., Guinée, J. B., … and Goedkoop, M., 2021. Principles for the application of life cycle sustainability assessment. *The International Journal of Life Cycle Assessment*, 26(9), 1900–1905.

van der Giesen, C., Cucurachi, S., Guinée, J., Kramer, G.J. and Tukker, A., 2020. A critical view on the current application of LCA for new technologies and recommendations for improved practice. *Journal of Cleaner Production*, 259, 120904.

Vilches, A., Garcia-Martinez, A. and Sanchez-Montanes, B., 2017. Life cycle assessment (LCA) of building refurbishment: A literature review. *Energy and Buildings*, 135, 286–301.

Vitale, P., Arena, N., Di Gregorio, F. and Arena, U., 2017. Life cycle assessment of the end-of-life phase of a residential building. *Waste Management*, 60, 311–321.

Weerasinghe, A.S., Ramachandra, T. and Rotimi, J.O., 2021. Comparative life-cycle cost (LCC) study of green and traditional industrial buildings in Sri Lanka. *Energy and Buildings*, 234, 110732.

Wen, T.J., Siong, H.C. and Noor, Z.Z., 2015. Assessment of embodied energy and global warming potential of building construction using life cycle analysis approach: Case studies of residential buildings in Iskandar Malaysia. *Energy and Buildings*, 93, 295–302.

Werner, F. and Richter, K., 2007. Wooden building products in comparative LCA. *The International Journal of Life Cycle Assessment*, 12(7), 470–479.

Yang, Y., 2017. Does hybrid LCA with a complete system boundary yield adequate results for product promotion? *The International Journal of Life Cycle Assessment*, 22(3), 456–460.

Yang, X., Zhang, S. and Wang, K., 2021. Quantitative study of life cycle carbon emissions from 7 timber buildings in China. *The International Journal of Life Cycle Assessment*, 26(9), 1721–1734.

Yılmaz, Y. and Seyis, S., 2021. Mapping the scientific research of the life cycle assessment in the construction industry: A scientometric analysis. *Building and Environment*, 204, 108086.

Zhang, Y., Luo, W., Wang, J., Wang, Y., Xu, Y. and Xiao, J., 2019. A review of life cycle assessment of recycled aggregate concrete. *Construction and Building Materials*, 209, 115–125.

Zhu, J., Zhang, C., Wang, S., Yuan, J. and Li, Q., 2022. Evolutionary game analysis of construction workers' unsafe behaviors based on incentive and punishment mechanisms. *Frontiers in Psychology*, 13, 907382.

Part IV

Social Transition towards Creating a Sustainable and Resilient Built Environment

10 Social Transition towards Creating a Sustainable and Resilient Built Environment

Thomas S.K. Tang

10.1 What Is a Sustainable and Resilient Built Environment?

The world's needs are changing. Driven by growing population expansion and urbanization, will we ultimately reach a disaster-ridden Malthusian conclusion as our growth exceeds the capacity of the planet to feed and support us? Science and technology have sustained us so far; the steam revolution in the 1740s was pivotal in the harnessing of energy for industrialization and urban development, and the green revolution back in the 1960s helped us through the use of pesticides and fertilizers to enable farmers to produce sufficient food to feed the hungry urban masses that were springing up in cities across the world. Today – achieved through good urban design and critical infrastructure – the *success* of cities as urban centres has been celebrated throughout the world as, freed from the worries of energy, water and food security, businesses and trade have flourished enriching and benefiting societies (Hollis 2013).

But in an urban built environment, there is a price to pay. The question is: can the built environment cope with the challenges that face us tomorrow?

The built environment is defined as the human-made surroundings that provide the setting for human activity, ranging in scale from buildings and parks or green space to neighbourhoods and cities that can often include their supporting infrastructure like roads and other facilities (Baker 2021). The built environment has not only to be functional enough to support the city in which people live but at the same time be ready for future changes. Successful cities will grow as they tend to draw in migrant populations in search of economic opportunity (Montgomery 2013). This means that accommodating a larger population will increase the demand for places to live and the corresponding demand for energy, water supply and treatment, transportation and solid waste management. A further factor is in changing demographics. As the population gets older and lives longer, homes need to be repurposed and changes made to public health and medical care, all of which involve changes to the built environment (World Economic Forum 2022).

The built environment must furthermore be designed with users in mind. The quality of life has to be taken into account (Glaeser 2012; Montgomery 2013). Consideration must therefore be given to the location of highways or railways which

DOI: 10.1201/9781003317890-14

bisect communities dividing the areas and generating unpopular noise and traffic impacts. This can be seen in rapidly expanding places like Bangkok or Singapore where neighbourhoods have been disrupted leading to the loss of community spirit and local culture. Another example is the increase in the number of tall buildings that are necessary to house a large population in a compact area, common in dense cities like Hong Kong. Such buildings can be intrusive and mask daylight and forcing people to live in cramped uncomfortable conditions. On the other hand, NIMBY or "not in my backyard" is a common expression used by residents who do not want the less popular infrastructure in their neighbourhoods like waste incinerators or electricity sub-stations. All of these factors should be addressed as part of developing the built environment.

The Sustainable Development Goals (United Nations, 2015) or SDGs offer rudimentary guidelines on how to design a sustainable urban centre. If followed robustly, the SDGs provide 169 targets for authorities and businesses to aspire towards. But to what extent are the Goals appropriate for the communities that live there? Do the Goals cater for affordability and equal access to infrastructure and utilities? Are the benefits distributed equitably?

Apart from the SDGs, future readiness for trending issues is another consideration. Some of the key issues are listed below.

- In 20–30 years, sea levels may have risen and temperatures may have breached the 1.5 degree rise warned by the Intergovernmental Panel on Climate Change leading to extreme heatwaves and weather dislocations.
- Societies will have changed too; by 2050. According to the World Health Organization (2016), one person in six will be over 65 years old having lived longer due to better medicine and lifestyles. This will entail changes in healthcare, housing and transportation to cater for the needs of a frailer and less mobile society.
- Cities will become more cosmopolitan as societies become more diverse. The future of work will be redefined as workers can choose whether to go into offices or work remotely in a metaverse virtual environment. Driven by pandemic concerns, studies by the World Economic Forum (2022) suggest that 20% of full work days will be completed from home in the future – compared to just 5% in the past.

Moving forward, it is clear that the design of the built environment will have to take social changes into account. As stated by the architect Janet Jacobs (1961), "cities have the capability of providing something for everybody, only because, and only when, they are created by everybody."

10.2 Why Social Transition of the Built Environment?

In the past, the built environment was designed for a purpose, whether to shelter us against inclement weather, to deliver water supplies and energy, to take away our waste or to transport us to our destinations. Living standards have evolved markedly as have expectations. Today the built environment not just serves us functionally but has to be aligned from a user's perspective. Designers have found that societies are more compliant to use facilities if they are designed properly. For example, people will use more public transport if bus stops are located appropriately in catchment areas and bus arrival times are known; similarly, waste recycling rates will be higher if sorting can be done at public places in an easy and convenient fashion.

Fundamental to understanding user needs is the comprehension of human behaviour. Maslow's hierarchy of needs (1943) is helpful to understand such needs.

Maslow's model comprises five tiers of needs starting with basic physiological or hygiene needs all the way up to the need to grow through self-actualization. If we consider each level, it becomes clear as to how social needs within cities change through the ascendency of the hierarchy.

- **Level 1.** Physiological – the basic hygiene needs of people must be met at the lowest level; food, water, housing, public health and waste removal are included (Taormina and Gao 2013).
- **Level 2.** Safety – comprises the need for public security, emergency response and disaster resilience (Poston 2009).
- **Level 3.** Love and belonging – communities and neighbourhoods are the foundation of any urban centre; people from families and communities retain close relationships and common beliefs to foster social harmony and mutual cooperation (Pinkus 2017).
- **Level 4.** Esteem – this component is paramount as government and businesses have to involve citizens or customers in decision-making usually through engagement and consultation – authorities that fail to do so often end up with disgruntled and unhappy populations who feel that respect and recognition of their rights are missing (Følstad and Krogstie 2004).
- **Level 5.** Self-actualization – the apex of the model is often referred to as the growth element of an individual or individuals and can only take place once the preceding four needs are satisfied (Krems et al. 2017).

Designing the built environment usually gets as far as the first two levels and sometimes up to the third level. The fourth level is usually the most difficult as there is, in the majority of cases, a political agenda behind how much engagement or involvement is considered allowable by government or corporate organizations (van Vuuren and Wicke 2022).

But with expanding populations and a burgeoning global awareness fuelled by social media and other channels of information (Wang et al. 2013), the attitudes and behaviour of societies have changed radically – leading to high expectations on authorities to do what is best for the population. Social concerns have also been elevated by NGOs and activists (Mhlanga et al. 2018) who regard worldwide issues such as poverty, climate change and pandemic outbreaks as challenges that leading authorities and businesses should and must address if we are to achieve sustainable development in the long run.

10.3 What Are the Social Transitions?

The SDGs mentioned earlier set a 15-year timeframe for the achievement of the goals. Let us assume that by 2030, the world has seen fit – whether driven by crises or other events – to place enough attention and effort towards these targets.

Whilst all of the SDGs are related to mankind's future, this chapter selected and reviewed the SDGs that have a pertinent bearing as a result of the necessities to humans, e.g. food, water, energy, economic industry, community living, sustainable consumption and the ability to live peacefully with one another, which are related to social transition. Possible scenarios achieved by 2030 are listed below to show what might be beneficial to people.

- **SDG 2** (Zero Hunger) – food supply chains may have improved by 2030 allowing the flow of food to needy countries from sources which are well-positioned to supply plentiful food beyond their own needs.
- **SDG 6** (Clean Water) – water security issues will still dominate but, as a result of better and more efficient ways of recycling water, cities will adapt (sometimes out of forced choice as countries controlling upstream river resources constraint the fresh water supply to downstream countries).
- **SDG 7** (Clean Energy) – there will be an abundance of green market opportunities in 2030 as companies and governments pursue renewable energy as a clean option shifting away from fossil fuels; alternative fuels like hydrogen will emerge.
- **SDG 9** (Industry, Innovation and Infrastructure) – driven by climate change, pandemic situations, digital advancements and other factors, businesses will come up with innovative ideas that don't just generate profits but social and environmental benefits as well.
- **SDG 11** (Sustainable Cities and Communities) – a growing awareness among millennials and later generations will force governments to be more socially accountable for their policies and actions with a longer term lens on services beyond just short-term needs and aimed at addressing resiliency within communities.
- **SDG 12** (Responsible Consumption and Production) – people will relate better to the consequences of their consumption patterns; an increase in the application of the circular economy will take place with a growing tendency for secondary goods, reverse supply chains and products designed for recycling together with cost models that favour recovered materials over virgin ones.
- **SDG 16** (Justice and Peace) – there will most likely be political shifts in superpowers by 2030 and the advent of regional trading blocs based on the EU model. China will undoubtedly lead the Asian trading bloc. Should this be the case, social cause and justice will be served at the regional level taking into account cultural and equity aspects.

Bissio (2019)'s reports that to achieve the 2030 goals laid out in the SDGs, "more holistic and more sweeping shifts in how and where power is vested, including through institutional, legal, social, economic and political commitments to realizing human rights" is required. Hence, if we look at the growth of urbanization alongside patterns of climate change, ageing populations, future work and pandemics together with the opportunities of digital technology and connectivity, the need for social transition of the built environment has never been higher.

In order to build for social needs, there has to be a paradigm shift in how we design and construct infrastructure and other facilities. But this is not without challenges. The Royal Institute of Chartered Surveyors (2020) states, "currently, the infrastructure sector does not consistently define or measure social value. It lacks a robust set of standards and objectives to guide the delivery and assessment of social value, and there is little capacity for social value measurement." It is true that defining social value let alone social transitions is fraught with loose and abstract notions. However, we can identify six transitions that are visible and implementable that can deliver value back to societies:

- Bringing nature back to the people
- Mobility for all
- People-friendly buildings
- Digital societies
- Co-living
- Equity and social justice

10.3.1 Social Transition #1 – Bringing Nature Back to the People

According to Frank Lloyd Wright (1939), architects and designers should "study nature, love nature, stay close to nature. It will never fail you." As a proponent of the built environment, Wright's words may seem disconsonant with his profession but in his work, he often experimented with the newest technologies and practices in the spirit of innovation and appreciation of nature and the environment.

Research in urban greening (Davern and Farrar 2017) has shown that found that people adopt more positive and happier behaviour living in areas with more birdlife and increased neighbourhood tree coverage, and there are lower rates of antidepression medical prescriptions in neighbourhoods close to woodlands. These and other findings in research on the correlation between wellbeing and natural environments (Rojas-Rueda et al. 2019; Templeton 2019) point to the importance of incorporating nature into the built environment.

Use of space is the first priority. People need spaces. Spaces empower people through engagement, interaction, transactions and recreation. When people have space, they have freedom to express themselves and at the same time, there is sufficient area to avoid being intrusive whilst maintaining a sociable distance. A community space breaks down barriers of age, class, religion and other differences. People, especially children, mingle once they have a common area of interest like a makeshift football match or a space for swings and slides. Older folks congregate in town squares or around public benches and tables. So, it is important to keep sufficient land for public spaces even in adverse times. The urban fabric must be maintained in order to keep spaces vibrant and relevant.

A vital component is parks. Public parks are an expression of society and they play a role as economic catalysts. Cities all over the world are revitalizing old parks to create places where people feel they belong and where they can connect with others. Parks thus must constantly evolve to continue serving communities. The flow of ideas about how to make a great public place should not cease once the park is built. Successful ideas include:

- A variety of destinations within the park
- Easy pedestrian access
- Connections to surrounding neighbourhoods and public transit
- Grass, walkways and trees that invite people to enjoy themselves
- Night-time movie screenings
- Places for chess and other games
- Outdoor musical performances
- Art markets

Another feature of space utilization is blue-green infrastructure; often recognized as an alternate model to stormwater management that seeks to avoid or minimize development impacts on the natural environment. Examples include bioswales, green roofs and porous paving. An added advantage of blue-green infrastructure over concrete hardscape is the social value it brings to communities that can utilize this aesthetic approach to beautify neighbourhoods. Related to connecting with nature, urban farming is a growing initiative not just for food security imperatives but also for the therapeutic opportunities this activity gives to people, especially the aged and marginalized.

Lastly, water is an essential element for reconnecting people to nature. Waterfront locations always command a premium in cities for views and access to water-based activities. But another way of linking the water connection to people is the revitalization of rivers. There are many examples of how river clean-up and beautification of rivers have helped to improve social cohesion. One of the most famous cases is the Cheongyecheon River in Seoul, South Korea. A two-year project costing nearly US$1 billion was carried out in 2003 to re-expose the covered river as part of a larger urban renewal effort. In the years following the project's completion, the area attracted more than 60,000 visitors daily Landscape Architecture Foundation (2011), and investment in the downtown area increased dramatically. The reduction in traffic resulted in a significant improvement in air quality and a decrease in local temperatures during the Seoul summer months. According to the Worldwide Fund for Nature (2102), these improvements in environmental conditions were accompanied by the return of several species of birds, fish and insects to the restored river.

10.3.2 Social Transition #2 – Mobility for All

The growth of the built environment has raised the question of urban mobility. The presence of commercial buildings provides work opportunities but these tend to be located far away from homes and houses. What are we doing to get people from their homes to workplaces and vice versa? How are we able to connect communities to maintain the social integrity of urban dwellings? At the same time, there are implications for our choices of how we travel from home to work. With the predominance of motorized vehicles or automobiles used for commuting, the course that many cities are taking is leading to traffic congestion, deteriorating air quality and loss of productive time due to hours spent in traffic jams.

At an individual level, in general, the built environment should strike a balance between automobile and pedestrian needs. Making walking a pleasure should be a priority. This means that the built environment should be designed to cater for all personal walks of life. These factors include:

- Beautiful spaces even if the buildings are mediocre
- Focusing on the bottom 10 m of the building to ensure a functional and aesthetic relationship with the sidewalk
- Putting up sun and wind shelters
- Planting trees
- Social interaction between dwellers and pedestrians
- Maintaining things to make it look like someone cares.

Designing the built environment with pedestrians in mind includes street typologies that account for pedestrian needs in different land use contexts, minimum dimensional requirements for sidewalks (including edges, furnishings, throughway and frontage zones), minimizing pedestrian crossing distances, setting standards for distances between pedestrian crossings and standards for how driveways cross sidewalks, with the sidewalk taking priority.

Examples include:

- Vegetation buffers act as grass areas between the street and the sidewalk to make sidewalks safer and also absorb the carbon dioxide from automobile emissions and assist with water drainage.
- On streets with parking, curb extensions allow pedestrians to see oncoming traffic better where they otherwise would be forced to walk into the street to see past parked cars.
- Cities undertaking pedestrian projects for better traffic flow by closing automobile access and only allowing pedestrians to travel. Projects such as the High Line in New York increase walkability by connecting neighbourhoods, using landscape architectural elements to create visually aesthetic green space and allowing for physical activity.

Still on individual transportation, according to Healthline (2021), cycling is probably the most sustainable form of transport, offering efficiency, low pollution, convenience, lower cost and better health. Factors to encourage cycling are:

- Numbers – the more cyclists that take to the roads, the more attention that will be paid by motorists to good driving behaviour in order to avoid hitting cyclists.

- Safety – well-designed dedicated and shared lanes will provide protection for cyclists and ensure better road safety when riding.
- Distances – cycling is best when the distances are of a length sufficient to travel with optimum effort yet not so short that one could walk the distance. Typically commuting distances of 3–4 km are preferred.
- Socializing – cycling can be a social exercise whether for recreation (with cycling clubs), as a family activity or when commuting. Cycle lanes that can accommodate two cyclists in both directions can encourage social interaction between riders.
- Portability – bicycles can be used for transporting goods using baskets and trailers. Bicycles themselves can be transportable through folding mechanisms.
- Comfort – part of the value proposition of cycling is the physical activity and the enjoyment of fresh air. This can have varying effects in different climates but on the whole, they form part of the attractiveness of cycling. It is useful to have shower facilities at the workplace for washing and freshening up which add to the sense of wellbeing after the journey.

Mobility of citizens whether for work, pleasure or socializing however will still be dominated by the use of private automobiles. On a mass scale, vehicles require roads and highways, which run the risk of cutting off and marginalizing communities. Vehicles also emit air pollutants causing health problems. To manage private vehicles and avoid the need to build obtrusive roads, measures like road tolls, parking charges and vehicle taxes, should be imposed to eliminate transport problems including congestion, high carbon emissions, deteriorating air quality and road safety.

But ultimately, as stated by the American Public Transportation Association (2012), public transit is the key to social mobility for all. There are many supporting arguments for this:

- In urban areas it is more cost-effective to invest in public transit to create roadway capacity from the existing infrastructure than to invest in adding more roadways.
- Switching to public transit reduces greenhouse gas emissions and air pollution.
- Transit is an equitable means of transport.
- Improved health, as there is some level of walking involved in getting to and from stations and the ultimate destinations.

Buses offer flexibility and cost advantages over rail (Hallisey 2021). They can provide a high level of trunk or major corridor service in emerging transit markets and essential feeder services in more mature markets. However, in general, the challenge for buses lies in obtaining the right of way and siting of bus stop locations. When these factors are not managed properly, buses become a secondary option to private vehicles, particularly in suburban areas. Buses also suffer from an unfair reputation of unreliability, poor cleanliness and sub-par maintenance (Byrne and Pharo 2018).

On the other hand, mass transit rail has an advantage over buses in terms of capacity which can reduce costs in high-demand corridors. As well as offering a smoother ride, rail tends to be quieter and cleaner than bus transit. The downside is the high capital costs, although the rail property model (Leong 2016) which allows railway operators to become property landlords helps alleviate this problem. For mass transit rail to work, stations must be accessible and conveniently located. If location is a problem, then good affordable feeder systems must support the stations.

In the future, as we steadily move away from dependence on private vehicles, we will be able to take advantage of shared transport to complement public transit systems; the built environment of the future may well become "smart" with better systems to better manage traffic flow and more efficient ways of discouraging private car ownership through car sharing, parking charges and road usage pricing. Time, cost and quality of travel will determine which method proves to be most impactful. Transit on demand could well become popular in the future whereby people "order" their transportation Uber-style when needed minimizing the need to own an automobile. When "driverless" cars become more than just a demo-type pilot, it is possible that this form of transport will become prevalent to complement public transport in first- and last-mile transport to homes and destinations.

10.3.3 Social Transition #3 – People-Friendly Buildings

We spend typically up to 90% of our time in buildings (Roberts 2016). It is engrained in our lifestyles that our work and living have to be enclosed to protect us against the weather and to maintain a degree of comfort and functionality in our lives. Small wonder then that we are susceptible to many physical and mental stresses as a result of this time spent inside buildings.

Buildings are designed to be energy efficient. But the price of this can be deleterious to our health. It is better and less demanding to recirculate chilled air rather than introduce new air which has to be cooled down or heated according to the building needs. The downside is poor air quality if the recirculated air is not properly treated; bacteria, chemicals, dust and particulates are the main agents of damage. In addition, air quality issues can diminish work productivity and lead to sick building syndrome, where no disease or cause can be identified, yet acute health effects are linked to time spent inside a building (Gawande et al. 2018). In general, this can be alleviated with overall hygiene measures like cleaning and ventilation. But other factors can be introduced such as:

- Banning of smoking
- Reduced use of Volatile Organic Compounds
- Air filtration
- Microbe and mould control
- Pest control

Our working environment is important as well. If people are unable to focus at work, prone to recurring illnesses or spend too much time sitting, they are liable to leave their jobs to find ones that offer better working conditions. According to the World Green Building Council (2014), a range of factors – from air quality and lighting to views of nature and interior layout – can affect the health and job performance of office workers. Office design can hence significantly impact not just the health and wellbeing of staff but also their productivity.

There are a number of measures which can be taken to address these challenges:

- Good lighting is crucial for occupant satisfaction. Studies have shown that productivity increases as a result of proximity to windows, especially if the view offers a connection to nature (Augustin 2021).
- Personal control over thermal comfort can result in improvements in productivity (Luo et al. 2016).
- Biophilic design, i.e. integration of the built space with greening like planters or green walls, has a positive impact on working conditions as well as reducing indoor air pollution (Angermann 2022).
- Noise distraction is related to interior layout and can be ameliorated by configuration of work space, breakout space and workstation density (Cyril 2023).
- The look and feel of the office has a potential impact on wellbeing and mindset – both for occupier and visiting clients. This is though highly subjective depending on age, gender and culture (Olson 2015).
- Access to exercise through the provision of gyms, bicycle storage and green space gives a positive release of energy; some of these facilities may be inside the office building or office grounds, or in the local vicinity (Kohll 2019).

Mental wellbeing, involving movement and relaxation, is also key to people-oriented buildings. As well as the personal ability to control temperature, noise levels and air quality, workers need breakout areas, ergonomic furniture, plants and even sleeping space pods where they can rest during their lunch break. Personality types, such as introverts and extroverts, need to be considered; open-planned offices favour extroverts, while different design solutions are needed for introverts. The International Well Building Institute or WELL (2013) uses the "nudge theory" to encourage people to make healthier choices, through recommendations such as providing healthier food for staff and making water easily available.

There are many sound arguments for people-oriented buildings. For one, people thrive in such buildings; the work atmosphere is healthy and relaxed, workers can socialize in unimposing surroundings and special areas can be designated for wellness so that occupants can maintain a balance between work pressures and physical and mental wellbeing. According to the Beam (2019), it is not surprising that firms which offer buildings with greenery, natural lighting and well-designed workspaces are the ones attracting staff and recording low levels of absenteeism.

10.3.4 Social Transition #4 – Digital Societies

Mention digital societies and automatically we think of technology and smart cities and how it is enabling us to lead faster, smarter and more convenient lives. Across the world we are seeing the built environment being upgraded or rebuilt, using digital technologies to solve social, economic and environmental problems brought on by urbanization. Digital is inexorably linked to the built environment so that services can be more efficient and public systems like healthcare and education can be utilized more effectively. However, digital is also about information, harvesting the right data and using this in a way that benefits cities to be resilient, competitive and successful.

But there is a higher purpose to harness digital technology other than just to improve efficiency in cities. Digital is part of the social transition of the built environment.

Digital feedback loops make sure that the impact of the built environment is providing the right types of benefits to the right communities according to UNESCO (2019) so that social and economic opportunities are optimized across the spectrum of society. As cities evolve and grow, it can often be seen that marginalized groups do not have access to the same opportunities as others in terms of quality of housing, jobs, neighbourhoods, community services, utilities and more.

This is a dilemma that authorities face because, even in prosperous cities, wealth and privileges may not be equally shared among the citizens. Local governments are called upon to make difficult decisions. Do investments in digital technologies benefit all or only a handful? Who gets access to these privileges and who gets left behind? It is incumbent to seek greater equality of outcome and to use infrastructure – digital and physical – to alleviate economic and social distress, as well as foster opportunities for citizens to leave a positive legacy for generations to come. The digital built environment should address these challenges – and hopefully not exacerbate them.

To some extent, this is already happening. In Helsinki, for example, the adoption of digital infrastructure and open data has produced a collaborative and inclusive ecosystem that benefits government, businesses and citizens alike. Helsinki (2021)'s city strategy is being driven by a local government with a defined purpose: to improve the quality of life of its citizens through digital solutions and to achieve the city's goal of becoming carbon neutral by 2035.

Similarly in Singapore – which ranked first on the Institute for Management Development Smart City Index (2020) – the government is utilizing digital innovation to deliver community-focused projects across the built environment such as smart mobility, healthcare, safety and governance, resulting in overall improved liveability for all its citizens. This Smart Nation initiative (see Table 10.1 below) was launched in 2016 with the aim of using technology to create new opportunities and to better the lives of its people.

Table 10.1 City Smart City Rank 2020 (Source: Institute for Management Development)

1	Singapore	11	Munich
2	Helsinki	12	Washington, DC
3	Zurich	13	Dusseldorf
4	Auckland	14	Brisbane
5	Oslo	15	London
6	Copenhagen	16	Stockholm
7	Geneva	17	Manchester
8	Taipei City	18	Sydney
9	Amsterdam	19	Vancouver
10	New York	20	Melbourne

The lessons from Helsinki, Singapore and other aspiring cities are that people-centred planning strategies are at the heart of creating shared value across communities for digital solutions. At the same time, we are beginning to see how this shared value, i.e. the investment in infrastructure and the societal returns to the community – and in particular, a shared higher purpose – can help cities be more resilient to urban challenges, both expected and unexpected.

To obtain the best value from digital technologies incorporated in the built environment, we must:

- Be clear on what outcomes are expected, whether it is to protect and empower marginalized groups or to build overall resilience.
- Collect information and data that can be strategically analysed to detect needs and gaps in the community.
- Ensure that people have access to digital services and open platforms so that the views of users of infrastructure are heard.
- Collaborate with key stakeholders in the public and private sectors, through interconnectivity and the use of digital arenas for the exchange of ideas.

Digital solutions are constantly evolving to help people manage and apply data to enrich their lives. However, it should be noted that there is a dark side to digital in the exploitation of data at the expense of individual privacy rights. The sharing of data must thus be respected and this is the common principle of digital enablement. The shared values of digital societies therefore must not only encompass transparency and openness but also ethical reasoning on what is right and what is wrong to be a key part of social transitioning.

10.3.5 Social Transition #5 – Co-Living

Whether owning or renting a home in the built environment, the aspiration of every inhabitant of a city is to have their own property. But the costs of purchasing or leasing property are prohibitive in many urban centres. How can the social transition of the built environment take place to address this need?

Co-living is a means of offsetting sky-high rents by sharing the costs as well as the facilities. With pressures on space and shifts in lifestyle among the under 40s, co-living or shared housing is becoming increasingly attractive, combining private space with communal facilities. During the 1960s, to deal with a shortage of inner city dwellings, Copenhagen authorities introduced this method using social housing blocks in the built environment dedicated to young families. The success of this experiment has led to an established culture of shared homes in that city that prevails today (Katz and Noring 2017).

Another aspect of modern living is that many young people choose to remain single but, despite the rise of social media, encounter record levels of loneliness through the inability to form real-life relationships. Unsurprisingly, research has found that some are interested in the co-living model as it creates new ways of socializing. According to Lewis (2020), the introduction of a wide range of co-living models could not only help cope with rising levels of loneliness but also the increasing need for assisted living in ageing populations.

From a commercial perspective, COVID has forced many companies to insist that employees work from home. The social precautions aligned with good connectivity showed that not only can this method of work be effective but also high office rental costs could be averted. With these conditions likely to last post-COVID, it is time for us to rethink what living and working in our homes can look like together with saving costs through co-sharing in the built environment. In this paradigm, the addition of technology hubs in housing estates where access to IT is difficult is an important feature to bridge the digital divide in low-income households.

To make co-living work, living space must be generous and forthcoming. In compact living quarters, facilities have to be versatile such as beds doubling up as furniture when not in use and Lego-style furniture that can be assembled into tables and chairs when needed and stowed away when not. It helps to employ modular design as well to maximize efficiency and use of materials. Whilst compactness will no doubt predominate in urban centres, the use of green spaces located within bustling areas is important to relieve household stress. According to the World Health Organization (2016), green spaces can contribute to addressing major public health issues related to non-communicable diseases, such as mental illness, obesity, cardiovascular diseases, diabetes and cancer.

Lastly co-living engenders a community spirit. NGOs are working on incentives like good neighbour credit systems whereby community activities like minor repairs, escorting elders or looking after others' children can be rewarded with shopping tokens or services in kind – but more importantly it serves for neighbours to get to know one another. Helping the aged and the vulnerable is always a must in any society, so shaping the built environment for the future should not that onerous to include the leap to co-live together. With innovative thinking, co-living can not only bring about better and more caring societies but also create affordable housing solutions for all.

10.3.6 Social Transition #6 – Equity and Social Justice

The development of the human race has not always taken place in a consistent manner. Geopolitical factors have led to the rise and decline of empires and societies with the result that there is an uneven distribution of power and wealth across the globe. Within nations, there are similar disparities between certain parts of society. The Gini coefficient is used to represent the income disparity or the wealth inequality within a nation or a social group (see Table 10.2). A Gini coefficient of 0 expresses perfect equality, where all values are the same, while a Gini coefficient of 1 (or 100%) expresses maximal inequality among values. For example, if everyone has the same income the Gini coefficient will be 0, while if for a large number of people only one person has all the income or consumption and all others have none, the Gini coefficient will be nearly one.

What the above table states is that there is income inequality across the world, and one can infer that this exists within cities and societies as well as nations. According to the book, Jared Diamond ("Guns, Germs and Steel" 1999), it is stated that that

Table 10.2 Gini coefficient of countries (Source: World Population Review, 2022)

Countries with the highest Gini coefficient	*Countries with the lowest coefficient*
South Africa – 2014 – 0.63	Slovenia – 2018 – 0.24
Namibia – 2015 – 0.59	Czech Republic – 2018 – 0.25
Suriname – 1999 – 0.58	Slovakia – 2018 – 0.25
Zambia – 2015 – 0.57	Belarus – 2019 – 0.25
Sao Tome and Principe – 2017 – 0.56	Moldova – 2018 – 0.25
Central African Republic – 2008 – 0.56	United Arab Emirates – 2018 – 0.26
Eswatini – 2016 – 0.55	Iceland – 2017 – 0.26
Mozambique – 2014 – 0.54	Azerbaijan – 2005 – 0.267
Brazil – 2019 – 0.53	Ukraine – 2019 – 0.27
Belize – 1999 – 0.53	Belgium – 2018 – 0.27

"much of human history has consisted of unequal conflicts between the haves and have-nots." Underlying this problem is the dearth of social equity and justice.

SDG 10 aims to reduce inequality within and among countries with a target of empowering and promoting the social, economic and political inclusion of all, irrespective of age, sex, disability, race, ethnicity, origin, religion or economic or other status. This intent makes sound sense in promoting a society of even opportunity but, in reality, failure occurs due to human bias, prejudice and greed. Social equity is a means of addressing this conundrum where labour laws, policies and corporate responsibility attempt to make companies and societies comply with the basic principles of fairness and abhorrence of exploitation of fellow human beings (Movendi International 2019).

Social justice on the other hand deals with human rights as a fundamental principle. These range from the compensation of indigenous villagers to make way for new infrastructure like dams and railways to acknowledging the angst of aged workers being replaced by robots.

How does this concern the built environment? On ethical grounds, this means that infrastructure must acknowledge the interests and sustainable development of communities even if the latter themselves are unable to express this view. There are methods and tools to capture data on what is termed social impact, most of which serve ultimately to monetize the impact into an easily understandable economic value (Biden 2020). The challenge to this process however is the quantification of intangible outcomes like loss of heritage or removal of cultural values. In cases like this, designing or delivering the built environment must relate to the creation of social value that is positive and where the benefits accruing outweigh the disbenefits to the societies affected.

This is straightforward enough when delivering what is termed "social infrastructure," which comprises community facilities like schools, hospitals and recreation grounds. However, how would a road or a bridge create social value? This is usually measured by using a combination of socio-economic cost-benefit methods and engagement with stakeholders to ensure that the negative impacts are mitigated and the positive returns are optimized. There are principles such as the Equator Principles (2003) and UN Principles for Responsible Investment (2020) that put forward ethical investor instructions – like minimizing environmental damage and increasing benefits to society whilst at the same time maintaining robust living standards – which are helpful as guidelines in setting up new projects. In this way, social value can be created so that investors obtain social as well as economic returns on their investments. Digital applications are also playing an important role in the form of advanced visual technologies such as AR/VR, which are shaping new forms of engagement with stakeholders who can access visual scenarios of the changes affecting them and become part of the decision-making process in how the changes will be rolled out and the scenarios can be tailored to their needs.

10.4 Discussion

When we talk of social transition, there is much that can be done to change the way we design the built environment to better serve the community. The six transitions previously discussed are just examples of how new and not-so-new ways of thinking can be applied. But importantly, in combination, these transitions help to build resilience in society that leads to sustainable living. For instance, buildings that allow healthy wellbeing and work to prosper help people make a living in comfortable surroundings. Also, transport choices like cycling to work offer healthy options. It is the combination of such choices and the freedom to express oneself that is paramount. The built environment plays a significant role in the choices to be offered. As an example, suppression of the human spirit through forbidding architecture will dampen any society's aspirations, hence the allocation of open space in the midst of densely packed cities has to be respected.

Man is a sociable animal; yet loneliness prevails in many urban centres. That is why co-living offers a panacea for this predicament. With ageing populations, there is a growing need to foster kinder and more caring housing for the vulnerable; this can be achieved through intergenerational living and re-establishing the family at the core. This in turn leads to social equity whereby every member of society gets the right to receive benefits equably and justly.

In this coming age of advanced technology, digital applications will become more ubiquitous as we allow the use of data to run our lives. Digital applications have undoubtedly made our lives easier and choices wider; nonetheless, it is worth remembering that personal rights and boundaries should be acknowledged and the benefits of digital should be the entitlement of everyone and not just a few.

Important lessons that we can glean from these observations are:

- The need to rethink infrastructure from a social perspective
- How simple facilities like buildings and workplaces have to be designed with the wellbeing of users in mind
- There are economic savings as well as social benefits through a well-designed and economically structured built environment that meets specific user needs
- Making sure that the benefits of social design are obvious and measurable
- Accepting that the responsibility for change to be successful lies not just in the hands of a few but amongst many, whose roles are to respect the rights of all and to help when needed

- Urgency for change in social behaviour; to be less profligate and more responsible as the consumption of resources outpaces our ability to replenish the latter.

10.5 Conclusions

So, what conclusions can we draw from the social transitions described above? The need to include societies as an integral part of the design of the built environment is clearly vital to meet their needs at different levels. We can summarize these as follows:

- Meeting society's needs is not just about physical needs, there are mental and emotional aspirations that can be uplifted by the built environment.
- Connecting people with nature has many advantages – apart from conservation and fostering green resilience, people have a sense of wellbeing and show positive signs of health when in natural surroundings.
- Connecting communities is also important whether these are social communities or work-related communities. As a species, we are gregarious and love to interact. Isolation can be negative and cause dysfunction and discrimination.
- Workplaces and future homes will be designed to encourage co-living and sharing. Already with a focus on wellbeing, space allocation and interior design will be green, communal, pandemic-ready and efficient taking advantage of the latest digital technology and modular design.
- The built environment has to take social justice and equity into account so that the benefits of new and existing infrastructure are evenly distributed.
- Lastly, the built environment has not only to be functional enough to support the city but also be future-ready for changes most likely in energy, water supply and treatment, transportation and solid waste management for escalating population needs and quality of life.

References

Angermann, M. (August 2022) Biophilic Architecture and Interior Design: Benefits for You and the Planet, Utopia Newsletter.

Augustin, S. (June 2021) How Lighting Choices Can Affect Your Mood, Psychology Today.

Biden, A. (2020) 5 Tools and Resources for Social Impact Assessment, EvalCareers.

Bissio, R., (2019) Civil society reports show conflicting priorities and trade-offs in SDG implementation, Spotlight on Sustainable Development 2019, Global Civil Society Report on the 2030 Agenda and the SDGs (pp 20–26)

Byrne, J., Pharo, E., (May 2018) Don't Forget Buses: Six Rules for Improving City Bus Services, The Conversation UK.

Cyril, H. (June 2023) Silencing the Chaos: Effective Strategies to Block Out Noise at Work, Tech Announcer.

Davern, M., Farrar, A. (May 2017) Higher-Density Cities Need Greening to Stay Healthy and Liveable, The Conversation.

Equator Principles (2003) - A financial industry benchmark for determining, assessing and managing environmental and social risk in projects. https://equator-principles.com/about-the-equator-principles/#

Gawande, S., Tiwari, R., Narayanan, P., Bhadri, G. (2020 Jan–Apr) Indoor Air Quality and Sick Building Syndrome: Are Green Buildings Better than Conventional Buildings? Indian Journal of Occupational Environmental Medicine 24 (1): 30–32.

Hallisey, K. (2021) How Riding the Bus Benefits Everyone, Public Transit UCLA.

Katz, B., Noring, L., (2017) The Secret Copenhagen Model for Regenerating Cities, Next City, Solutions for Liberated Cities.

Kohll, A. (January 2019) Why We Pay Our Employees To Exercise At Work, Forbes.

Krems, J.A., Kenrick, D.T., Neel, R. (2017) Individual Perceptions of Self-Actualization: What Functional Motives Are Linked to Fulfilling One's Full Potential? Personality and Social Psychology Bulletin 43 (9): 1337–1352.

Leong, L. (2016) The 'Rail Plus Property' Model: Hong Kong's Successful Self-Financing Formula Commentary. McKinsey & Company.

Lewis, P. (March 2020) How to Make Co-Living Serve People Before Profit, RIBA Journal.

Maslow, A. (1943) A Theory of Human Motivation, Psychological Review, American Psychological Association.

Mhlanga, R., Gneiting, U., Agarwal, N. (September 2018) Walking the Talk: Assessing Companies' Progress from SDG Rhetoric to Action, Oxfam Discussion Paper.

Olson, C. (April 2015) Wellness, Wellbeing and Productivity in the Office, Buildings Magazine.

Poston, R. (August 2009) An Exercise in Personal Exploration: Maslow's Hierarchy of Needs, The Surgical Technologist p.347–353.

Roberts, T., (2016) We Spend 90% of Our Time Indoors. Says Who? Building Green Blog Post.

Rojas-Rueda, D., Nieuwenhuijsen, M., Gascon, M., Perez-Leon, D., Mudu, P. (2019) Green Spaces and Mortality: A Systematic Review and Meta-Analysis of Cohort Studies, Lancet Planet Health 3: 469–477.

Taormina, R., Gao, J. (2013) Maslow and the Motivation Hierarchy: Measuring Satisfaction of the Needs, The American Journal of Psychology 126 (2): 155–177.

Templeton, L. (2019) Green Spaces in Cities Can Help People Live Longer, Medical News Today.

van Vuuren, D., Wicke, B. (2022) Scientific Evidence on the Political Impact of the Sustainable Development Goals. Nature Sustainability 5, pp 795–800.

Conferences

Følstad, A., Krogstie, J. (2004) User Involvement in e-Government Development Projects. Proceedings of NordiCHI '04, (pp. 217–224) Tampere, Finland Norwegian University of Science and Technology.

Wang, F., Yan, W., Song, P. (2013) Conceptualization of the public's expectation on government micro-blogs, ICEGOV '13: Proceedings of the 7th International Conference on Theory and Practice of Electronic Governance (October) pp 272–279

Books

Baker, K. (2021), Chapter Ten – Power, Buildings, and Other Critical Networks: Integrated Multisystem Operation, New Technologies for Power System Operation and Analysis Pages 319–358, Academic Press

Glaeser, E. (2012) Triumph of the City. Macmillan.

Hollis, L. (2013) Cities Are Good for You. Bloomsbury Publishing.

Jacobs, J., (1961) The Death and Life of Great American Cities, Random House.

Lloyd, F.L. (1939) An Organic Architecture: The Architecture of Democracy. Lund Humphries Publishers.

Luo, M., Cao, B., Ji, W., Ouyang, Q., Lin, B., Zhu, Y. (January 2016) The Underlying Linkage between Personal Control and Thermal Comfort: Psychological or Physical Effects? Energy and Buildings 111: 56–63, Elsevier.

Montgomery, C. (2013) Happy City. Penguin Books.

Pinkus, R.T. (2017) Love and Belongingness Needs. In: Zeigler-Hill, V., Shackelford, T. (eds) Encyclopaedia of Personality and Individual Differences. Springer.

Corporate Author

American Public Transportation Association (2012) Benefits of Public Transportation, American Public Transportation Association.

Healthline (2021) 12 Benefits of Cycling, Plus Safety Tips, Healthline Media.

Helsinki (2021) Helsinki City Strategy, City of Helsinki.

Institute for Management Development Smart City Index (2020) Singapore, Helsinki and Zurich triumph in global smart city index Singapore, Helsinki and Zurich triumph in global smart city index. https://www.imd.org/news/competitiveness/updates-singapore-helsinki-zurich-triumph-global-smart-city-index/

International WELL Building Institute (2013) WELL Building Standard, New York: International WELL Building Institute.

Landscape Architecture Foundation (2011) Landscape Performance Series: Cheonggyecheon Stream Restoration Project, Washington: Landscape Architecture Foundation.

Movendi International (2019) World Failing to Achieve SDGs, published in Democracy, Human Rights, Obstacle To Development, Policy, Sustainable Development, Movendi International.

Royal Institute of Chartered Surveyors (2020), Measuring Social Value in Infrastructure: Lessons from the Public Sector, Royal Institute of Chartered Surveyors, London.

The Beam Magazine (2019) How Do Green Buildings Attract, Engage, & Retain Talent? Clean Technica.

UN Principles for Responsible Investment (2020) What Are the Principles for Responsible Investment? UN Principles for Responsible Investment

UNESCO (2019) Digital Technologies for the Inclusion of Disadvantaged Communities and Marginalized Groups Supported by IFAP, UNESCO.

United Nations (2015) Sustainable Development Goals, United Nations.

World Economic Forum (2022) An Ageing Population Needs a Different Approach to Housing and Care. This Is How to Provide It, World Economic Forum.

World Green Building Council (2014) Health, Wellbeing & Productivity in Office, World Green Building Council.

World Health Organization (2016) Urban Green Spaces and Health, World Health Organization.

World Population Review (2022) Gini Coefficient by Country, World Population Review.

11 Stakeholders' Transformative Adaptation in Sustainable Design and Construction

Cen-Ying Lee, Bruno Lot Tanko, Chris Harty and Martin Skitmore

11.1 Introduction

The 2015 United Nations Sustainability Development Goals (Kawakubo et al., 2018) have provided a blueprint for the built environment sector to foster innovation in promoting buildings and infrastructure that are more resilient and sustainable as well as taking urgent actions to reduce the impacts of climate change. In the built environment sector, stakeholders can no longer rely solely on promoting the use of green practices, but they must also equip themselves with the adaptive capacity to mitigate against climate change and embed more environmental sustainability elements into their work practices. Adaptive capacity is the skill to reconfigure existing processes and translate them as a reaction to anticipated changes (Pahl-Wostl, 2020). While *adaptation* outlines a process of altering the changing environment but maintaining a similar configuration of the same set of structures, *transformation* is a process of transitioning to a more diversified structure of the societal system and environment (Jozaei et al., 2022). The concept of transformative adaptation is multifaceted, depending on the context of change. Hence, identifying its attributes is crucial to ensure a strategy works. According to Few et al. (2017), the differences between "transformational" and "transformative adaptation" are the influences of the terms. "Transformational" is a phase of change, whilst the activity of "transformative adaptation" generates a sustained alteration to an object or matter that develops wider transformation. "Transformative adaptation," on the other hand, enables stakeholders to plan, formulate, initiate and implement change (Wolfram, 2016). It is also considered to be a fundamental process that results in a change in a system from one type or function to another (Mendizabal et al., 2018; Dilling et al., 2023). If the adaptive capacity of stakeholders in transformative adaptation could be enhanced, a more sustainable and resilient built environment could be realised. However, there is a lack of research aimed at identifying the ability of stakeholders to become accustomed to the changes involved in enhancing sustainable practices (Mensah et al., 2021).

Fedele et al. (2019) identified several transformative adaptation features for sustainable agricultural systems, whereas Asadzadeh et al. (2022) revealed four domains that can drive existing urban governance towards transformative resilience. Little is known about the transformative adaptation that could enhance the capacity

DOI: 10.1201/9781003317890-15

of construction stakeholders in adopting environmentally sustainable practices effectively. This study refers to such construction stakeholders as project owners, architects, engineers, surveyors, contractors, subcontractors and suppliers. The project owners could be private or public entities. The construction literature heavily emphasises the key traits or strategies of construction professionals to improve their sustainability practices (Al Amri et al., 2021; Munyasya and Chileshe, 2018).

Given that global warming is happening, which has numerous impacts on relatively complicated human civilisations and misconfiguration between sustainability planning and climate crisis predictions, it is a vital that efficient adaptation takes place (Filho et al., 2022). Despite involving technological advancement, material innovation, knowledge and expertise, sustainable construction projects are not always successful (Maqbool and Amaechi, 2022). Effective adaptation includes enhancing environmental sustainability, minimising vulnerability and risks, improving monetary resources, strengthening governance and creating resilient community systems (Owen, 2020). In this sense, the best means of adaptation should be transformative to sustainably decrease vulnerability due to the impact created by construction projects, by acknowledging and addressing the fundamental causes and creating a sustainability competence (Probst et al., 2019). It is imperative, therefore, to investigate the transformative adaptation attributes that construction stakeholders should have, so the industry can adopt a useful approach to improve the adaptive capacity of stakeholders to enhance their capability of maintaining sustainable design and construction practices.

In this chapter, we first review the existing sustainable design and construction practices of stakeholders and the social challenges encountered in their adoption. We then explore the relevant transformation adaptive theories that could be used to enhance the adaptive capacity of stakeholders in adopting sustainable design and construction practices more effectively. Through reviewing the attributes of transformative adaptation strategies contained in the literature thematically, a conceptual framework is proposed that could be used to enhance the adaptive capacity of construction stakeholders and to serve as a blueprint to advance sustainable design and construction practices.

11.2 Sustainable Design and Construction Practices

11.2.1 Sustainable Development Goal 2015

As countries continue to develop, the idea of sustainable development has gained widespread acceptance worldwide and evolved into a shared construction objective tied to the future of humanity because of the ongoing degradation of the world's natural environment. The United Nations Member States' leaders and governments adopted the *2030 Agenda for Sustainable Development* in 2015 (Zhu et al., 2020), which is a political resolution that calls for the implementation of a "comprehensive, far-reaching, and people-centred set of universal and transformative goals and targets" by the year 2030 (Johnsson et al., 2020; Opoku, 2016). The *Agenda* provides a shared framework for peace and prosperity for people and the planet now and in the future (Zhu et al., 2020). Developed and developing countries can use it to address climate change and promote sustainable cities and communities. The *Agenda* incorporates 17 sustainable development goals (SDGs) and 169 related targets. The SDGs have been referred to as "the 'five Ps' plan of action for people, prosperity, planet, peace, and partnership" (Opoku, 2016). All individuals (People) will be able to live prosperous and fulfilling lives (Prosperity) by addressing climate change head-on to prevent the planet's destruction (Planet), and a society that is peaceful, just and inclusive (Peace) will be fostered through international cooperation and solidarity with all stakeholders (Partnership) for sustainable development. The goals are intended to ensure that all people can realise their full potential with pride, fairness and a healthy environment (United Nations, 2022).

A sustainable built environment can help achieve the SDGs in various ways, including developing sustainable communities and smart cities, installing solar technology on constructed assets (Opoku, 2016), and designing and building infrastructure using sustainable materials. According to the Circle Economy (2020), of the 50.8 billion tonnes of minerals consumed annually, 38.8% are used to construct homes. Along with these facts, the Waste Framework Directive 2008/98/EC mandates that, by 2020, all European Union (EU) members must achieve a minimum of 70% non-hazardous construction and demolition waste (CDW) reuse, recycling or other recovery activities (Horry et al., 2021). In response, most nations have responded to the requirement for sustainable development by carefully studying environmental sustainability issues, with a special emphasis on energy efficiency, renewable energy and asset life cycle management. Recently, many countries have begun implementing green building regulations to promote sustainability within the construction industry (Mattinzioli et al., 2021). Extensive research is being conducted into sustainable construction, and both developed and developing nations are actively adopting sustainable building assessment systems. Several countries serve as notable examples, such as the United States with its LEED (Leadership in Energy and Environmental Design) programme, Australia with Green Star, Canada with LEED Canada, the United Kingdom with BREEAM (Building Research Establishment Environmental Assessment Method), Germany with DGNB (German Sustainable Building Council) and Singapore with Green Mark (Cordero et al., 2019; Edwards et al. 2019, Alhilli and Burhan 2021, Mahyuddin et al. 2022). Moreover, numerous other nations, including France, Sweden, Japan and the Netherlands, have also introduced their own systems to advance sustainable construction practices.

Sustainable design and construction practices are closely related to the effective utilisation of resources and energy (Yin et al., 2018; Tafazzoli et al., 2020; Omopariola et al., 2022). The use of sustainable design and construction techniques can make buildings more energy efficient. To attain a net-zero carbon building stock by 2050, according to the International Energy Agency, building CO_2 emissions must be reduced by 50% (Müller-Puch et al., 2021). Furthermore, using renewable energy (Omopariola et al., 2022), recycled materials (Yin et al., 2018), green building materials (Müller-Puch et al., 2021) and robust and adaptive materials can help boost the application of sustainable design and construction practices. Environmental, economic, social and other advantages of using sustainable construction techniques, sometimes known as "green building practices," include reducing CO_2 emissions and energy usage (Alsanad, 2015), resource conversation and improved indoor environmental quality (Tokbolat et al., 2020), reduced environmental impact (Yanita and Mochtar, 2021), enhanced resilience (Ling and Chiang, 2018) and long-term financial benefits. Sustainable construction practices can also improve environmental performance through water and insulation protection (Yin et al., 2018). The next section examines stakeholder involvement in sustainable design and construction practices.

11.2.2 Stakeholder Involvement in Adopting Sustainable Design and Construction

Stakeholders are groups or individuals that may impact on, or be impacted by, the project. According to Li et al. (2022), developers, clients, designers, suppliers, contractors, end users, the government and the public are all considered construction project stakeholders. A study by Kongela (2021) considered "engineers, architects, property managers, property owners, property users and academicians" as stakeholders in the sustainable built environment, while Berawi et al. (2019) regard developers, project managers, consultants, architects, contractor, researcher institutions, construction management organisations and government institutions as the most important stakeholders in sustainable construction. However, facility managers, energy service providers and specialists in green accreditation are also involved in sustainable projects because of the excessive demands of green buildings on technical and managerial resources (Li et al., 2022). Figure 11.1 depicts the categories of stakeholders involved in sustainable design and construction practices.

According to Li et al. (2018), the government, with its ability to consider the interests of multiple stakeholders, plays a primary role in advocating a more inclusive and participatory decision-making and evaluation process for sustainable building. However, Akotia et al. (2020) conducted a study revealing that legal requirements and laws by governments were not the primary motivators for practitioners to promote sustainability principles in their projects; instead, some have simply embraced sustainability to meet specific "green construction" standards and regulations set by awarding bodies or clients.

It is certainly the case that construction stakeholders are making more effort to address environmental, social and economic goals by seeking certifications to ensure that sustainability targets have been met (Herazo and Lizarralde, 2016). Recent studies have shown that stakeholder involvement and actions are essential for the effective development and operation of green buildings (Li et al., 2022). According to a study on stakeholders' viewpoints on green building rating in Indonesia (Berawi et al., 2019), energy efficiency and conservation are the most significant sustainability criteria, and environmental protection is the primary goal for sustainable design and construction. However, most UK organisations and practitioners working in the built environment advocate socio-economic sustainability drivers because they think it would help achieve their corporate objectives (Akotia et al., 2020). In their research, Mashwama et al. (2020) found that lean

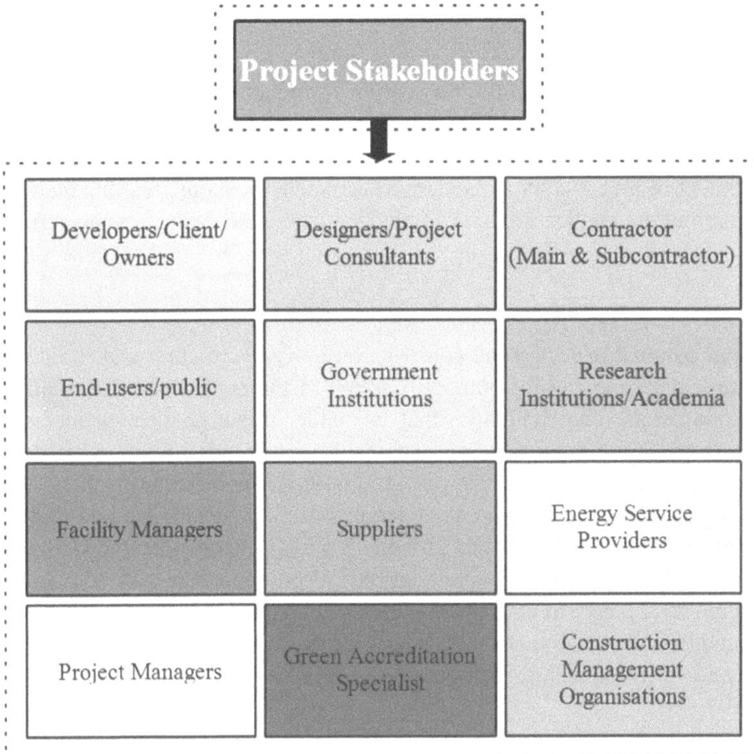

Figure 11.1 Categories of stakeholders involved in sustainable design and construction practices

construction, life cycle costing, construction ecology and industrialisation building systems were the most favoured practices in sustainable construction according to most stakeholders. However, as highlighted by AlWaer et al. (2008), the challenge lies in interpreting and judging these practices due to the subjective viewpoints and varying stakeholder perspectives of assessment.

Nevertheless, Herazo and Lizarralde (2016) found that involving stakeholders, project managers and professionals can lead to several positive outcomes. These include anticipating potential conflicts, promoting sustainable agendas, increasing participation from civil society groups and facilitating collaboration among decision-makers. These stakeholders can develop improved projects that safeguard the environment by working together.

Stakeholders can directly influence the diverse variables that support sustainable design and construction processes by promoting the essential needs of sustainable construction, such as social, economic, environmental, technical, managerial and legal norms. However, some hurdles must be overcome before stakeholders can adopt sustainable design and construction.

11.2.3 Challenges of Sustainable Design and Construction

Understanding the causal structure of vulnerability is critical in transformative adaptation because it serves as the foundation for adaptation planning (Mendizabal et al., 2018). The challenges of sustainable design and construction practices comprise four groups of knowledge and information-related factors, attitudinal factors, management-related factors and external factors. These constitute the main obstacles hindering the application of stakeholder sustainable design and construction.

11.2.3.1 Knowledge and Information-Related Factors

Applying sustainable design and construction practices requires stakeholders with adequate knowledge and information relating to the performance of quality sustainable materials. The challenges that fall under this classification include lack of knowledge and being aware of sustainable construction and design practices (Ametepey et al., 2015; Aghimien et al., 2019; Ali and Alkayed, 2019; Ahmed and El-Sayegh, 2022); limited experience in selecting sustainable building procedures and techniques (Ahmed and El-Sayegh, 2022); inadequate knowledge of the performance quality of sustainable materials (Aghimien et al., 2019; Ahmed and El-Sayegh, 2022); construction professionals and stakeholders not being involved early enough in the design phase (Ali and Alkayed, 2019; Ahmed and El-Sayegh, 2022); and insufficient information on the handling and storage of sustainable materials.

11.2.3.2 Attitudinal Factors

Attitudinal factors influence the adoption or application of sustainable practices in construction projects. Low demand by potential buyers (Abidin et al., 2013), weak scope definitions of sustainable construction requirements, lack of interest among clients (Ershadi et al., 2021; Kalyana Chakravarthy et al., 2022), general resistance to change within the sector (Ametepey et al., 2015; Aghimien et al., 2019) and the frequency of client or architect design changes (Ahmed and El-Sayegh, 2022) are attitudinal factors that affect the incorporation of sustainable design and construction practices.

11.2.3.3 Management-Related Factors

Management and stakeholders' support and active involvement are vital for implementing sustainable practices in construction projects. Some of the challenges include a lack of consensus over sustainability values among stakeholders and inadequate top management commitment to sustainability (Ershadi et al., 2021), inadequate sustainability measurement tools and difficulty in working with green codes compliance or certification (Aghimien et al., 2019), fragmented guidelines for green procurement and the contracting parties' failure to communicate and coordinate with each other (Ershadi et al., 2021; Ahmed and El-Sayegh, 2022).

11.2.3.4 External Factors

External factors consist of the conditions in which stakeholders operate to allow the adoption of effective sustainable design and construction practices. This classification has the potential to drive the project team to achieve the needs of sustainable design and construction within the constraints of available resources. The challenges include additional costs, as green materials might need to be imported (Abidin et al., 2013; Ametepey et al., 2015; Aghimien et al., 2019; Ali and Alkayed, 2019; Ahmed and El-Sayegh, 2022; Kalyana Chakravarthy et al., 2022); lengthy production times for sustainable materials (Saleh and Alalouch, 2015; Ahmed and El-Sayegh, 2022); limited options for sustainable materials on the market or in market circumstances (Aghimien et al., 2019); lack of laws, regulations and standards (Ametepey et al., 2015; Kalyana Chakravarthy et al., 2022); and price inflation of sustainable materials.

11.3 Enhancing Adaptive Capacity of Stakeholders in Sustainable Design and Construction

11.3.1 Adaptation of Stakeholders

Climate change has been accelerating, increasing the need for built environment stakeholders to adopt effective environmentally sustainable design and construction practices. While the focus should be on promoting such practices, we also need to emphasise the construction stakeholders' adaptability to deal with sustainability requirements, invent green practices and cope with the changes. Adaptation measures the ability to mitigate the ramifications arising from construction activities and deploy strategies for achieving positive effects. However, there is no consensus on the definition of "adaptation" in the built environment context, as it has been used in various contexts based on the extent to which adaptation applies (Askar et al., 2021). For instance, adaptable housing means adapting a property to change the users' physical needs when necessary (Till and Schneider, 2005). For urban water transformation, the existing system requires adaptation to ensure it is safe and provide an adequate water supply in the future (Dilling et al., 2023). There is also an argument surrounding the terms "adaptation" and "coping." While coping involves processes that require short-term attention, adaptation refers to long-term plans (Adegun, 2023). The adaptation process requires the acceptance of diverse stakeholders who coordinate project activities and complete the project goal (Dolla et al., 2023). Institutions form a set of rules and norms for stakeholders to improve their adaptive capacity for implementing sustainability practices. Moreover, the adaptation process that involves the active engagement of stakeholders facilitates their awareness, communication with other stakeholders and maintaining collaborative governance, which is of overarching importance in developing trust in institutions (Dilling et al., 2023). By observing the definitions used in various studies, we identified three essential elements that influence the adaptive capacity of built environment stakeholders: (1) adaptation as a process, (2) adaptation requires stakeholders as actors for social capital development and (3) adaptation to the institution.

11.3.1.1 Adaptation as a Process

The adaptation of stakeholders in sustainable design and construction should be considered a process to reduce environmental impacts and improve resiliency against the environmental crisis. Adaptation in the built environment context is a procedure of creating, devising and executing an approach and plan to reduce the impact of environmental issues and become familiar with them (Mimura, 2021). It also involves the process or capacity to navigate change continually to quantify a social and technical system's resilience when considering the stakeholders' adaptive capacity (Doorn et al., 2019). The output of social adaptation is similar to that of resilience, whereby the built environment community familiarises itself with the environmental crisis by fundamentally restructuring itself to shed its former path (Lyon, 2014). Some claim that adaptation is primarily concerned with "who regulates who and what" and "what oversees who and what" when reflecting on the adaptation process and its influences on the built environment society (Hutcheon, 2012). This means that actors and governance should also be considered in the stakeholder adaptation process.

11.3.1.2 Actors of Adaptation Process

Stakeholders are the actors in the adaptation process for social capital development. Transformation of a built environment object requires the adaptive capacity of stakeholders to be involved and engage proactively. Social networks that define individual and inter-organisational relationships are vital for developing the stakeholders' adaptive capacity (Lee et al., 2018). It is important to improve knowledge-sharing, exchange resources and facilitate opportunities to develop social capital and improve decision-making effectiveness in responding to disruption and alteration (Barnes et al., 2017). Similarly, it can reduce transaction costs by coordinating responses and developing more effective strategies to cope with alterations (Henry and Vollan, 2014). However, a higher adaptive capacity cannot be developed if stakeholders have no self-reflection or have not reviewed their fundamental principles and beliefs (Pahl-Wostl, 2020). Reducing project risks and vulnerability, creating social resilience systems, enhancing environmental and economic resources and improving governance and institutions are considered effective adaptations developed from social capital. Other essential elements that can be found through social capital to improve the adaptive capacity of stakeholders include trust, shared understanding, acknowledgement of diversity, the ability to learn and self-organisation (Filho et al., 2022).

11.3.1.3 Institutions

An institution can be defined as a set of regulations that constitute social interactions in specific practices (Knight, 1992). It consists of the common rules utilised to manage, organise and coordinate activities in repetitive circumstances (Hodgson, 2007; Ostrom, 2015). The institution that influences interactions among built environment stakeholders is developed from the stakeholders' organisations, government, and professional bodies and projects. Examples of formal institutions that apply to the built environment sector include government policies and regulations, while informal institutions mean norms, customs, values etc. (Lakoff, 2010).

To enhance stakeholder capacity in adopting built environment sustainability, actors, adaptation processes and institutions should be closely connected as they depend on each other to work effectively. For instance, the adaptation process is a procedure involving designing, planning and operating the plans to minimise environmental impact and enables actors to work on attributes to ensure the success of projects. Nevertheless, the adaptation process would not be successful unless the actors are equipped with the transformative adaptation attributes needed to initiate and execute the process effectively. The actors, which include public actors from the institutions that govern the formal and informal institutions and project actors involved in construction projects, are the key to ensuring a successful adaptation process.

11.3.2 Transformative Adaptation

The literature review identifies several transformative adaptation attributes with which stakeholders should be equipped to enhance their adaptive capacity in deploying sustainable design and construction practices in the built environment sector. An important adaptive behaviour is whether the stakeholder is flexible to change (Barnes et al., 2020). *Flexibility* in this study refers to the capability of stakeholders to be trained and become accustomed to innovative ways of doing things. The actors' *flexibility* in a governance system is needed to offer insights into the stakeholders' capability of learning and inform the planning theories and practices of adaptive capacity (Schmitt and Wiechmann, 2018). *Restructuring* is another important transformative adaptation attribute that stakeholders should possess as it enables a major shift in essential functions and interactions within each project discipline (Fedele et al., 2019). For instance, owners who implement sustainable practices can alter existing methods and work with their designers, consultants and contractors to plan the project delivery structure capable of adapting to sustainability requirements. Another example of sustainable technological inventions is restructuring supply chain delivery to one that requires skilled workers to perform knowledge-intensive jobs (Al Amri et al., 2021).

Reorientation reconfigures social beliefs and social interactions (Few et al., 2017). It is used to deliver the concept of alteration in actors' beliefs, abilities and attitudes at numerous levels varying from individual actors to official authorities (Colloff et al., 2017). It also refers to alterations that influence behaviours and judgements concerning adaptation and relates to a developing set of principles in social learning as a route to transformation in planning and governance (Asadzadeh et al., 2022). *Social learning* enables us to understand how people recognise environmental alteration, the factors that affect alleviation and changed behaviours and how the alteration will influence people's well-being (Clayton and Myers, 2009). Recent evidence suggests that psychological variables are important in influencing stakeholder adaptive capacity in adopting sustainability practices (Mendizabal et al., 2021). The theory of planned behaviour should be considered to improve social learning, as it has often been used to suggest drivers that influence certain practices. This theory could connect the engagement process closely with behavioural outcomes at an individual level (Eaton et al., 2021), which include personal attitudes, socially informed subjective norms and perceived behavioural control (Ajzen, 2020).

To improve the adaptive capacity of stakeholders, *participation* from all stakeholders is imperative. *Inclusiveness* is the key element that highlights the transparency of how a sustainable built environment is developed, which includes any relevant opportunities for stakeholders to contribute to the decision-making process to lead the existing governance towards transformation. *Participation* from stakeholders should be connected with planning elements from top to bottom or from bottom to top (Asadzadeh et al., 2022). The *inclusiveness* of all stakeholders in sustainability evaluation is important to enhance the clarity of enquiry by the public and reduce conflicts arising from collaboration (Berardi, 2012).

Innovation is an attribute of transformative adaptation as sustainable practices emerge due to innovative ideas created by stakeholders. It is asserted that companies have no option but to invent sustainable technologies to enhance their value chain, particularly in offering their services or products to remain competitive (Nidumolu et al., 2009). Sustainably oriented innovation can strategise different areas of built environment services, procedures, products and practices to achieve targeted environmental sustainability objectives (Klewitz and Hansen, 2014; Ghassim and Bogers, 2019).

Compared with stakeholder collaboration, "*orchestration*" refers to more intensive coordination to mitigate the adverse ramifications arising from fragmented practices. It is used to resolve complicated systems requiring collective ideas or solutions from multi-disciplinary stakeholders and promote communication mechanisms for enhancing knowledge-sharing, nurturing synergies, reducing transaction costs and disputes and developing shared plans and actions (Asadzadeh et al., 2022). While orchestration is important, transformative *leadership* is also significant for transitioning, despite centralised decision-making and localised decisions not reaching a consensus (Lonsdale et al., 2015). Humans power every built environment project, and it is extremely important for experts with strong *leadership*, knowledge and skills to coordinate the project for sustainability (Maqbool and Amaechi, 2022).

11.4 Development of a Stakeholder Transformative Adaptation Framework for Sustainable Design and Construction

11.4.1 Methodology

Thematic analysis has been selected as the main method to identify the transformative adaptation attributes of stakeholders in adopting sustainable design and construction practices. This method allows researchers to examine the data to identify common themes for interpreting and analysing the collected data. A deductive approach has been used to analyse data based on perceived themes and existing concepts of transformative adaptation attributes we have identified (Guest et al., 2014). The main data source has been obtained from journal articles in the University of Reading electronic database search connected to various electronic databases such as the Web of Science or Scopus. A set of keywords involving construction AND industry AND sustainability AND social AND transformation and construction AND stakeholder AND sustainability AND transformation have been used to identify relevant peer-reviewed journal articles from 2018 to 2022. All data were retrieved from the most recent five years to identify the latest developments of construction stakeholder transformative adaptation attributes. After using more focused search criteria involving environmental science and engineering disciplines, we found 78 results and 36 results from each set of keywords. Each article was screened to remove the publications unrelated to our research topics. Some were not chosen because their topics were more technically oriented around green practices with relatively little discussion about transformative adaptation attributes and no informative statements that could support the claims of transformative adaptation. After a detailed filtering process, only 20 peer-reviewed journal articles were identified as relevant. Figures 11.2 and 11.3 show the types of journals and their year of publication, respectively. *Sustainability*, *Ambio* and the *Journal of Cleaner Production* recorded the most publications. The literature was analysed based on the themes of transformative adaptation attributes identified in Section 11.3. Most publications (eight) were published in 2020.

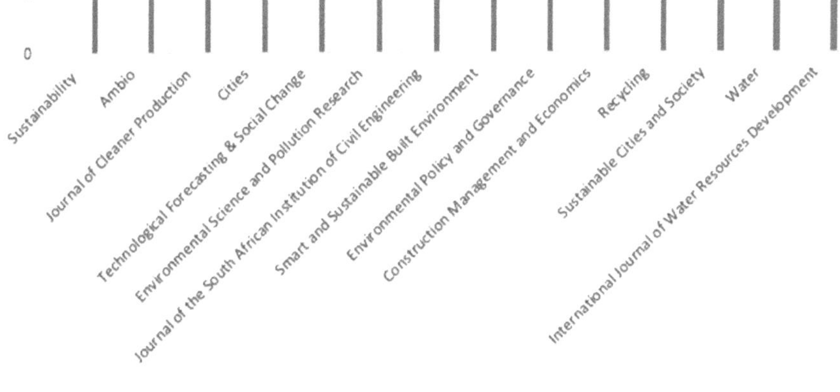

Figure 11.2 Types of publications

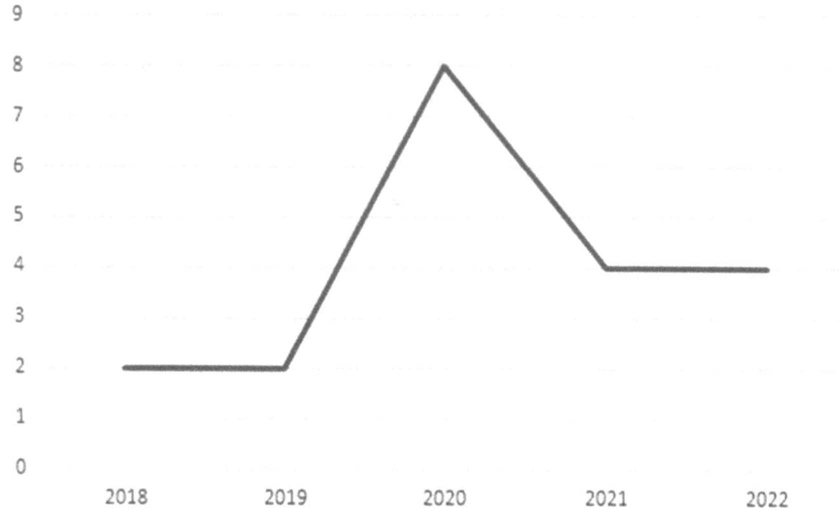

Figure 11.3 Years of publications

11.4.2 Thematic Analysis

We identified seven main categories of transformative adaptation from the 20 journal articles. Each category is grouped by its natural attributes and purpose of transformational activities as a catalyst for transformative adaptation to build environment stakeholders (Table 11.1).

11.4.2.1 Flexibility and Restructuring

We obtained various forms of organising restructuring from the literature analysis, such as multiform governance, decentralisation, polycentrism, sustained intermediaries and hybridisation. These types of governance have the common feature of suggesting that, to make sustainable projects effective, centralised governance should be reorganised to improve its governance arrangement. The sustainable schemes should be flexible enough to adapt to local requirements, be circumstantial and constantly enhanced through accepted practice to recognise their limitations and prospects (Padovezi et al., 2022). As such, stakeholders should be able to evaluate their managerial flexibility and reduce any inflexible elements (Mensah et al., 2021). For instance, codes of design should emphasise environmental aspects and provide more flexibility to designers in selecting materials leading to the improved usage of products that could be recycled (Shooshtarian et al., 2020). The governance context should be reconfigured, processed and allowed opportunities with essential modifications and incentives (Yasmin et al., 2020). The existing governance could also be reformatted to organise new units to oversee sustainability programmes: for example, by establishing an ecosystem stewardship unit to deal with unforeseen circumstances and realign social-economic systems where their future

Table 11.1 References of transformative adaptation attributes

Transformative adaptation attributes	*References*
Flexibility and restructuring	Brodnik and Brown (2018); Soma et al. (2018); Castán Broto et al. (2019); Wolfram (2019); Fu et al. (2020); Murtagh et al. (2020); Newton and Rogers (2020); Rodger et al. (2020); Shooshtarian et al. (2020); Triyanti et al. (2020); Yasmin et al. (2020); Al Amri et al. (2021); Marsh et al. (2021); Mensah et al. (2021); Padovezi et al. (2022); Shahani et al. (2022)
Participation and inclusiveness	Brodnik and Brown (2018); Soma et al. (2018); Castán Broto et al. (2019); Wolfram (2019); Fu et al. (2020); Newton and Rogers (2020); Triyanti et al. (2020); Yasmin et al. (2020); Moosavi and Brown (2021); Padovezi et al. (2022); Shahani et al. (2022)
Innovation	Brodnik and Brown (2018); Munyasya and Chileshe (2018); Castán Broto et al. (2019); Wolfram (2019); Newton and Rogers (2020); Shooshtarian et al. (2020); Triyanti et al. (2020); Al Amri et al. (2021); Marsh et al. (2021); Shahani et al. (2022)
Leadership and orchestration	Brodnik and Brown (2018); Soma et al. (2018); Castán Broto et al. (2019); Wolfram (2019); Fu et al. (2020); Murtagh et al. (2020); Newton and Rogers (2020); Triyanti et al. (2020); Yasmin et al. (2020); Al Amri et al. (2021); Marsh et al. (2021); Mensah et al. (2021); Moosavi and Brown (2021); Maqbool and Amaechi (2022); Padovezi et al. (2022); Shahani et al. (2022)
Reorientation and social learning	Brodnik and Brown (2018); Castán Broto et al. (2019); Wolfram (2019); Fu et al. (2020); Newton and Rogers (2020); Shooshtarian et al. (2020); Triyanti et al. (2020); Yasmin et al. (2020); Al Amri et al. (2021); Maqbool and Amaechi (2022); Padovezi et al. (2022); Shahani et al. (2022)
A collective vision for change and defined goals	Brodnik and Brown (2018); Castán Broto et al. (2019); Litt et al. (2022); Newton and Rogers (2020); Padovezi et al. (2022); Shahani et al. (2022); Triyanti et al. (2020); Wolfram (2019)
Knowledge co-production	Wolfram (2016); Brodnik and Brown (2018); Castán Broto et al. (2019); Newton and Rogers (2020); Triyanti et al. (2020); Marsh et al. (2021); Moosavi and Brown (2021); Litt et al. (2022); Padovezi et al. (2022), Shahani et al. (2022)

is often disputable. This promotes the voluntary participation of residents who know their rights to access local environmental resources (Padovezi et al., 2022). Multiform governance that links powerful and diverse disciplines could develop strategic planning and shared decision-making among stakeholders (Brodnik and Brown, 2018). The decentralised system implementation should incorporate plans to increase trust among stakeholders and reliability of the government and allow shared knowledge for making decisions (Soma et al., 2018). Plans to facilitate the participation of such intermediaries as independent funding units could be integrated into project governance (Castán Broto et al., 2019). Intermediaries and hybridisation units could work with a wider responsibility to allow for autonomous and reliable intervention, which could promote effective stakeholder engagement (Wolfram, 2019).

11.4.2.2 Participation and Inclusiveness

The involvement of key stakeholders in planning and implementing sustainable programmes and policies is the key to a successful project that involves green practices. One study reveals that active public participation in transformative urban planning, particularly its sustainability vision, is the most appropriate (Soma et al., 2018). Inclusiveness in the context of this study refers to the involvement and connections among various elements inherent in the concept of sustainability; for instance, in evaluating the need for an urban transformation, the local and individual perspectives of householders who require system change should be included. To achieve both participation and inclusiveness, a unit that provides essential support to promote public participation could be created. The support centre is evident in energy retrofitting for urban regeneration projects, which promotes public participation and initiatives (Wolfram, 2019). Another example is the transformation of Australia's existing urban water and energy systems to reduce their carbon footprint, involving many stakeholders, from residents to policymakers. More than 250 policymakers from various fields of environmental sustainability areas, such as planners and designers as well as industry practitioners, were invited to attend a series of workshops across six cities to discuss and envisage the priority values of sustainable development in their cities in the next 50 years (Newton and Rogers, 2020).

11.4.2.3 Innovation

Universities and independent research centres could be the key contributors to a sustainable and resilient built environment (Marsh et al., 2021). Innovation drives proactive formulation to enhance sustainable built environment transformation programmes such as those projects that have been evident in resource management, water conservation and innovative renewable energy (Munyasya and Chileshe, 2018). For instance, a small-scale innovation experiment could be conducted to provide an important platform for actors to embed common values in creating a new way of transforming and managing urban stormwater drainage systems (Brodnik and Brown, 2018). Larger scale innovations, such as energy and water systems guided by the Australian Government's Cooperative Research Centre (CRC), could be established to provide sustainability foresight and introduce advances and superior systems to transform inadequately performing existing systems. The centre could develop smart design tools and methods to scope, plan, evaluate and build the enormous pipeline systems in urban development projects that are predicted to arise in 21st-century cities (Newton and Rogers, 2020). Another innovative strategy is forming a boundary organisation that operates in multi-disciplinary fields on various scales and with various stakeholders. One study found that boundary organisations allow better collaborative forms of governance through the revolutionary invention, research and cooperation of multi-disciplinary teams and create a priority list of needs (Triyanti et al., 2020).

11.4.2.4 Leadership and Orchestration

Most of the literature describes the stakeholders' transformative leadership and intensive coordination as the key attributes of transformative adaptation. While *leadership* is essential to determine ownership of an issue and its components at various levels (Murtagh et al., 2020), *orchestration* involves deliberate actions by stakeholders to initiate and manage collaborations within a network (Hurmelinna-Laukkanen and Natti, 2022). Stakeholders with strong leadership skills head sustainable practices and motivate passion by articulating common goals (Fu et al., 2020). Project owner or developers should lead the green network in built environment projects. Their sustainability approaches are essential to realising green-built facilities by managing every aspect of green construction. Project managers unfamiliar with green building practices can rarely develop a clear plan for green projects, leading to over-budget and low sustainability performance (Maqbool and Amaechi, 2022).

One approach to developing better green governance is by organising a green leadership forum to determine appropriate actions and enhance stakeholder business decisions. Leadership demonstrated by the construction company can uphold green practices and influence green principles for employees to plan and implement green practices (Mensah et al., 2021). Another essential leadership attribute that the industry and government stakeholders demonstrate is organising a structured discussion that amalgamates their implicit knowledge with external technical knowledge and taking the lead to orient strategic implementations (Moosavi and Brown, 2021). Also, stakeholders' leadership should involve creating an opportunity for research innovation at a particular scale, which enables stakeholders to manage risks effectively. This assesses the limits of existing governance arrangements to identify the transformation required and establishes solid relationships among important stakeholders who jointly develop new inventions for alterations *via* visions and experimentation (Newton and Rogers, 2020).

Leaders with an inclusive vision and appropriate transformation strategies would receive enormous support from most stakeholders and make the reform successful (Wolfram, 2019). The leaders sensitive to complex issues and systems in green strategy implementation and the various interests of stakeholders affected by the strategies could engage in the joint decision-making with stakeholders. Take urban regeneration, for example: although higher level leadership is essential for the alterations and incentives needed, the innovations demonstrated by local leaders are also crucial for boosting the transition of other stakeholders and hence improving their resilience. As such, the local leaders could play a role as a broker to bridge the gaps where a higher level of governance could not perform effectively for some factors, including capabilities, source of funding, access and legality. Leadership could be from stakeholders from any level who share common goals and enhance coordination with other stakeholders to achieve successful collective strategies and actions (Yasmin et al., 2020).

11.4.2.5 Reorientation and Social Learning

Other stakeholders can easily influence the environmental sustainability behaviours of suppliers and designers. However, they normally rely heavily on incentives given by the government; hence, reorienting sustainability values among stakeholders is important to enhance their capacity to adopt sustainable practices (Fu et al., 2020). Various events and campaigns from non-governmental organisations have been organised to increase awareness of sustainability among stakeholders (Maqbool and Amaechi, 2022). Research centres could develop sustainability guiding activities to encourage the deployment of sustainable practices among stakeholders who would voluntarily shift their existing practices to higher sustainable levels. This includes increased stakeholder preparedness to accept such new sustainability challenges as population growth, development density and climate crisis (Newton and Rogers, 2020). Stakeholders could also be oriented to access green information and receive education from construction experts to increase their desire to use sustainable building materials. Therefore, substantial efforts are necessary to notify stakeholders about the properties of sustainable materials and their impact on environmental performance (Shooshtarian et al., 2020).

In addition to reorientation, reflexivity, such as recording and assessing green initiatives to reflect their implementation and how their visions were achieved *via* open discussions and open dialogues with stakeholders, is an effective means of social learning (Castán Broto et al., 2019). For instance, the innovative delta programme is considered successful through reflexive monitoring to predict modifications. Reflexive governance is useful for examining the types of changes that are required and under which conditions. It addresses the loopholes of feedback and mobilises the knowledge for reflection to propose possible ways of forwarding to enhance sustainable practices further (Triyanti et al., 2020). Social learning in the formal and informal networks is identified as an essential element in reorientation to increase the capacity of a system to transform (Yasmin et al., 2020). It ensures stakeholders are equipped with the necessary knowledge of sustainable practices by altering circumstances and overcoming barriers. These could be achieved through open discussions with other stakeholders in project development, examining and informing project outcomes, and assessing applied technologies and methods (Shahani et al., 2022).

11.4.2.6 A Collective Vision for Change and Defined Goals

This attribute was not identified in Section 11.3, but we observed it as an important feature to improve the adaptive capacity of stakeholders in the transformation of built environment projects. This attribute is also repeatedly emphasised in the literature identified in Table 11.1. Stakeholders need to be provided with a clear and achievable vision which frames a prosperous future. This allows better sustainability foresight to frame societal well-being and predict future uncertainties (Brodnik and Brown, 2018). Visions must be devised, targeted for a profound shift and endorsed by stakeholders from various levels. Exercises of envisioning should include achievable outcomes linked with governed policies and environmental circumstances (Castán Broto et al., 2019). Goals should be defined clearly with medium- and long-term strategic visions, with the detailed identification of the steps to follow and a systematic introduction of the logic behind the steps to strengthen stakeholder capacity for real transition (Litt et al., 2022). It is also essential to convert approaches of visions and goals into tangible activities in a corporate context that is operative, spatial and organisational (Newton and Rogers, 2020). Without a clear vision and defined goals collectively agreed by construction stakeholders, it is difficult for them to enhance their adaptive capacity in applying sustainable design and construction practices.

11.4.2.7 Knowledge Co-Production

Similar to a collective vision and defined goal, knowledge co-production, which was not identified in the literature review section, has been considered an important attribute to enhance stakeholder adaptive capacity in adopting sustainable practices. It has also been repeatedly discussed in the literature identified in Table 11.1. When knowledge of complex sustainable practices is only consulted by one discipline, conflicts among other project team members could occur if a shared understanding or a collective solution cannot be developed (Moosavi and Brown, 2021). Interdisciplinary collaboration creates knowledge that can be shared with various stakeholders (Brodnik and Brown, 2018). Strategic knowledge management could analyse existing governance structures and environmental conditions to develop innovations and plan interventions (Castán Broto et al., 2019). Knowledge co-production, which various stakeholders develop, could provide useful knowledge concerning project goals and develop activities consistent with the sustainability practices context that are innovative and replicable (Litt et al., 2022). Co-production allows a combination of ground-breaking and gradual low-carbon inventions that co-form the pathway to profoundly alternate futures. Sustainability foresight is one of the important components in co-production as it enables knowledge and resources to be shared from various levels, simplifying the procedure and approaches, offering a holistic view, and altering the policies and strategies *via* stakeholder initiatives (Shahani et al., 2022).

11.4.3 Conceptual Framework of Transformative Adaptation for Built Environment Stakeholders

Few et al. (2017) have suggested three elements that could be used to evaluate how an adaptation activity exists in connection with transformation concepts. These include objects of transformation, methods of transformation and expected outcomes.

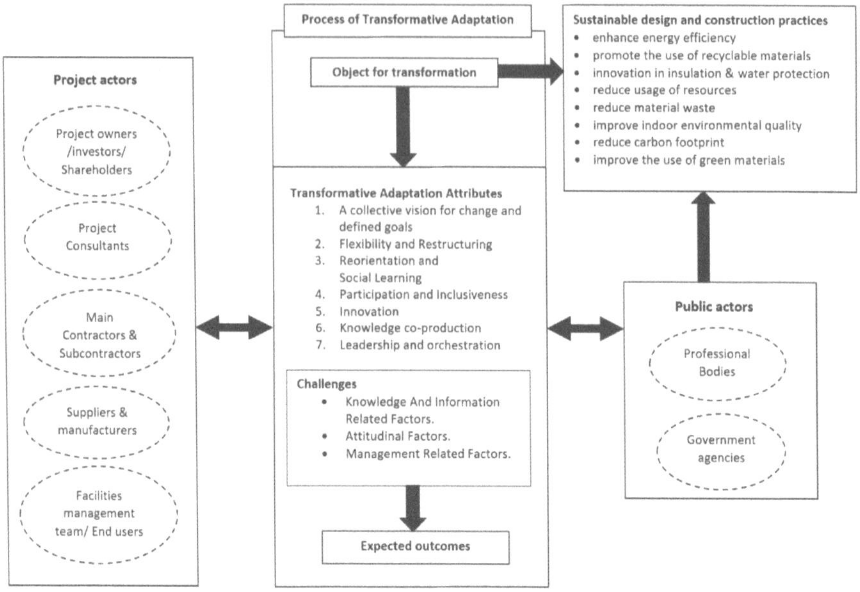

Figure 11.4 Conceptual framework for stakeholder transformative adaptation in sustainable design and construction practices

From the outset, the main actors, such as project owners, should work with public actors to define the object and requirement for built environment transformation. The object of transformation should include the sustainable design and construction practices highlighted in Figure 11.4. The object of transformation in the context of this study includes any built facilities, such as buildings and infrastructure systems, as well as coastal, rural and urban developments that require change. As stated in Figure 11.4, the scope of the objects needs to be clearly defined as they would influence the direction of other project actors, such as project consultants and main contractors, in taking the necessary steps to innovate and adopt sustainability design and construction practices. When defining the object and scope of the transformation, we suggest that stakeholders align the object of transformation with the SDG and consider the applicability of the sustainable design and construction practices discussed in Section 11.2 and Figure 11.4. With a clear object and scope for transformation identified in the project initiative and planning stage, the transformation process could be operated easier by overcoming challenges through the seven transformative adaptation attributes and effective connections with the public and project actors in Figure 11.4.

As Figure 11.4 demonstrates, both project and public actors should actively participate in the overall adaptation process. Public actors such as professional bodies and government agencies who aim to promote a sustainable built environment should work together with project actors to provide education and training and bridge the knowledge gap by providing sustainable foresight, green technical knowledge and necessary transformative adaptation skills from the attributes of project actors identified in Figure 11.4. This could help address such challenges as the lack of knowledge and awareness of sustainable design and construction practices, limited experience in the selection of sustainable construction procedures and techniques, insufficient information about the performance quality of sustainable materials and inadequate information about the handling and storage of sustainable materials and thereby improve their adaptation capacity in adopting sustainable practices. The expected outcomes will only be achieved when the project and public actors engage and collaborate to determine the objects and sustainable practices they would like to apply, plus the attributes of the stakeholder adaptive capacity specified in Figure 11.4.

The adaptation process should involve the seven main transformative adaptation attributes identified from the thematic analysis. Each attribute makes an important contribution towards a sustainable and resilient built environment. Without *clear and defined visions and goals*, the transformation becomes meaningless as there is no way of knowing whether the transformation outcomes have been achieved. Project owners involved in built environment transformation initiatives should define their scope and requirements clearly by informing other project actors of the sustainability outcomes and enabling them to plan and innovate the sustainability practices needed.

Likewise, the main project actors, such as project owners and public actors, should provide *flexibility* to reorganise various sustained forms of units to implement sustainability practices; this will enable stakeholders to adopt more approachable and workable sustainable practices. Another key characteristic that is extremely important and identified by most of the literature concerns *orchestration and leadership*. These synergise fragmented interests and challenges arising from implementation, enable better information sharing and develop innovative solutions to address lack of knowledge and information and management-related issues, enhancing stakeholders' adaptive capacity to adopt sustainable practices. These attributes could be used to address such management-related challenges as the lack of consensus on sustainability values among stakeholders and inadequate top management commitment to sustainability, inadequate sustainability measurement tools and difficulty in compliance with certification and green codes, fragmented guidelines for green procurement, and lack of communication and coordination among the contracting parties. Although the roles of project actors in a construction project are fragmented, each should have the attributes of orchestration and leadership to enable a better built environment transformation.

Another attribute which is of importance is providing *orientation* to all project actors as well as public actors on the importance of sustainability values. Any implemented practices should be reflected consistently to provide *social learning* to stakeholders, so they know how to apply practice change. These could help address attitudinal issues from stakeholders, such as low demand by potential buyers, weak scope definitions of sustainable construction requirements, lack of interest among clients, resistance to change within the construction industry and frequent design changes by the client or architect. The end users are the last stakeholders who use the built facility; hence, their *participation* should be *inclusive* and considered in the decision-making process. Participation and inclusiveness should also apply to all other project and public actors as they contribute to the decision-making process, further enhancing stakeholder collaboration and adaptive capacity in effectively dealing with the sustainable built environment. As discussed earlier, to address attitudinal challenges, psychological factors that influence green behaviours and planned behaviour theory should be explored to determine the adaptive capacity that could affect adopting sustainable practices. One of the significant characteristics of transformative adaptation is *innovation*, as adopting sustainable design and construction practices requires frequent inventions, and it requires adapting to new policies, knowledge and technologies. Innovations are influenced by location, societal values, transformational leadership and numerous actors involved with exploratory and practical experiences (Wolfram, 2016). The stakeholders could strengthen their adaptive capacity in developing new sustainable inventions. *Co-production knowledge* created by multi-disciplinary stakeholders and sustainability foresight allows stakeholders to develop their capacity to innovate and produce sustainable technologies and practices that benefit the built environment.

11.5 Conclusion

Transformative adaptation should be treated as a process that requires commitment and proactive participation for each stakeholder, including the end users who use the built facility. In the transition process towards improving sustainable built environment practices, stakeholders should be clear about the types of transformation required. The adaptation process should be coupled with all seven categories of transformative adaptation characteristics to ensure better social transition among stakeholders towards a sustainable and resilient built environment. This study is limited to identifying transformative adaptation attributes of stakeholders in general. An in-depth study is required to determine how the stakeholders could deploy each attribute and how each element discussed in the framework could be connected to improve the adaptive capacity of stakeholders more effectively. The attribute of transformative adaptation could be influenced by the context of the objects of transformation, expected outcomes, existing conditions etc. Also, all categories of adaptive capacity – social, physical and economic – are important across cases (Dilling et al., 2023). The present study focuses on social capacity only. Future studies should consider the social implications of each transformative adaptation attribute for various objects of transformation and the expected outcomes in the built environment sector. Comparative studies between two different objects of transformation could also be carried out to determine the transformative adaptation attribute effectiveness. For example, transformative adaptation attributes of brownfield coastal areas could be compared and distinguished from those used to transform energy and water infrastructure systems. The conceptual framework developed in this chapter could serve as a guiding framework for researchers and practitioners to formulate strategies for implementing sustainability practices and increasing their capability in transforming built environment objects to being more environmentally sustainable.

References

Abidin, N.Z., Yusof, N., and Othman, A.A.E., 2013. Enablers and challenges of a sustainable housing industry in Malaysia. *Construction Innovation*, 13 (1), 10–25.

Adegun, O.B., 2023. Climatic disasters within a flood-prone coastal slum in Lagos: coping capacities and adaptation prospects. International journal of disaster resilience in the built environment, 14 (2), 212–228.

Aghimien, D.O., Aigbavboa, C.O., and Thwala, W.D., 2019. Microscoping the challenges of sustainable construction in developing countries. *Journal of Engineering, Design and Technology*, 17 (6), 1110–1128.

Ahmed, S. and El-Sayegh, S., 2022. The challenges of sustainable construction projects delivery–evidence from the UAE. *Architectural Engineering and Design Management*, 18 (3), 299–312.

Ajzen, I., 2020. The theory of planned behavior: Frequently asked questions. *Human Behavior and Emerging Technologies*, 2 (4), 314–324.

Akotia, J., Manu, E., Opoku, A., and Sackey, E., 2020. The role of built environment practitioners in driving the socio-economic sustainability aspects of sustainable regeneration. *Construction Economics and Building*, 20 (2), 89–108.

Alhilli, H.K. and Burhan, A.M., 2021. Developing a system for assessing the sustainability in school building projects. *Environment, Development and Sustainability*, 23 (12), 17483–17502.

Ali, H.H. and Alkayed, A.A., 2019. Constrains and barriers of implementing sustainability into architectural professional practice in Jordan. *Alexandria Engineering Journal*, 58 (3).

Alsanad, S., 2015. Awareness, Drivers, Actions, and Barriers of Sustainable Construction in Kuwait. In: *Procedia Engineering*. Elsevier Ltd, 969–983.

Ametepey, O., Aigbavboa, C., and Ansah, K., 2015. Barriers to successful implementation of sustainable construction in the Ghanaian construction industry. *Procedia Manufacturing*, 3, 1682–1689.

Al Amri, T., Khetani, K.P., and Marey-Perez, M., 2021. Towards sustainable i4.0: Key skill areas for project managers in GCC construction industry. *Sustainability (Switzerland)*, 13 (15), 1–17.

AlWaer, H., Sibley, M., and Lewis, J., 2008. Different stakeholder perceptions of sustainability assessment. *Architectural Science Review*, 51 (1), 48–59.

Asadzadeh, A., Kötter, T., Fekete, A., Moghadas, M., Alizadeh, M., Zebardast, E., Weiss, D., Basirat, M., and Hutter, G., 2022. Urbanization, migration, and the challenges of resilience thinking in urban planning: Insights from two contrasting planning systems in Germany and Iran. *Cities*, 125, 1036421.

Askar, R., Bragança, L., and Gervásio, H., 2021. Adaptability of buildings: A critical review on the concept evolution. *Applied Sciences (Switzerland)*, 11, 4483.

Barnes, M.L., Bodin, Ö., Guerrero, A.M., McAllister, R.R.J., Alexander, S.M., and Robins, G., 2017. The social structural foundations of adaptation and transformation in social–ecological systems. *Ecology and Society*, 22 (4):16.

Barnes, M.L., Wang, P., Cinner, J.E., Graham, N.A.J., Guerrero, A.M., Jasny, L., Lau, J., Sutcliffe, S.R., and Zamborain-Mason, J., 2020. Social determinants of adaptive and transformative responses to climate change. *Nature Climate Change*, 10, 823–828.

Berardi, U., 2012. Sustainability assessment in the construction sector: rating systems and rated buildings. *Sustainable Development*, 20 (6), 411–424.

Berawi, M.A., Miraj, P., Windrayani, R., and Berawi, A.R.B., 2019. Stakeholders' perspectives on green building rating: A case study in Indonesia. *Heliyon*, 5 (3), 01328.

Brodnik, C. and Brown, R., 2018. Strategies for developing transformative capacity in urban water management sectors: The case of Melbourne, Australia. *Technological Forecasting and Social Change*, 137, 147–159

Castán Broto, V., Trencher, G., Iwaszuk, E., and Westman, L., 2019. Transformative capacity and local action for urban sustainability. *Ambio*, (48), 449–462.

Clayton, S. and Myers, G., 2009. *Conservation Psychology: Understanding and Promoting Human Care for Nature*. Applied Signal Processing.

Colloff, M.J., Martín-López, B., Lavorel, S., Locatelli, B., Gorddard, R., Longaretti, P.Y., Walters, G., van Kerkhoff, L., Wyborn, C., Coreau, A., Wise, R.M., Dunlop, M., Degeorges, P., Grantham, H., Overton, I.C., Williams, R.D., Doherty, M.D., Capon, T., Sanderson, T., and Murphy, H.T., 2017. An integrative research framework for enabling transformative adaptation. *Environmental Science and Policy*, 68, 87–96.

Cordero, A.S., Melgar, S.G., and Márquez, J.M.A., 2019. Green building rating systems and the new framework level(s): A critical review of sustainability certification within Europe. *Energies*, 13 (66), 1–25.

Dilling, L., Daly, M.E., Travis, W.R., Ray, A.J., and Wilhelmi, O.V., 2023. The role of adaptive capacity in incremental and transformative adaptation in three large US Urban water systems. *Global Environmental Change*, 79, 102649, 1–12.

Dolla, T., Jain, K., and Delhi, V.S.K., 2023. Strategies for digital transformation in construction projects: Stakeholders' perceptions and actor dynamics for industry 4.0. *Journal of Information Technology in Construction*, 28, 151–175.

Doorn, N., Gardoni, P., and Murphy, C., 2019. A multidisciplinary definition and evaluation of resilience: the role of social justice in defining resilience. *Sustainable and Resilient Infrastructure*, 4 (3), 112–123.

Eaton, W.M., Brasier, K.J., Burbach, M.E., Whitmer, W., Engle, E.W., Burnham, M., Quimby, B., Kumar Chaudhary, A., Whitley, H., Delozier, J., Fowler, L.B., Wutich, A., Bausch, J.C., Beresford, M., Hinrichs, C.C., Burkhart-Kriesel, C., Preisendanz, H.E., Williams, C., Watson, J., and Weigle, J., 2021. A Conceptual Framework for Social, Behavioral, and Environmental Change through Stakeholder Engagement in Water Resource Management. *Society and Natural Resources*, 34 (8), 1111–1132.

Edwards, R.E., Lou, E., Bataw, A., Kamaruzzaman, S.N., and Johnson, C., 2019. Sustainability-led design: Feasibility of incorporating whole-life cycle energy assessment into BIM for refurbishment projects. *Journal of Building Engineering*, 24, 100697.

Ershadi, M., Jefferies, M., Davis, P., and Mojtahedi, M., 2021. Barriers to achieving sustainable construction project procurement in the private sector. *Cleaner Engineering and Technology*, 3, 100125.

Fedele, G., Donatti, C.I., Harvey, C.A., Hannah, L., and Hole, D.G., 2019. Transformative adaptation to climate change for sustainable social-ecological systems. *Environmental Science and Policy*, 101, 116–125.

Few, R., Morchain, D., Spear, D., Mensah, A., and Bendapudi, R., 2017. Transformation, adaptation and development: relating concepts to practice. *Palgrave Communications*, 3(17092).

Filho, W.L., Wolf, F., Moncada, S., Salvia, A.L., Balogun, A.L.B., Skanavis, C., Kounani, A., and Nunn, P.D., 2022. Transformative adaptation as a sustainable response to climate change: Insights from large-scale case studies. *Mitigation and Adaptation Strategies for Global Change*, 27 (20), 1–26.

Fu, Y., Dong, N., Ge, Q., Xiong, F., and Gong, C., 2020. Driving-paths of green buildings industry (GBI) from stakeholders' green behavior based on the network analysis. *Journal of Cleaner Production*, 273, 122883.

Ghassim, B. and Bogers, M., 2019. Linking stakeholder engagement to profitability through sustainability-oriented innovation: A quantitative study of the minerals industry. *Journal of Cleaner Production*, 224, 905–919.

Guest, G., MacQueen, K., and Namey, E., 2014. *Applied Thematic Analysis*. Applied Thematic Analysis.

Henry, A.D. and Vollan, B., 2014. Networks and the challenge of sustainable development. *Annual Review of Environment and Resources*, 39, 583–610.

Herazo, B. and Lizarralde, G., 2016. Understanding stakeholders' approaches to sustainability in building projects. *Sustainable Cities and Society*, 26, 240–254.

Hodgson, G., 2007. What are institutions? *Voprosy Ekonomiki*, 2007 (8), 28–48.

Horry, R., Booth, C.A., Mahamadu, A.–M, Manu, P., and Georgakis, P., 2021. Environmental management systems in the architectural, engineering and construction sectors: a roadmap to aid the delivery of the sustainable development goals. *Environment, Development and Sustainability*, 24, 10585–10615.

Hurmelinna-Laukkanen, P., Möller, K. and Nätti, S., 2022. Orchestrating innovation networks: Alignment and orchestration profile approach. Journal of Business Research, 140, 170–188.

Hutcheon, L., 2012. *A Theory of Adaptation*. Routledge.

Johnsson, F., Karlsson, I., Rootzén, J., Ahlbäck, A., and Gustavsson, M., 2020. The framing of a sustainable development goals assessment in decarbonizing the construction industry – Avoiding "Greenwashing". *Renewable and Sustainable Energy Reviews*, 131, 110029.

Jozaei, J., Chuang, W.-C., Allen, C.R., and Garmestani, A., 2022. Social vulnerability, social-ecological resilience and coastal governance. *Global Sustainability*, 5, e12, 1–9.

Kalyana Chakravarthy, P.R., Suganya, R., Nivedhitha, M., Parthiban, A., and Sivaganesan, S., 2022. Barriers and project management practices in green buildings. *Materials Today: Proceedings*, 52, 2248–2254.

Kawakubo, S., Murakami, S., Ikaga, T., and Asami, Y., 2018. Sustainability assessment of cities: SDGs and GHG emissions. *Building Research and Information*, 46 (5), 528–539.

Klewitz, J. and Hansen, E.G., 2014. Sustainability-oriented innovation of SMEs: A systematic review. *Journal of Cleaner Production*, 65, 57–75.

Knight, J., 1992. *Institutions and Social Conflict*. Institutions and Social Conflict.

Kongela, S.M., 2021. Sustainability potential awareness among built environment stakeholders: experience from Tanzania. *International Journal of Building Pathology and Adaptation*, 41 (2), 301–319.

Lakoff, G., 2010. Why it matters how we frame the environment. *Environmental Communication*, 4 (1), 70–81.

Lee, C.-Y., Chong, H.-Y., Liao, P.-C., and Wang, X., 2018. Critical review of social network analysis applications in complex project management. *Journal of Management in Engineering*, 34 (2), 04017061.

Li, H., Zhang, X., Ng, S.T., and Skitmore, M., 2018. Quantifying stakeholder influence in decision/evaluations relating to sustainable construction in China – A Delphi approach. *Journal of Cleaner Production*, 173, 160–170.

Li, Y., Li, M., Sang, P., Chen, P.H., and Li, C., 2022. Stakeholder studies of green buildings: A literature review. *Journal of Building Engineering*, 54, 104667.

Ling, T.Y. and Chiang, Y.C., 2018. Well-being, health and urban coherence-advancing vertical greening approach toward resilience: A design practice consideration. *Journal of Cleaner Production*, 182, 187–197.

Litt, G., Ferraioli, E., Magni, F., Lucertini, G., and Musco, F., 2022. Inter-municipal methodology for climate transition strategies: The first case in Italy. *Sustainability (Switzerland)*, 14 (5), 2926.

Lonsdale, K., Pringle, P., and Turner, B., 2015. *Transformative adaptation: what it is, why it matters and what is needed*. Oxford.

Lyon, D., 2014. Surveillance, Snowden, and Big Data: Capacities, consequences, critique. *Big Data and Society*, 1 (2).

Mahyuddin, N., Samzadeh, M., Zaid, S.M., and Ab Ghafar, N., 2022. Towards nearly zero energy building concept – Visual comfort and energy efficiency assessments in a classroom. *Open House International*, 47 (1), 167–187.

Maqbool, R. and Amaechi, I.E., 2022. A systematic managerial perspective on the environmentally sustainable construction practices of UK. *Environmental Science and Pollution Research*, 29, 64132–64149.

Marsh, R.J., Brent, A.C., and de Kock, I.H., 2021. Understanding the barriers and drivers of sustainable construction adoption and implementation in South Africa: A quantitative study using the Theoretical Domains Framework and COM-B Model. *Journal of the South African Institution of Civil Engineering*, 63 (4), 52–62.

Mashwama, N.X., Thwala, D., Aigbavboa, C., and Hamma-Adama, M., 2020. A stakeholder perspective on the implementation of sustainable construction practices. *Proceedings of International Structural Engineering and Construction*, 7 (1), 1–6.

Mattinzioli, T., Sol-Sánchez, M., Moreno, B., Alegre, J., and Martínez, G., 2021. Sustainable building rating systems: A critical review for achieving a common consensus. *Critical Reviews in Environmental Science and Technology*, 51 (5), 512–534.

Mendizabal, M., Feliu, E., Tapia, C., Rajaeifar, M.A., Tiwary, A., Sepúlveda, J., and Heidrich, O., 2021. Triggers of change to achieve sustainable, resilient, and adaptive cities. *City and Environment Interactions*, 12, 100071.

Mendizabal, M., Heidrich, O., Feliu, E., García-Blanco, G., and Mendizabal, A., 2018. Stimulating urban transition and transformation to achieve sustainable and resilient cities. *Renewable and Sustainable Energy Reviews*, 94, 410–418.

Mensah, S., Nani, G., and Ayarkwa, J., 2021. Contractors' adaptation to environmentally sustainable construction: a micro-level implementation framework. *Smart and Sustainable Built Environment*, 11 (4), 832–851.

Mimura, N., 2021. Rising seas and subsiding cities. *Nature Climate Change*, 11, 296–297.

Moosavi, S. and Brown, G.R., 2021. Advancing the Adaptive, Participatory and Transdisciplinary decision-making framework: The case of a coastal brownfield transformation. *Cities*, 111, 103106.

Müller-Puch, A., Hegre, E., and Innerarity, M., 2021. 2 in 1 – A playful approach to the sustainable use of building materials. *Plan Journal*, 6 (2), 323–336.

Munyasya, B.M. and Chileshe, N., 2018. Towards sustainable infrastructure development: Drivers, barriers, strategies, and coping mechanisms. *Sustainability (Switzerland)*, 10 (12).

Murtagh, N., Scott, L., and Fan, J., 2020. Sustainable and resilient construction: Current status and future challenges. *Journal of Cleaner Production*, 268, 122264.

Newton, P.W. and Rogers, B.C., 2020. Transforming built environments: Towards carbon neutral and blue-green cities. *Sustainability (Switzerland)*, 12 (11), 4745.

Nidumolu, R., Prahalad, C.K., and Rangaswami, M.R., 2009. Why sustainability is now the key driver of innovation. *Harvard Business Review*, 87 (9), 57–64.

Omopariola, E.D., Olanrewaju, O.I., Albert, I., Oke, A.E., and Ibiyemi, S.B., 2022. Sustainable construction in the Nigerian construction industry: Unsustainable practices, barriers and strategies. *Journal of Engineering, Design and Technology*, 22 (4), 1158–1184.

Opoku, A., 2016. SDG2030: A sustainable built environment's role in achieving the post-2015 United Nations sustainable development goals. *In*: *Proceedings of the 32nd Annual ARCOM Conference, ARCOM 2016*.

Ostrom, E., 2015. *Governing the commons: The evolution of institutions for collective action*. Governing the Commons: The Evolution of Institutions for Collective Action.

Owen, G., 2020. What makes climate change adaptation effective? A systematic review of the literature. *Global Environmental Change*, 62, 102071.

Padovezi, A., Secco, L., Adams, C., and Chazdon, R.L., 2022. Bridging social innovation with forest and landscape restoration. *Environmental Policy and Governance*, 32(6), 520–531.

Pahl-Wostl, C., 2020. Adaptive and sustainable water management: from improved conceptual foundations to transformative change. *International Journal of Water Resources Development*, 36 (2–3).

Probst, L., Bardach, L., Kamusingize, D., Templer, N., Ogwali, H., Owamani, A., Mulumba, L., Onwonga, R., and Adugna, B.T., 2019. A transformative university learning experience contributes to sustainability attitudes, skills and agency. *Journal of Cleaner Production*, 232, 648–656.

Rodger, D., Callaghan, N., and Thomson, C.S., 2020. The loosening control of social housing: Creating a holistic retrofit system for an ageing population through the lens of governmentality. *Construction Management and Economics*, 38 (12), 1101–1121.

Saleh, M.S. and Alalouch, C., 2015. Towards Sustainable Construction in Oman: Challenges & Opportunities. In: *Procedia Engineering*, (118), 177–184.

Schmitt, P. and Wiechmann, T., 2018. Unpacking spatial planning as the governance of place: Extracting potentials for future advancements in planning research. *DISP*, 54 (4), 21–33.

Shahani, F., Pineda-Pinto, M., and Frantzeskaki, N., 2022. Transformative low-carbon urban innovations: Operationalizing transformative capacity for urban planning. *Ambio*, 51, 1179–1198.

Shooshtarian, S., Caldera, S., Maqsood, T., and Ryley, T., 2020. Using recycled construction and demolition waste products: A review of stakeholders' perceptions, decisions, and motivations. Recycling, 5 (4), 31.

Soma, K., Dijkshoorn-Dekker, M.W.C., and Polman, N.B.P., 2018. Stakeholder contributions through transitions towards urban sustainability. *Sustainable Cities and Society*, 37, 438–450.

Tafazzoli, M., Mousavi, E., and Kermanshachi, S., 2020. Opportunities and challenges of green-lean: An integrated system for sustainable construction. *Sustainability (Switzerland)*, 12 (11), 4460.

Till, J. and Schneider, T., 2005. Flexible housing: The means to the end. *ARQ: Architectural Research Quarterly*, 9 (3–4), 287–296.

Tokbolat, S., Karaca, F., Durdyev, S., and Calay, R.K., 2020. Construction professionals' perspectives on drivers and barriers of sustainable construction. *Environment, Development and Sustainability*, 22, 4361–4378. doi.org/10.1007/s10668-019-00388-3

Triyanti, A., Hegger, D.L.T., and Driessen, P.P.J., 2020. Water and climate governance in deltas: On the relevance of anticipatory, interactive, and transformative modes of governance. *Water (Switzerland)*, 12 (12), 3391.

United Nations, 2022. Transforming our world: The 2030 Agenda for Sustainable Development [online].

Wolfram, M., 2016. Conceptualizing urban transformative capacity: A framework for research and policy. *Cities*, 51, 121–130.

Wolfram, M., 2019. Assessing transformative capacity for sustainable urban regeneration: A comparative study of three South Korean cities. *Ambio*, 48 (5), 478–493.

Yanita, R. and Mochtar, K., 2021. Legal aspect of value engineering implementation in Jakarta (Indonesia) construction projects. *International Journal of Construction Management*, 21 (2), 131–139.

Yasmin, T., Farrelly, M., and Rogers, B.C., 2020. Adaptive governance: a catalyst for advancing sustainable urban transformation in the global South. *International Journal of Water Resources Development*, 36 (5), 818–838.

Yin, B.C.L., Laing, R., Leon, M., and Mabon, L., 2018. An evaluation of sustainable construction perceptions and practices in Singapore. *Sustainable Cities and Society*, 39, 613–620.

Zhu, B., Zhu, C., and Dewancker, B., 2020. A study of development mode in green campus to realize the sustainable development goals. *International Journal of Sustainability in Higher Education*, 21 (4), 799–818.

12 Building Back Better

Residential Retrofit Options for Older Australians in Bushfire Prone Areas

Alison Wain, Saeed Banihashemi, Sara Wilkinson, Lois Towart, Aso Haji Rasouli, and Iina Lohi

12.1 Introduction

As climate change impacts escalate in frequency and severity, it is essential that society explores and implements resilience measures for vulnerable communities or groups. In Australia, bushfires have occurred for centuries; however, increased urbanisation and population growth, coupled with increasing frequency and intensity of bushfire, means more people are now affected. The 2019/2020 Black Summer Bushfires were unmatched in terms of scale, intensity and impacts (Damany-Pearce et al., 2022). The extreme drought preceding the bushfires caused the exceedance of critical flammability thresholds for the most prolonged period in the last 30 years, enabling uncontrolled fire spread. Lucas et al. (2007) in their Bushfire Report stated that bushfires are inevitable in Australia, and whilst they are most common in the north, the south-east, where most people live, is also susceptible to fires which threaten people and property. Climate projections undertaken in 2007 indicate south-east Australia will become hotter and drier: estimates indicate that 'very high' and 'extreme' fire danger days will increase by 4–25% by 2020 and by 15–70% by 2050.

The five most deadly bushfires in Australia were Black Saturday 2009 in Victoria with 173 people killed, and 2000 homes lost; Ash Wednesday in 1983 in Victoria and South Australia with 75 dead and nearly 1900 homes lost; Black Friday 1939 in Victoria with 71 people killed and 650 homes lost, Black Tuesday 1967 in Tasmania with 62 people killed and almost 1300 houses lost and the Gippsland fires and Black Sunday of 1926 in Victoria with 60 people killed. Other major bushfires were the 1851 Black Thursday bushfires, the 2006 December bushfires, the 1974–1975 fires that burnt 15% of Australia (Australian Disaster Knowledge Resilience Hub, 2023) and the 2019–2020 Black Summer Bushfires (Damany-Pearce et al., 2022).

Housing is the single most expensive investment most people make during their lifetime. Typically, buildings and peoples' possessions are damaged or destroyed in bushfires. The extent of the damage depends on the pace and direction of the fire and the building design and materials used in the construction of peoples' homes. Many buildings predate current design thinking on bushfire protection, and therefore retrofitting is a good option to improve the resilience of housing.

Older people (aged 65 and older) are disproportionately affected by bushfires and other disasters, but there is limited research into the details of their ability to prepare

DOI: 10.1201/9781003317890-16

for these events, and their resilience following them (Tuohy, Stephens and Johnson, 2014). They are not a homogenous group and have varying levels of physical, financial and cognitive capacity. The 'flight or fight' decision is often different for this group than for younger people due to their different physical and cognitive capacities (Horton, Hanna and Kelly, 2010), and while many are 'asset rich' because they own their own home, their ability to retrofit their homes for bushfire resilience is often limited by being 'cash poor' through dependence on the age pension (Korda et al., 2014).

This chapter begins with a discussion of the extent of bushfire risk in Australia and the damage caused in the 2019/2020 Black Summer Bushfires, followed by an outline of the demographics of older Austalians in two case study local government areas (LGAs) (temperate Bega Valley in New South Wales (NSW) and sub-tropical Noosa in Queensland), and a review of the housing options most commonly used by older Australians. The chapter then reviews housing typologies typically found in bushfire-prone regions in NSW and Queensland and discusses the issue of bushfire resilience in the context of the multi-hazard disaster scenarios increasingly being experienced in these states. The issues surrounding personal heritage, its importance to individuals and its role in underpinning identity, connection and emotional recovery is then reviewed, along with current approaches to heritage protection in bushfire events. Bushfire retrofit benefits and measures are then highlighted to illustrate the potential resilience options available, with preliminary notes on research to develop a toolkit to assist older Australians in making practical and informed retrofitting decisions.

12.2 Bushfire risk in Australia and the 2019 Black Summer Bushfires

Natural disasters are significant risks in Australia. The total value of losses from natural disasters for the 50-year period ending at 2016 was more than AUD $180 billion, including more than $20 billion in the cost of death and injuries. Bushfires, which cause 17% of these losses, are a natural part of the Australian disaster landscape but they have devastating consequences on people and communities (Handmer, Ladds and Magee, 2016).

As one of the most recent ecological catastrophes in Australia, the Black Summer Bushfires of 2019–2020 changed the local face of the Earth. The scale, breadth and severity of the fires drastically reflected the impacts of global warming and climate change, with the lives of more than 80% of Australian people being impacted by this disaster, either directly or indirectly (Biddle et al., 2020). More than 400 people lost their lives (Arriagada et al., 2020), more than 3000 houses, 7000 buildings, urban blocks and over 20 million hectares of land were burned, and the loss of local flora and fauna was devastating. Many Australian cities were covered with the smoke and haze. Canberra, in the Australian Capital Territory, had the worst air quality in the world. As a rule of thumb, air quality index readings above 200 are considered hazardous (Banihashemi and Zarepour Sohi, 2022) while one of the Canberran monitoring sites peaked at 7700 in 2020 (iqair, 2020). Particulates and chemicals in the smoke affected untold numbers of people, both in Australia and beyond.

The Earth is now about 1.1°C warmer than it was during the 1800s (United Nations Climate Action, 2023). Among the many impacts of this climate change are: heat stress, which impacts people's health negatively, with younger and older people being more vulnerable; increasing drought due to a lack of rainfall, which reduces the ability to fight fire and increases the likelihood of fire occurring; and increasing incidence of fire itself, which clearly affects both physical property and people's health, through burning and poor air quality.

These facts are evidence of the significance of the scale and extent of the bushfire damage in the Australian natural and built environments. Bushfires damage housing and infrastructure (including communications, energy and water) and disrupt supply chains for food and building materials. Hence, it is incumbent on the sector to take effective action to increase the resilience and sustainability of our built environment.

12.3 Vulnerable Groups – Older Australians

Bushfires, along with other climate-related hazards, are more likely to affect regional locations and these regional locations generally have a higher proportion of older people (Australian Bureau of Statistics, 2021; Paddam, Liu and Philip, 2022). The main reasons for this ageing profile are residents ageing in place, as well as inward migration. Those ageing in place are often 'born and bred' and have lived in their region for decades. Inward migration by retirees, referred to as amenity retirement migration, is a feature of many Australian regions, and this group has often sold their previous home to fund the relocation. These different drivers compound the existing heterogeneity of older Australians.

This study focuses on two regional LGAs that were affected by the 2019/2020 bushfires: Bega Valley Shire in NSW and Noosa Shire in Queensland. Both LGAs are coastal, with a surrounding hinterland comprising rural land uses, state forests and national parks, and urban development ranging from rural/residential subdivisions to small townships to medium-density housing in coastal suburbs. Bega Valley LGA in NSW is on Aboriginal Yuin lands and is situated 350 km south of the Sydney CBD; Noosa LGA in Queensland is on Kabi Kabi and Gubbi Gubbi lands and is located 120 km north of the Brisbane CBD. Bega Valley comprises 6278 km^2 with a population of 29,016 people; Noosa is smaller and more heavily populated with an area of 870 km^2 and a population of 56,298 people (Australian Bureau of Statistics, 2021).

Bega Valley and Noosa LGAs have an ageing population profile with median ages of 52 and 50, respectively. In contrast Australia as a whole has a median age of 38. Both LGAs are established retiree destinations with growth rates greater than predicted from ageing in place (Australian Bureau of Statistics, 2021). There is a diversity of older people within each LGA. In Bega Valley the towns of Bega and Eden have a greater proportion of pensioners and part pensioners, while the coastal suburbs of Merimbula, Tura Beach, Burma Narooma and Bermagui have a higher proportion of self-funded retirees, as these suburbs are popular with retirees from Canberra, many with Commonwealth public service superannuation benefits.

In Noosa the hinterland areas of Tewantin, Peregian Beach and Noosa Hinterland have a greater proportion of pensioners and part pensioners and are closer to the bushland that comprises the greatest fire hazard. The coastal suburbs of Noosa Heads and Sunshine Beach, which are noted for newer luxury residential developments with a lower fire hazard, have a greater proportion of self-funded retirees.

Compared to younger Australians, the ability of older Australians to defend themselves and their property in the event of a fire is lower, and for those whose income is fixed to a relatively low pension, their capacity to pay for home alterations to provide additional fire protection is limited. However, while older people may have mobility and health issues, most older people living independently in the community have high levels of functionality. For instance in Bega Valley and Noosa LGAs, the proportion of this group reporting that they need assistance with daily activities ranges from just 9% up to 15%, with much of this care being supplied by a family member (Australian Bureau of Statistics, 2021).

Locations with greater proportions of older people renting tend to be those with existing disadvantage, including lack of health facilities such as hospitals, and low employment and low wages (Australian Bureau of Statistics, 2021), although in both Bega Valley and Noosa older people are more likely to live in detached houses which they own outright. The housing in both LGAs displays a similar profile, ranging from older workers' cottages in more established non-coastal locations, to recently constructed luxury houses (of over five bedrooms) in predominantly new coastal subdivisions, and everything in between. This heterogeneity is prevalent throughout many Australian locations which feature an established 'born and bred' population along with incoming sea- and tree-changers, and it makes 'one size fits all' retrofit options problematic for older people. Communicating the nature of bushfire hazards and ways to retrofit housing for greater resilience to this diverse population requires multiple channels and points of connection, and diverse methods of presenting and engaging people with information.

12.4 Ageing in Place versus Aged Care and Other Options

Australian government policy has been focused on enabling older Australians to age in place in their community, either in their own home or in specialised retirement accommodation. Most older Australians do remain living independently in the community, with assistance when needed from a family member or home support, and/or home care services (which may be Commonwealth funded). Just 185,000 older Australians (4% of the older Australian age cohort) live in residential aged care, mostly in major cities (69%) and regional centres (21%) as increasing regulation of residential aged care has resulted in a consolidation of operators and of facilities (Australian Institute of Health and Welfare, 2021). Residential aged care facilities are required to conform to high building standards and operators are responsible for the well-being of their residents. In the event of bushfire or other hazard, they are required by law to have appropriate procedures to ensure the well-being of their residents.

Older people living in the community who are owner-occupiers, are responsible for their own well-being and safety and that of their property and are therefore responsible for their own bushfire retrofitting actions. Those living in rental properties, however, must rely on landlords to undertake this retrofitting. Specialised retirement housing is more complicated: in retirement villages the dwelling is the responsibility of the operator, but in land lease communities (manufactured home estates and caravan parks) the responsibility for retrofitting depends on who owns the dwelling, which can be either the resident or the operator. In the latter type of housing, any retrofitting may not be feasible due to the type of construction.

12.5 Housing Typologies

This section of the chapter provides a brief review of the major housing styles used in Australia, and their vulnerabilities to bushfire risk. Australia is large place, and its different climate zones and market conditions, as well as personal taste, have influenced the use of materials and the uptake of different styles in different parts of the country through the decades. Housing styles in different areas over time have also been affected by the skills of the local labour force, and the techniques and materials they are familiar with. It is worth noting also that even though new homes may meet building codes when built, owner-builders may conduct undocumented alterations which may impact fire resistance.

Australia's housing begins with Indigneous, pre-colonial architecture, which ranged from temporary shelters to sophisticated dome-like shelters and included elaborately decorated stone dwellings and caves, in temporary, semi-permanent and permanent forms (Memmott and Go-Sam, 1999).

Following colonisation from 1788 to 1840, a variety of settlement architecture developed. Convicts used outdoor spaces or preliminary sheds and shelters, free settlers used readily available tools, materials and techniques to construct their own homes, and the higher echelons of society used local sources and imported material to construct Georgian architecture, reflecting the style of houses in the United Kingdom. Commonly used materials in the construction of such buildings were lime plaster, reeds, clay, wood, brick and sandstone (Willis and Goad, 2008). These materials could have varying degrees of effectiveness against fire. The parts of the buildings that used more fire-resistant materials such as brick and stone, including foundations, chimneys and loadbearing walls are more resilient against fire effects.

Early Victorian architecture followed from 1840 to 1860, representing and replicating vernacular UK architecture during the same period. In the construction of early Victorian houses timber shingles, brick, weatherboards, natural slate, wooden picket fencing and corrugated iron were commonly used. Early Victorian was followed by Gothic Victorian around 1850–1880, which was inspired by the European Gothic style of the 12th and 15th centuries. This style of housing was known for its use of post-Gothic romanticised features and elements such as steeply pitched gables and asymmetrical design with larger windows. Commonly available material on site would determine the materiality of such buildings: for example, sandstone was used in Sydney with very rich timber detailing, and bluestone was used frequently in Melbourne. Brick was also used in this typology of housing, however, this was usually rendered and lined to mimic a masonry style (Kelly, 1996;

MacMahon, 2001; Smith, 1969). Similar to Colonial Georgian buildings, the degree of fire-resiliency of this typology of buildings depends on the construction methods, materials used in the building and any successive updates and alterations made through the years.

Queenslander homes originated in the 1840s and are an iconic and distinctive local style designed to respond to the climate of the north and north-eastern part of Australia. Queenslander homes are set high on timber stumps, or stilts, to allow for the free movement of floodwater and humid air. They are single-storey, detached buildings built out of timber weatherboards, with steeply pitched roofs constructed from tiles, slate and corrugated steel and with wide wrap-around verandas to provide shading and passive cooling (Bajracharya, Kumarasuriyar and Demirbilek, 2003; King, 2010; Naylor, 2010). As timber is the main construction material in Queenslanders, the fire risk for this housing type is very high.

The mid-Victorian style, established during 1860–1875, was another version of the Victorian home style, but with more decorative brickwork. Verandas were constructed with timber, with cast iron lacework ornamentation and decoration. Corrugated iron or terracotta tiles were mainly used in the construction of roofs (Kelly, 1996; MacMahon, 2001; Smith, 1969). Similar to the early Victorian styles, the fire-resiliency of this type of building is determined by the type of materials and construction methods that were used in the construction of particular houses.

The late-Victorian style of architecture featured more ornate, decorated architecture in both freestanding and terrace homes, with square-edged weatherboards and rendered or textured brickwork used for construction. Verandas were highly decorated with cast iron lacework, filigrees and complex tiling, but towards the end of this era the decoration moved towards timber fretwork (Kelly, 1996; MacMahon, 2001; Smith, 1969). Both weatherboards and the decorative wooden fretwork contribute to the fire risk for these houses.

Homes in the Queen Anne style (1895–1910) were constructed using weatherboard or brickwork, with steeply pitched roofs of terracotta tiles. The facades of these homes were highly decorated with gargoyles, finials, post-brackets, veranda posts and chimney cornices (Saunders, 1969). The high level of woodwork and decoration in this type of building creates a high level of fire risk.

The Federation architectural style, dating to post-federation of Australia in 1901, includes elements from Victorian and Queen Anne typologies, with predominantly brick walls and steeply pitched roofs of terracotta tiles, slate or corrugated iron, and verandahs with wooden fretwork (Saunders, 1969; Williamson, 2019). The use of ornate timber elements increases the risk of fire for these homes.

A type of housing which became very popular during the 1920s–1930s was the Californian Bungalow. These buildings were usually single-storey and were constructed with different materials depending on where they were built. For example, brick was common in Sydney and Melbourne, limestone in South Australia and weatherboards in Queensland. Asbestos was commonly used in these homes as a fire retardant, which ironically creates additional hazards in the event of fire, with asbestos fibres complicating fire suppression and cleanup operations. The roofs were sloped and usually had a dormer window as the focal point. A wide veranda with a winding path towards the door is a feature of this style. The windows and doors were made of timber and usually detailed in contrasting colours (Butler, 1992; Clare, 1986). The degree of fire resilience in this type of architecture is dependant on design features, such as the decorative embellishments and woodwork, and the style and materials of the roof, which in Australia is commonly ceramic tile or galvinised iron.

Interwar architecture was influenced by the financial consequences of World War I and the effects of the Great Depression. After the end of World War II a lack of materials, equipment and skills meant that houses were mainly manufactured with cheap materials and on a mass scale to respond to the high demand of housing. During these periods, as well as the Californian bungalow style, several other architectural styles such as Streamline Moderne, Art Deco and Spanish Mission became popular. These homes were generally single-storey and had a heavy and solid appearance, using non-organic and geometric forms and shapes. Red brick was the primary construction material for such buildings; however, weatherboard, rendered and roughcast variants can also be seen. They had a double-hung windows and hipped roofs (Edquist, 2008). These houses commonly had an L plan shape and were constructed using double, or triple, brick veneer. During this time garages were implemented into the housing plan (Hogben, 2012). During the 1960s, and with the growing influence of modernism, the entire concept of house and home was reinvented. Common features of housing from this period are flat roofs, sharp angular lines and floor-to-ceiling windows, with reinforced concrete, concrete blocks, brick, steel and metal frequently being used in their construction (Hogben, 2012; Smith, 2008). These materials are generally non-combustible, but the large amount of glass and exposed steelwork increases the risk of fire.

With increased urban density many Australians now live in apartments in high-rise city towers and low-rise suburban developments. Although there is more diversity of architectural style in the Australian housing market to suit different lifestyles and environments, more recent trends in design have been for smooth finish building which may be brick with a timber frame behind, or concrete. These are typically all rendered so look the same, but they have different fire responses. In the 1990s oversized suburban homes became popular (Nastasie, 2020), however, more recently they have been criticised, often due to their inherent energy inefficiency. Australian homes from the 1990s onwards are typically large double-storey buildings with separate living zones for parents and children and open plan design prevailing (Milson and Milson, 2018). Housing lots in this period shrank and less outside space is provided, potentially enabling easier spread of fire from one property to another.

So-called Classical Style properties and the Pavilion-style home, popular since 2000, also typically have rendered or concrete external finishes with large windows (Soho, 2022). The Pavilion-style home has a primary rectangular, box-shaped style and an open plan layout with glass replacing much of the external wall. Windows are often steel-framed, and due to the transparency of the walls, this design is popular for providing privacy or a view. Roofs are frequently low-pitched with a skillion. Again the large amount of glass increases fire risk, and the different wall materials behind the rendering have differing fire responses.

Increasing opportunities to travel and the ability to promote both architectural plans and images of new housing styles across the country have homogenised Australian housing in recent decades, with local architectural adaptations to climate being replaced by powered heating and air conditioning. More attention has been paid to materials in recent years as it is known that poorly designed and constructed homes result in inefficient energy use contributing to climate change (Milson and Milson, 2018), however, fire and flammability qualities of the designs are rarely mentioned in design sources.

12.6 Other Climate Change Risks to Housing

For many regions in Australia which have experienced catastrophic bushfire, other climate change-related hazards are also contributing to weakening housing resilience, such as increased frequency and intensity of flooding, cyclones, storm events and heat waves, and these cannot be seen in isolation from one another. Between 2004 and 2014, however, bushfires have been found to be the most common disasters in the state of NSW, being responsible for 108 disaster declarations, followed by 55 storm declarations and 44 flood declarations (Sewell et al., 2016). Heat waves and drought often trigger the fear of bushfire, and it is both anecdotally and scientifically true that these events often precede a high-risk bushfire season (Jyoteeshkumar reddy et al., 2021a, 2021b).

When it comes to recommending retrofits that aim to address human comfort and safety, there is a tension between trying to simultaneously counteract the effects of heat and to protect a house from fire or flood. For example, a significant amount of research recommends the reduction of the urban heat island effect by introducing green infrastructure close to buildings. Bushfire building protection approaches, however, recommend the reduction of trees and vegetation near buildings. This is observed in the development of new suburbs where bushfire asset protection standards lead developers to seek lowered Bushfire Attack Level (BAL) ratings through a reduction in vegetated areas instead of integrating green areas within and around housing. Complications thus arise in determining measures to counteract heat, drought and floods at the same time as implementing bushfire protection measures for housing.

To explain this multi-hazard building conundrum further, it is useful to view the case study of the Black Summer Bushfires in terms of what happened before and afterwards. The Black Summer Bushfires were caused by an extended period of heat and drought associated with a combination of global weather pattern factors which caused extremely dry and hotter than average ground conditions, leading to loss of soil moisture. Immediately following the Black Summer Bushfires, a period of extreme rain events and flooding in 2021 and 2022 caused some of the areas that experienced bushfires to then experience flood disasters. Due to the atmosphere being capable of holding more water in warmer temperatures, in the words of Bureau of Meteorology sources: 'when it does rain, we're seeing more high-intensity, extreme rain events – the ones that are associated with flash flooding' (ABC News Story Lab, 2020). This has led to the realisation that an 'all-hazards approach' is appropriate (NSW Government, 2022).

This, compounded with the ageing post-war housing stock, means that homes built prior to the introduction of bushfire, heat wave and cyclone-resistant building requirements by the Building Code of Australia are at higher risk of having multiple hazards compromise housing safety and resilience. A higher number of older Australians are likely to be living in this older housing stock and to have mobility and chronic health issues. Roughly 50% of people who died outside their home in bushfires were within 50 m of their home (Leonard, 2022), indicating that it is vital to design the surroundings of homes to improve the safety of residents in the event of bushfire and to provide safe places for shelter near to the home. Evidence shows that the broad vegetation types located within neighbourhood proximity can have a significant impact on the likelihood of housing resilience to bushfire attack (Tolhurst, 2022).

To compound this problem, bushfires and other hazards are statistically more prevalent in areas of disadvantage (Paddam, Liu and Philip, 2022) and within these areas disadvantaged groups are more adversely affected by these hazards (ACOSS, 2020). For example, using subsidised housing as an indicator of socioeconomic disadvantage, Gabbe and Pierce (2020) found that subsidised housing has the most sensitive populations and barriers to adaptation in regard to extreme heat when it comes to housing types. But it is not only the socioeconomically disadvantaged who face such challenges. Some areas that have previously *not* been areas of disadvantage but, on the contrary, have been attractive to those who move when of older age – coastal and warmer climates – are now seen to be more prone to increasing natural disaster issues related to climate change (Bukvic et al., 2018; Molinksy and Forsyth, 2023). To help communities prepare for disasters, the causes and effects of disadvantage need to be addressed (Dominey-Howes et al., 2016).

Molinsky and Forsyth (2023) explain how, within the home setting, 'vulnerability to climate change is shaped by exposure to climate-related stressors, the sensitivity of a group or individual to those particular stressors, and people's capacity to adapt and respond'. Susceptible regions may be increasingly facing exposure to health-threatening incidents caused by stressors such as heat, insect- and water-borne illness and lack of access to safe food and water. Research from the United States where such effects have already been reported supports these statements (Balbus et al., 2016; Crimmins et al., 2016; Gabbe and Pierce, 2020; Molinksy and Forsyth, 2023). As Costello et al. state (2009), the effects of climate change 'are changing patterns of disease and mortality, extreme events, food, water, shelter, and population'. The possibility of needing temporary shelter as housing becomes more vulnerable to climate-related disasters may be particularly problematic for older people, with the possibility of overcrowding (increasing the chance for spreading infectious diseases), and being located in structures that are problematic for people with mobility difficulties, or which may separate older adults from their social support networks (WHO, 2008). People's sensitivity and ability to adapt to these types of stressors will vary with their circumstances and personal characteristics. For instance, if their home has in-built bushfire-resistant features (a feature they may be able to control), then exposure to bushfire can be reduced. The surrounding and regional vegetation biome is, however, unlikely to be something they can control, and if it consists of flammable vegetation and is not managed to control fuel loads and continuous fire paths, then exposure to risk may be higher. In terms of personal characteristics, older people may have more health-related hindrances, but they can also often demonstrate a high capacity for resilience by drawing on previous experience. Conversely, successful adaptation in the past can lead them to underestimate their vulnerabilities and fail to take action to mitigate risk.

12.7 Personal Heritage

Current insurance and building practices assume that people want merely to set the clock back to before the disaster, and that this can best be done by just replacing affected material. But the fabric of people's homes and possessions embodies their personal heritage and identity, and that of their families, and merely replacing it does not erase the impact of the disaster. In fact, it may erase things that can help them come to terms with the disaster (Kousa and Pottgiesser, 2020), as well as embrace change both on a societal and personal level.

The ABC series 'People's Republic of Mallacoota' (2022), which documented the recovery process for a small community affected by the Black Summer Bushfires, revealed that although painful, the process of sorting through the rubble of their burnt houses to find items of personal significance became a transformational experience for those who had lost everything, helping them to move on to a new stage of their lives. Items that carry a record of damage caused by an event can act as witnesses to the past (a concept embedded in Article 7 of the Venice Charter (ICOMOS, 1964), and tangible items from the past can become touchstones for processing change and negotiating decisions – what to keep, what to let go, what to renew or to replicate (Jones, 2010). In the community of Nerriga, also burnt in the Black Summer Bushfires, processes of cleaning and repair in homes and community buildings that had been damaged rather than destroyed, brought people together and helped them to regain a sense of control over their lives and environment (personal communication, Sarah Horobin, Nerriga, 18.02.2022). These activities of tending and sharing are empowering and stand in opposition to feelings of despair and helplessness felt by people who see the fabric of their lives recategorised as rubbish and carted away (Denhart, 2009).

In both disaster planning and disaster recovery, there is also huge potential for the circular economy to enter into dialogue with the historic environment and built forms that people inhabit. One pressing issue is the need for heritage professionals to engage with 'more recent material legacy – previously deemed to have little or no heritage value' (Ross and Angel, 2020), which includes the ordinary fabric of people's homes. Another issue is the need to acknowledge the benefits of partial preservation and recontextualisation – the reuse of building materials in new ways and places, a practice that was traditional in earlier times, but which has become negatively recast in formal heritage doctrines as parasitic actions that reduce authenticity (Huuhka and Vestergaard, 2020). Partial preservation is particularly relevant to post-disaster rebuilding, both where materials can be preserved in situ, and even where homes are reduced to rubble, as the saving and reuse of materials can create jobs (with flow-on benefits in skills development) and engender pride and a sense of healing through the sale of, or gifting, of materials to others (Arlotta, 2020; Denhart, 2009).

When it comes to personal implementation of practices of material retention, however, dealing with legislative, bureaucratic and insurance regulations can be obstructive and demoralising (Dastgerdi et al., 2020) and can also be compounded by decision fatigue. Therefore, adequate social support becomes a key factor influencing how well circular economy principles might be able to be implemented in practice, and how much they can contribute to processes of recovery and planning for future resilience (Bihari and Ryan, 2012). Understanding that 'heritage values can be located not just in a building or even a building fragment, but in the dynamic process of reuse and reconfiguration' (Arlotta, 2020) offers the opportunity to ground decisions about the future through positive valuations of the tangible remains of personal pasts.

12.8 Benefits of Retrofitting for Enhanced Sustainable Resilient Housing

This section defines the terms 'retrofit' and 'resilience' before outlining some of the features of sustainable housing and the concept of 'future proofing', which is the process of taking action in the present to prevent future damage.

Retrofitting is the practice of altering something after it has been manufactured or built: in this case, housing. There are different classifications of retrofit, from light to deep retrofit. Deep retrofit is associated with extensive energy upgrades to the building fabric and envelope and potentially offers an opportunity to consider bushfire resilience measures. Typically, only 1–2% is added to the total stock of buildings each year (Ravetz, 2008), so a focus entirely on new building will take decades to have an impact on the resilience of the total building stock. Building retrofits have the potential to enhance the resilience of existing building stock more quickly and effectively (Sayce et al., 2022).

Retrofitting housing involves changing its systems or structure after the initial construction and occupation (Dixon and Eames, 2013). Retrofitting can restore functionality, improve the amenities of the building and improve its technical performance environmentally and economically. As technology evolves over time, building retrofits can significantly reduce energy and water usage and improve sustainability performance. Further, given the impacts of climate change and rising temperatures, retrofitting may be essential if the building is to remain thermally functional. This has been brought sharply into focus, as weather patterns change more rapidly than previously predicted, making some buildings constructed only a few years ago already require technology upgrades to combat extreme heat. There are several terms used in different countries to describe retrofit; these include refurbishment, renovation, adaptation and tune-ups (Wilkinson and Sayce, 2022, in Sayce et al., 2022).

Resilience is the capacity to recover quickly from difficulties. Defining resilient housing is challenging as it is a comparatively new addition to the real estate lexicon. For some it can be interpreted using social, economic and environmental lenses (Nicol and Knoepfel, 2014) where the capacity of occupants is a focus. Others consider the physical building fabric and envelope most important (Attia et al., 2021), where the impacts of excessive heat and fire and how to reduce them is the focus. In essence, resilience embodies the concept of future proofing; that is to take action now to make the property less vulnerable to damage and to be more comfortable to occupy.

It is important to consider the characteristics of sustainable housing, which is viewed in terms of energy efficiency and/or water-saving technologies (Tahsildoost and Zomorodian, 2020). Other definitions focus on materials and embodied carbon to reduce the whole life footprint of the building (Sahagun and Moncaster, 2012). Further definitions of sustainable buildings extend to user behaviours which exercise considerable and variable impacts on actual building performance (Wilkinson and Zajelska-Jonsson, 2021). Whilst some aspects of 'sustainability' can be built in through design and original construction, the process of retrofitting is key to climate targets, given that most buildings were neither built with carbon neutrality in mind, nor are they future-proofed.

The benefits therefore of retrofitting for enhanced, sustainable, resilient housing is that the building is future-proofed against extreme weather events affecting its locality, including bushfire, sea level rise and/or flooding. The physical fabric and envelope of the building should be upgraded to enhance fire protection giving the occupants a sense of enhanced protection and safety (Bell, 2020; Penman et al., 2017). Whilst undertaking these improvements, it should be ensured that sustainability criteria are adopted to reduce environmental impacts. This includes the specification of environmentally friendly materials. The physical retrofit improvements may also reduce operating costs and greenhouse gas emissions from the building (Beccali et al., 2013). In this way economic, social and environmental benefits accrue to the occupants. Clearly the decisions made regarding the degree of vulnerability and risk to the property, what measures are possible to take, and the budget available to the owner all determine what measures, if any, are adopted.

12.9 Developing a Bushfire Retrofitting Toolkit

The research discussed above is now being used to develop a toolkit to assist older Australians to make decisions about retrofitting their homes to increase their ability to survive bushfire and to recover well afterwards. The project draws on initial work completed for the Canberra Regional Joint Organisation (2021), in collaboration with Resilience NSW, Minderoo Foundation and Edge Environment, and extends that early work to a wider range of climate environments (coastal temperature Bega Valley Shire and coastal sub-tropical Noosa Shire), with a specific focus on the needs of older Australians.

Titled *Retrofit Toolkit for Bushfire Resilience: Making it work for older people*, the new toolkit aims to provide users with a decision-making pathway that takes into account their lifestyles and preferences, their physical and cognitive capacities and their financial position. An understanding of these lifestyle aspects is critical to guiding people towards retrofitting actions that are compatible with their values, aspirations and capacities, and which are therefore mostly likely to gain their acceptance and be acted upon. Importantly, the toolkit recognises the many other resources that are being developed in the wake of the 2019 bushfires and other recent climate disasters in Australia and provides links to the resources that are emerging as most effective and most relevant to older Australians.

12.10 Conclusions

The risk, frequency and severity of bushfire events in Australia is increasing. Not only is the natural environment damaged during bushfires but also the built environment, including housing, and the fire damage inflicted on buildings is predicted to increase in scale and severity. Older people have been shown to be a particularly vulnerable group, not only in terms of their physical and cognitive ability to protect their property but also in terms of their financial capacity to pay for retrofits to enhance fire protection.

The capacity to recover from the physical and emotional trauma and loss associated with bushfires is correlated to people's levels of education about, and preparation for, bushfires, and their capacity to repair their properties. To maintain a habitable environment in the face of increased bushfire risk, preventative actions should be undertaken during planned maintenance and property upgrades to minimise future fire damage, or should be implemented during remediation following fire damage. Prioritising and paying for protective measures, however, can be a challenge, both because the ability of the homeowner to fund the protective measures varies, and because there is little information to help homeowners decide on the best options for their circumstances. Research is being undertaken to work with local contexts and stakeholders to develop and trial a practical toolkit, empowering people to take effective proactive steps to protect their homes, their loved ones and their personal heritage.

References

ABC. (2022). *People's Republic of Mallacoota*. iView: Australian Broadcasting Corporation. https://iview.abc.net.au/show/people-s-republic-of-mallacoota

ABC News Story Lab. (2020). *How Extreme Conditions Drove Australia's Record Bushfire Disaster*. Australian Broadcasting Corporation. https://www.abc.net.au/news/2020-02-19/australia-bushfires-how-heat-and-drought-created-a-tinderbox/11976134

ACOSS. (2020). Climate Change Disproportionately Impacts People Experiencing Poverty and Disadvantage. In *Presentation to the St Vincent DePaul NSW*. ACOSS, Sydney.

Arlotta, A. (2020). Locating Heritage Value in Building Material Reuse, *Journal of Cultural Heritage Management and Sustainable Development, 10*(1), 6–15, DOI: 10.1108/JCHMSD-06-2019-0076

Arriagada, N., Palmer, A., Bowman, D., Morgan, G., Jalaludin, B. and Johnston, F. (2020). Unprecedented Smoke-Related Health Burden Associated with the 2019–20 Bushfires in Eastern Australia, *Medical Journal of Australia 213*(6), 282–283.

Attia, S., Levinson, R., Ndongo, E., Holzer, P., Kazanci, O. B., Homaei, S., Zhang, C., Olesen, B. W., Qi, D., Hamdy, M. and Heiselberg, P. (2021). Resilient Cooling of Buildings to Protect against Heat Waves and Power Outages: Key Concepts and Definition, *Energy and Buildings*, *239*, 110869.

Australian Bureau of Statistics. (2021). *Census of Population and Housing*. ABS, Canberra.

Australian Disaster Knowledge Resilience Hub. (2023). *Australian Institute for Disaster Resilience*. Available at https://knowledge.aidr.org.au/

Australian Disaster Resilience Hub. (2023). *Bushfire New South Wales December 1974.* https://knowledge.aidr.org.au/resources/bushfire-new-south-wales-1974/

Australian Institute of Health and Welfare. (2021). *Older Australians.* Australian Government. DOI: 10.25816/5ec5bda5ed178

Bajracharya, B., Kumarasuriyar, A. and Demirbilek, N. (2003). Vernacular Architecture in Queensland: Current Planning Issues and Opportunities (pp. 412–422). Paper presented at *International Conference on Quality of Urban Life: Policy Versus Practice*, Istanbul Technical University Urban and Environmental Planning and Research Center, Istanbul, Turkey.

Balbus, J., Crimmins, A. and Gamble, J. (2016). *Climate Change and Human Health.* A Bold Blend of News and Ideas, Brewminate. https://brewminate.com/climate-change-and-human-health/

Banihashemi, S. and Zarepour Sohi, S. (2022). *Data-Centric Regenerative Built Environment: Big Data for Sustainable Regeneration*, London, Routledge. https://www.iqair.com/au/australia/act/canberra

Beccali, M., Cellura, M., Fontana, M., Longo, S. and Mistretta, M. (2013). Energy Retrofit of a Single-Family House: Life Cycle Net Energy Saving and Environmental Benefits, *Renewable and Sustainable Energy Reviews*, *27*, 283–293.

Bell, N. (2020). Beating Bushfire: Retrofitting for Safer Homes in Fire-Prone Areas, *Sanctuary: Modern Green Homes*, *53*, 68–72.

Biddle, N., Edwards, B., Herz, D. and Makkai, T. (2020). *Exposure and the Impact on Attitudes of the 2019–20 Australian Bushfires.* Australian National University Centre for Social Research and Methods. http://dx.doi.org/10.26193/S1S9I9

Bihari, M. and Ryan, R. (2012). Influence of Social Capital on Community Preparedness for Wildfires, *Landscape and Urban Planning*, *106*(3), 253–261, ISSN 0169-2046, https://doi.org/10.1016/j.landurbplan.2012.03.011

Bukvic, A., Gohlke, J., Borate, A. and Suggs, J. (2018). Aging in Flood-Prone Coastal Areas: Discerning the Health and Well-Being Risk for Older Residents, *International Journal of Environmental Research and Public Health*, *15*(12), 2900, DOI: 10.3390/ijerph15122900. PMID: 30567352; PMCID: PMC6313428.

Butler, G. (1992). *The Californian Bungalow in Australia: Origins, Revival, Source Ideas for Restoration.* Lothian Books.

Canberra Regional Joint Organisation. (2021). *Your Bushfire-Resilient Home. A Toolkit for Residents.* https://www.edgeimpact.global/case-studies/bushfire-resilient-homes-toolkit

Clare, J. (1986). The Californian Bungalow in Australia, *Historic Environment*, *5*(1), 19–39.

Costello, A., Abbas, M., Allen, A., Ball, S., Bell, S., Bellamy, R., Friel, S., Groce, N., Johnson, A., Kett, M., Lee, M., Levy, C., Maslin, M., McCoy, D., McGuire, W., Montgomery, H., Napier, D., Pagel, C., Patel, J., Antonio Puppim de Oliveira, J., Redclift, N., Rees, H., Rogger, D., Scott, J., Stephenson, J., Twigg, J., Wolff, J. and Patterson, C. (2009). Managing the Health Effects of Climate Change: Lancet and University College London Institute for Global Health Commission, *The Lancet*, *373*(9676), 1693–1733, https://doi.org/10.1016/S0140-6736(09)60935-1

Crimmins, A. et al., (Eds.) 2016. *The Impacts of Climate Change on Human Health in the United States: A Scientific Assessment A.* Washington, DC: U.S Global Change Research Program. http://health2016.globalchange.gov

Damany-Pearce, L., Johnson, B., Wells, A. *et al.* (2022). Australian Wildfires Cause the Largest Stratospheric Warming since Pinatubo and Extends the Lifetime of the Antarctic Ozone Hole, *Scientific Reports 12*, 12665, https://doi.org/10.1038/s41598-022-15794-3

Dastgerdi, A. S., Stimilli, F., Pisano, C., Sargolini, M. and De Luca, G. (2020). Heritage Waste Management: A Possible Paradigm Shift in the Post-Earthquake Reconstruction in Central Italy, *Journal of Cultural Heritage Management and Sustainable Development*, *10*(1), 76–89, DOI: 10.1108/JCHMSD-07-2019-0087

Denhart, H. (2009). Deconstructing Disaster: Psycho-Social Impact of Building Deconstruction in Post-Katrina New Orleans, *Cities*, *26*(4), 195–201, DOI: 10.1016/j.cities.2009.04.003

Dixon, T. and Eames, M. (2013). Scaling Up: the Challenges of Urban Retrofit, *Building Research & Information*, *41*(5),499–503.

Dominey-Howes, D., Bruce, E., Stephens, R. and Perkins-Kirkpatrick, S. (2016). *Natural Disasters Are Affecting Some of Australia's Most Disadvantaged Communities.* The Conversation, November 8, 2016. Accessed 1st October 2024 from https://theconversation.com/natural-disasters-are-affecting-some-of-australias-most-disadvantaged-communities-68165

Edquist, H. (2008). Reading Modernity: Architecture and Urbanism in the Interwar Australian Novel, *Fabrications*, *18*(2), 50–69.

Gabbe, C. and Pierce, G. (2020). Extreme Heat Vulnerability of Subsidized Housing Residents in California, *Housing Policy Debate*, *30*(5), 843–860, DOI: 10.1080/10511482.2020.1768574

Handmer, J., Ladds, M. and Magee, L. (2016). *Disaster Losses from Natural Hazards in Australia, 1967–2013.* RMIT University, Melbourne.

Hogben, P. (2012). Architecture and Arts and the Mediation of American Architecture in Post-War Australia, *Fabrications*, *22*(1), 30–57.

Horton, G., Hanna, L. and Kelly, B. (2010). Drought, Drying and Climate Change: Emerging Health Issues for Ageing Australians in Rural Areas, *Australasian Journal on Ageing*, *29*(1), 2–7.

Huuhka, S. and Vestergaard, I. (2020). Building Conservation and the Circular Economy: A Theoretical Consideration, *Journal of Cultural Heritage Management and Sustainable Development*, *10*(1), 29–40, https://doi.org/10.1108/JCHMSD-06-2019-0081

ICOMOS. (1964). *International Charter for the Conservation and Restoration of Monuments and Sites (The Venice Charter (1964).* ICOMOS – International Council on Monuments and Sites, Venice, https://www.icomos.org/images/DOCUMENTS/Charters/venice_e.pdf

iQair (2020). *Air Quality in Canberra.* https://www.iqair.com/au/australia/act/canberra

Jones, S. (2010). Negotiating Authentic Objects and Authentic Selves: Beyond the Deconstruction of Authenticity, *Journal of Material Culture*, *15*(2), 181–203, DOI: 10.1177/1359183510364074

Jyoteeshkumar reddy, P., Sharples, J., Lewis, S. and Perkins-Kirkpatrick, S. (2021a). Modulating Influence of Drought on the Synergy between Heatwaves and Dead Fine Fuel Moisture Content of Bushfire Fuels in the Southeast Australian Region, *Weather and Climate Extremes*, 31, 100300, https://doi.org/10.1016/j.wace.2020.100300

Jyoteeshkumar reddy, P., Sharples, J. and Perkins-Kirkpatrick, S. (2021b). How Heatwaves and Drought Combine to Produce the Perfect Firestorm. *The Conversation.* https://theconversation.com/how-heatwaves-and-drought-combine-to-produce-the-perfect-firestorm-153890

Kelly, I. (1996). A Tale of Two Cities: The Impact of Victorian Immigrants on the Architecture of Perth (1895–1905), *Fabrications*, *7*(1), 51–76.

King, S. A. (2010). *Colony and Climate: Positioning Public Architecture in Queensland, 1859-1909.* University of Melbourne.

Korda, R. J., Paige, E., Yiengprugsawan, V., Latz, I. and Friel, S. (2014). Income-Related Inequalities in Chronic Conditions, Physical Functioning and Psychological Distress among Older People in Australia: Cross-Sectional Findings from the 45 and Up Study, *BMC Public Health*, *14*, 1–10.

Kousa, C. and Pottgiesser, U. (2020). Post Syrian-War Material Recovery, Reuse and Transformation in the Old City of Aleppo, *Journal of Cultural Heritage Management and Sustainable Development*, *10*(1), 90–103, DOI: 10.1108/JCHMSD-07-2019-0085

Leonard, J. (2022). *Your Vegetation: What You Need to Know.* Bushfire Resilience Inc. webinar 10th August 2022, https://bushfireresilience.org.au/webinar-recordings/2022-webinar-2/

Lucas, C., Hennessy, K., Mills, G. and Bathols, J. (2007). *Bushfire Weather in Southeast Australia: Recent Trends and Projected Climate Change Impacts.* Bushfire Cooperative Research Centre, CSIRO, Australia.

MacMahon, B. (2001). *The Architecture of East Australia.* Edition Axel Menges: Fellbach, Germany.

Memmott, P. and Go-Sam, C. (1999). Australian Indigenous Architecture: Its forms and Evolution. *Thresholds: Papers of the Sixteenth Annual Conference of the Society of Architectural Historians Australia and New Zealand*, Launceston and Hobart, TAS, September.
Milson, C. and Milson, P. (2018). *The Evolution of Australian Home Building*. Accessed on 3rd January 2024 from https://www.architectureanddesign.com.au/features/comment/the-evolution-of-australian-home-building
Molinsky, J. and Forsyth, A. (2023). Climate Change, Aging, and Well-Being: How Residential Setting Matters, *Housing Policy Debate*, *33*(5), 1029–1054, DOI: 10.1080/10511482.2022.2109711
Nastasie, D. (2020). *The Evolution of Australian Housing*. Defence Housing Australia (DHA). Accessed on 1st October, 2024 from https://www.dha.gov.au/investing/property-news/property-news-and-events/2020/08/21/the-evolution-of-australian-housing
Naylor, S. (2010). The Ethos of 'the Queenslander': A Journey into the Art of Building in North Queensland, *ETropic: Electronic Journal of Studies in the Tropics*, *9*, https://doi.org/10.25120/etropic.9.0.2010.3424
Nicol, L. A. and Knoepfel, P. (2014). Resilient Housing: A New Resource-Oriented Approach, *Building Research & Information*, *42*(2), 229–239.
NSW Government. (2022). *Independent Flood Inquiry*. NSW Government, Sydney New South Wales. https://www.nsw.gov.au/nsw-government/projects-and-initiatives/floodinquiry
Paddam, S., Liu, C. and Philip, S. (2022). *Home Insurance Affordability and Socioeconomic Equity in a Changing Climate*. Actuaries Institute with Finity Consulting, Sydney.
Penman, T. D., Eriksen, C., Horsey, B., Green, A., Lemcke, D., Cooper, P. and Bradstock, R. A. (2017). Retrofitting for Wildfire Resilience: What Is the Cost?, *International Journal of Disaster Risk Reduction*, *21*, 1–10.
Ravetz, J. (2008). State of the Stock—What Do We Know about Existing Buildings and Their Future Prospects?, *Energy Policy*, *36*(12), 4462–4470.
Ross, S. and Angel, V. (2020). Heritage and Waste: Introduction, *Journal of Cultural Heritage Management and Sustainable Development*, *10*(1), 1–5, https://doi.org/10.1108/JCHMSD-02-2020-116
Sahagun, D. and Moncaster, A. (2012). How Much Do We Spend to Save? Calculating the Embodied Carbon Costs of Retrofit. In: *Retrofit2012*, 24–26 Jan 2012, Salford, Manchester.
Saunders, D. (1969). Domestic Styles of Australia's Federation Period: Queen Anne and the Balcony Style, *Architecture in Australia (A)*, 58 (August 1969), 662.
Sayce, S., Wilkinson, S. J., Armstrong, G. and Organ, S. eds. (2022). *Resilient Building Retrofits: Combating the Climate Crisis*. Routledge, UK.
Sewell, T., Stephens, R., Dominey-Howes, D., Bruce, E. and Perkins-Kirkpatrick, S. (2016). Disaster Declarations Associated with Bushfires, Floods and Storms in New South Wales, Australia between 2004 and 2014, *Scientific Reports*, *6*, 36369, https://doi.org/10.1038/srep36369
Smith, B. (1969). Architecture in Australia, *Australian Historical Studies*, *14*(53), 85–92.
Smith, T. (2008). Modernism & Australia: Documents on Art, Design and Architecture 1917-1967, *Modernism/modernity*, *15*(2), 393–396.
Soho. (2022). *Australian Architecture in Homes through the Decades*. Accessed on 3rd January 2024 from https://articles.soho.com.au/australian-architecture-in-homes-through-the-decades
Tahsildoost, M. and Zomorodian, Z. (2020). Energy, Carbon, and Cost Analysis of Rural Housing Retrofit in Different Climates, *Journal of Building Engineering*, *30*, 101277.
Tolhurst, K. (2022). *Your Vegetation: What You Need to Know*. Bushfire Resilience Inc. webinar 10th August 2022, https://bushfireresilience.org.au/webinar-recordings/2022-webinar-2/
Tuohy, R., Stephens, C. and Johnson, D. (2014). Qualitative Research Can Improve Understanding about Disaster Preparedness for Independent Older Adults in the Community, *Disaster Prevention and Management*, *23*(3), 296–308.
United Nations Climate Action. (2023). *Climate Action Fast Facts*. https://www.un.org/en/climatechange/science/key-findings. Accessed 18th December 2023.

Wilkinson, S. and Sayce, S. (2022). Chapter 2. The Philosophy and Definitions of Retrofitting for Resilience. In Sayce, S., Wilkinson, S. J. Armstrong, G. & Organ, S. eds. *Resilient Building Retrofits: Combating the Climate Crisis*. Routledge, UK.

Wilkinson, S. J. and Zajelska-Jonsson, A. (2021). Student Accommodation, Environmental Behaviour and Lessons for Property Managers, *Property Management*, DOI: 10.1108/PM-09-2020-0055

Williamson, E. (2019). Visualising the Architecture of Federation: Digital Media and Cultural Identity in Australian Architecture, *ARCC Conference Repository*, *1*(1), https://arcc-repository.org/index.php/repository/article/view/775

Willis, J. and Goad, P. (2008). A Bigger Picture: Reframing Australian Architectural History, *Fabrications*, *18*(1), 6–23.

World Health Organization (2008). *Older persons in emergencies: An active ageing perspective)*. https://extranet.who.int/agefriendlyworld/wp-content/uploads/2014/06/WHO-Older-Persons-in-Emergencies-An-Active-Ageing-Perspective.pdf

Part V

Roadmaps towards Sustainability

Goals, Policies and Legislations in the Building and Construction Sector

13 Green Finance

A Tool for Financing Green Building Projects

*Caleb Debrah**, *Albert P. C. Chan*, *Amos Darko**, *De-Graft Owusu-Manu*, *and Eric Ohene*

**Corresponding author.*

13.1 Introduction

Global buildings and construction output are projected to reach US$15.2 trillion by 2030 (Robinson *et al.* 2021). Yet, the operations and materials used in buildings and construction account for nearly 37 per cent of final energy consumption and global carbon emissions (United Nations Environment Programme [UNEP] 2022). While consuming about 50 per cent of all materials extracted, buildings- and construction-related processes also generate almost 50 per cent of annual solid waste (World Green Building Council [WorldGBC] 2021, Transparency Market Research 2022) – making the sector highly unsustainable. It is critical to reduce the negative impacts of buildings and construction on the climate and the environment to keep global warming under 1.5°C. Green building (GB) has emerged as a concept to promote sustainability in buildings and construction.

GB is "the practice of creating structures and using processes that are environmentally responsible and resource-efficient throughout a building's lifecycle" (US Environmental Protection Agency 2016). While current interest in GB is enormous, the money allocated to research and development is still minimal. Recent increase in GB and energy efficiency investment in buildings amount to just about US$237 billion of the global buildings construction sector value of US$6.3 trillion (UNEP 2022). This approximates to less than US$3 investment in GB for every US$100 spent on new construction (WorldGBC 2021). Again, it accounts for about 0.2 per cent (or US$193 million) of all federally funded research in the USA annually (Kubba 2012). This represents a potential for green finance (GF) investment for GB, which currently lacks the needed research and investment (Debrah *et al.* 2022a). GF has received increasing recent attention (Zhang *et al.* 2019) and provides opportunities to address the investment gap in GB (Debrah *et al.* 2022a).

GF refers to financial instruments that support the climate-resilient transition by funding activities that protect the environment such as emissions reductions, reduced energy use, and the development of climate-resilient infrastructure (International Capital Market Association [ICMA] 2020). It is reported that GF products such as green bonds linked to GB represent about 25 per cent of all the green bonds so far (Climate Bonds Initiative [CBI] 2015). Similarly, about 32 per cent of green loans are allocated to green real estate. With investment prospects of US$24.7 trillion

DOI: 10.1201/9781003317890-18

by 2030 across just emerging markets (International Finance Corporation [IFC] 2019), GF-for-GB has been touted as a multi-trillion investment opportunity in the next decade. This is due to the possibility that GF may increase demand for GB and building retrofits (UNEP 2021). To provide a comprehensive understanding of the concept of GF-for-GB, this chapter explores its development over time. The insights provided in this chapter will therefore inform key stakeholders of some critical issues pertaining to GF-for-GB and opportunities for future development.

13.2 What Is Green Building?

Conventional buildings consume considerable energy, produce a lot of indoor and outdoor air pollution, and waste significant amounts of water and materials (Kubba 2012). To overcome this, GB has emerged to make our buildings sustainable and environmentally friendly (Debrah *et al.* 2022a, 2022b). GB is also known as *sustainable or high-performance building* (US Environmental Protection Agency 2016). It describes a building that reduces or eliminates negative impacts in its design, construction, or operation and can create positive impacts on our climate and natural environment, economy, and society (Chan *et al.* 2018a, WorldGBC 2022a). GB improves design and building techniques, which leads to more affordable, durable, and productive buildings with better working conditions for occupants. More importantly, it protects our natural resources and improves the built environment so that the planet's ecosystems, people, enterprises, and communities can live a healthier and more prosperous life (Kubba 2012).

While GB is still growing, evolving practices and technologies vary across regions, countries, states, and cities. However, there are still a few core principles that hold true for all green projects, including siting, efficient structural design, efficient water and energy use, material selection, indoor environmental quality, efficient operations and maintenance, and minimisation of wastes and toxics (Kubba 2012). GB represents the commitment of the buildings and construction sector to transition to cleaner technology and building practices (Chegut *et al.* 2019). This is also consistent with the quest to keep global temperatures under 1.5°C through a 30 per cent reduction of energy intensity of buildings by 2030 (IFC 2019). Yet, inadequate finance and high upfront cost remain a critical challenge to the GB movement globally. These economic challenges of GB can be addressed with GF (He *et al.* 2022, Debrah *et al.* 2022a, 2022d). The succeeding section identifies some economic challenges encountered by GBs and how they could be resolved with GF.

13.3 Economic Barriers to Green Building Market; the Potentials for Green Finance

The barriers to the adoption of GB are widely studied. They include social and cognitive barriers, economic barriers, legislative and institutional barriers, technical and knowledge barriers, management and government-related barriers, and market barriers (Chan *et al.* 2017, Nguyen *et al.* 2017). Among the barriers, economic or cost-related barriers are known to highly influence GB adoption. Previous studies have shown that economic barriers include, but not limited to, higher cost of green technologies, high market prices, rental charges, long payback periods, lack of financing schemes (e.g., GF and bank loans), inadequate incentives for private investors, inadequate public funding, high inflation rate, and high approval cost due to high taxes and fees on developers (Chan *et al.* 2017, 2018a, Darko and Chan 2017, Adabre *et al.* 2020, Debrah *et al.* 2022a). For instance, due to climatic conditions, cost and availability of building materials, among others, the cost of building green varies across locations (IFC 2019). Existing reviews reveal that the additional cost of GB measures ranges between −4 and 21 per cent (Dwaikat and Ali 2016, Zhang *et al.* 2018), with varying payback periods. Others such as split incentives and short-termism challenge the potency of the economic viability of GB among developers and occupants (Zhang *et al.* 2018, Agyekum *et al.* 2020).

GF presents opportunities to overcome some of the economic challenges of GB identified above. For example, the long payback periods of GB align with the long-term investment principles of GF (Wang *et al.* 2021). In addition, GF also creates new investment opportunities or capital for environmentally conscious projects such as GB (Akomea-Frimpong *et al.* 2022). It also provides better credit terms for green projects (Volz 2018). Other GF drivers or success factors such as favourable economic (investment) returns, lower interest rates charges, access to global climate funds, availability of institutional investors, and government support are critical to overcoming the economic challenges of GB described above (Akomea-Frimpong *et al.* 2022, Debrah *et al.* 2022a, 2024). The subsequent section clearly distinguishes between conventional and GF for buildings and construction. The potentials and benefits of GF-for-GB are also discussed elsewhere in this chapter.

13.4 Difference between Conventional and Green Finance

All projects, especially buildings and construction, require funding and none can move forward without it (Merna and Njiru 2002). Financial instruments are tools that help raise money for projects. Debt, equity, and mezzanine financing are the traditional forms of long- or short-term financial instruments. *Debt instruments* refer to the raising of *term loans* from banks and other financial institutions (including commercial banks, merchant banks, investment banks, development agencies, pension funds, and insurance companies), debentures, and export credits. Term loans are arranged between the borrower (building and construction project developer, *in this case*) and financial institutions. There are two types of interest rates for loans: fixed and floating (Merna and Njiru 2002). Repayment of construction loans is relatively shorter, usually aligning with the duration of construction projects of three or more years for commercial construction projects (Ross *et al.* 2021). All other claims on the project's cash flow are subordinate to debt. On the other hand, common investors' ownership stake in the project is referred to as *ordinary equity*. *Mezzanine finance* sits in the middle of the equity and senior debt markets (Merna and Njiru 2002). Traditionally, bank loans and government appropriations are the major sources of finance for buildings and construction. Others include syndicated loans, asset securitisation, project revenue bonds, special local government bonds, and infrastructure equity funds (Zhang *et al.* 2021). In addition, self-financing is a major conventional source of finance for buildings and construction projects (Chiang and Cheng 2010). Generally, conventional/traditional finance provides short-term finance which solely focuses on return and risk (Schoenmaker 2017, Sachs *et al.* 2019).

By contrast, *GF* is structured to ensure a better environmental outcome (World Economic Forum 2020). It is a broad concept of *carbon financing*; to reduce GHG emissions, and *climate finance*; to mitigate and adapt to climate change (Noh 2019). Basically, GF funds green growth, and environmental cost prevention (Chien *et al.* 2021). It thus provides financial support for environment improvement, climate change mitigation, and more efficient resource utilisation (European Investment Bank [EIB] and Green Finance Committee [GFC] 2017). GF intends to increase level of financial flows (from banking, micro-credit, insurance, and investment) from the public, private, and not-for-profit sectors to sustainable development priorities (UNEP 2018). In addition, it focuses on operation and risk management for green projects such as GB (EIB and GFC 2017) and takes up opportunities that bring both a decent rate of return and environmental benefits and deliver greater accountability (UNEP 2018). With approximately US$2 trillion in volume to date, green/climate bonds, green loans, and other long-term markets have traditionally been the domain of GF (CBI 2022a). Other GF instruments are green investment funds, green insurance, green mortgage, green credit, and green securitisation which are described subsequently in this chapter. Generally, GF facilitates the formulation and implementation of green projects by providing investment, financing, operating funds, and other financial services (Ji and Zhang 2019). Comparatively, GF provides long-term financing for green projects which are considered riskier and in some cases with low returns (Sachs *et al.* 2019). In addition to the global bond/loan market, the estimated US$55 trillion global value of short-term debt markets provides a huge source of financing for the expansion of GF. These include green commercial paper, green revolving credit facilities, short-term bank loans for green projects, export letters of credit for green projects, green deposits, green structured notes, green repurchase agreements, green leases, green trade finance, green supply chain financing, stock/inventory financing, and green bank guarantees (CBI 2022a). Arguably, there are enormous potentials in both short- and long-term markets for GF.

13.5 Evolution of Green Finance for Green Building

Generally, GF-for-GB products range from climate-certified bonds or green bonds linked to green or low-carbon building, green commercial building loans, green construction loans, green insurance, green mortgages, green credit, and green securitisation for GB (IFC 2019, Noh 2019, Gholipour *et al.* 2022). After energy investments, GF-for-GB represents the second-largest use of proceeds. As shown in Table 13.1 and Figure 13.1, it accounts for US$449 billion (being 28 per cent) of the total volume of US$1.6 trillion by the end of 2021 (CBI 2022b). Table 13.1 illustrates a significant and continuous increase in the number of issuers and deals since 2014, indicating a growing interest in green investments for buildings. However, there are some interesting observations from these findings. First, as shown in

Table 13.1 Green finance for green building

Year	*Number of deals*	*Number of issuers*	*Amount issued (US$ billion)*
2014	122	31	8.2
2015	132	37	9.7
2016	256	60	18
2017	1335	84	49
2018	1316	133	49
2019	1403	216	81
2020	1164	245	80
2021	1305	433	147
Total	**7208**	**736**	**449**

Source: CBI (2022b).

Figure 13.1, there is a sharp spike in the number of deals and the amount of green bonds issued in 2017. This surge could be attributed to increasing investor interest in aligning their financial activities with the goals of the Paris Agreement Accord in 2015 and the UN Sustainable Development Goals. Conversely, the noticeable decline observed in 2020 could be attributed to the global COVID-19 pandemic. A substantial body of literature has examined how the pandemic triggered a worldwide economic and financial crisis, particularly in 2020, which had a negative impact on GF (Zeidan 2020, Pisani and Russo 2021, Yi *et al.* 2021). However, despite these short-term fluctuations, it is anticipated that green bonds associated

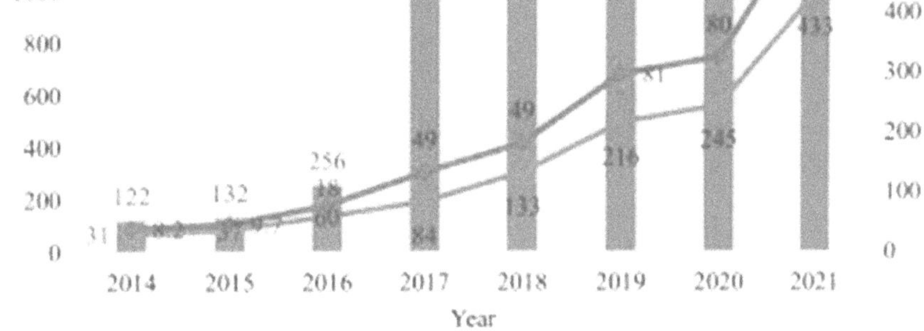

Figure 13.1 GF-for-GB from 2014 to 2021.

Source: Authors' elaboration based on CBI (2022b)

with low-emission or zero-carbon buildings will constitute a substantial portion, around 40 per cent, of the overall green bonds market in the long-term (CBI 2015).

This table reports the total annual issuance amount (in US$billion) as well as the number of GF deals linked to GB on annual basis from 2014 to 2021.

CBI's first-labelled green bond certified under the low-carbon (green) buildings criteria was issued by the ANZ Bank in May 2015, with proceeds of US$467 million allocated to GBs, wind energy, and solar loans in Australia, New Zealand, and parts of Asia (CBI 2015). Since then, GF-for-GB has been rising. As reported by CBI (2022b), emerging markets account for about 20 per cent of total green volume issued as of 2021. This indicates that there are untapped potentials in GF-for-GB in emerging market economies. Based on the CBI database, GF-for-GB-labelled instruments issued include green bonds, green loans, green repurchase agreements, green securitised bonds, and asset-backed securities – which could be sovereign, covered, municipal, or private placed bonds. The GF instruments issued so far are usually tenured from 2 to 47 years (CBI 2022c).

Figure 13.2 also presents a continental breakdown of GF-for-GB. According to CBI (2023)'s database, Europe alone accounts for about 73% (approx. US$57 billion) of the total volume of GF-in-GB of about US$79 billion issued as of December 2022. This is followed by Australia (12.4%), North America (6.9%), and South America (6.7%). Asia (1.1%) and Africa (0.1%) accounted for the least volume of GF-in-GB issued. The Netherlands, Norway, Australia, the USA, and Chile were the top five countries with the highest volume of GF-in-GB issued. On the contrary, Singapore, Hungary, Mexico, Ivory Coast, and Sweden had the least volume

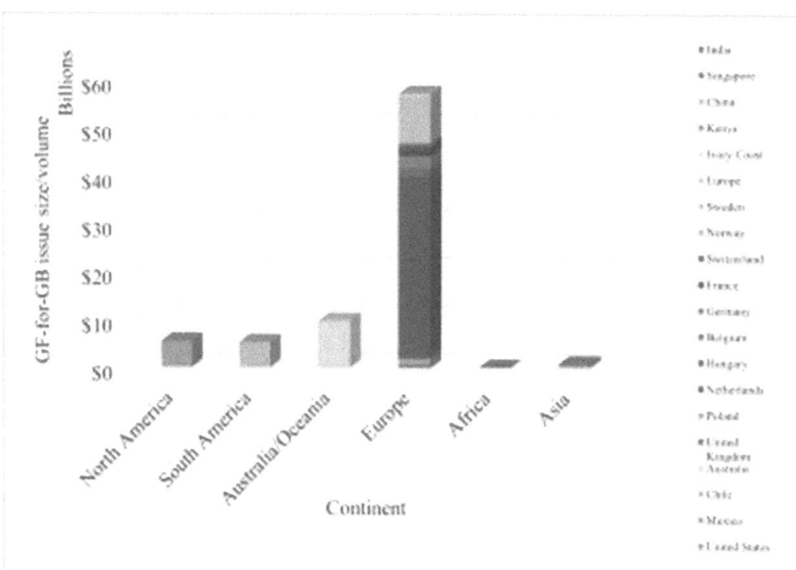

Figure 13.2 GF-for-GB from 2015 to 2022.

Source: Authors' elaboration based on CBI (2023)

of GF-in-GB issuance based on the CBI database. The high percentage of Europe contribution to GF globally is consistent with the EU's long-term commitment to be a market leader in sustainability, innovations, and greener economy. For instance, the European Investment Bank played a pioneering role in GF by issuing the first green bond in 2007 (Migliorelli and Dessertine 2019). Moreover, the increase in global buildings investment in energy efficiency primarily occurred among European countries with existing programmes of public investment in efficiency including Germany, the UK, and Italy (UNEP 2022). The next section describes the typology of GF-for-GB.

13.6 Typology of Green Finance for Green Building

There are several GF products or instruments available to finance the transition to GB portfolios. Verified GF products influential to GB projects (Akomea-Frimpong *et al.* 2022) have led to a number of universally accepted typologies of GF-for-GB as summarised in Figure 13.3 and described below. In addition, some implementation examples or success stories of the typologies are provided.

13.6.1 Green/climate Bonds for Green Building

Green bonds are "any type of bond instrument where the proceeds or an equivalent amount will be exclusively applied to finance or re-finance, in part or in full, new

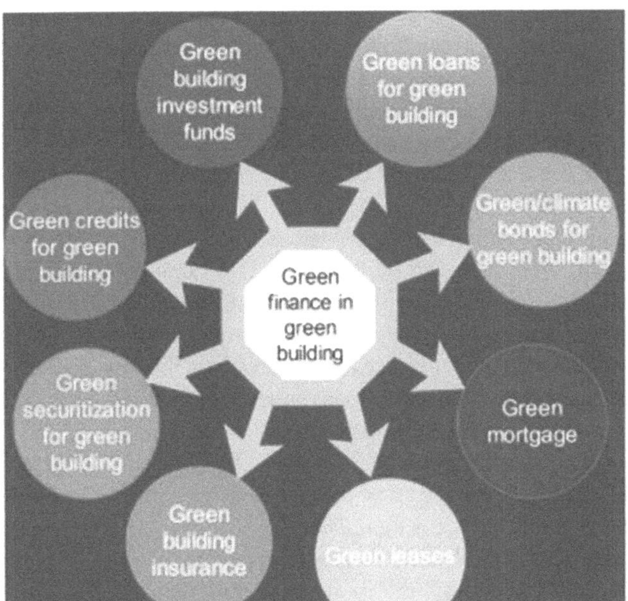

Figure 13.3 Typology of GF-for-GB

Source: Debrah (2024)

and/or existing green projects" (ICMA 2021). Eligible green projects include but not limited to renewable energy; energy efficiency; pollution prevention and control; environmentally sustainable management of living natural resources and land use; terrestrial and aquatic biodiversity; clean transportation; sustainable water and wastewater management; climate change adaptation; circular economy-adapted products; production technologies and processes and/or certified eco-efficient products; and *GBs* (ICMA 2021). Green bonds usually have a bigger volume, may have higher transaction costs, and could be listed on an exchange or placed privately (IFC 2021). As noted earlier, green (climate) bonds for GB have risen substantially over time.

Since 2010, the European Bank for Reconstruction and Development (EBRD) has dominated the markets for green bonds. It had issued 124 green bonds totalling €8 billion in 14 currencies by April 2022. A record €524 million was invested by the EBRD in 17 green bonds in 2021 (EBRD 2022). By purchasing green-labelled and climate-related bonds issued by customers, the bank also encourages capital market transactions. For example, it provided support for a €288 million bond issued by Rönesans Holding to fund a state-of-the-art public-private partnership hospital in Elazig. In addition, the bank provided €89 million as short-term funding to reduce the risks related to the development and maintenance of the green project (EBRD 2017). Its latest investment of €65 million in green-covered bonds from PKO Bank Hipoteczny will be used to finance low-carbon residential buildings in Poland (EBRD 2022). Duci et al. (2021) revealed €17.3 billion of green bonds have been issued by 29 European-listed property companies between 2014 and June 2021. According to CBI (2023), Australia, US, and the Netherlands are the largest issuers of green bonds for GB. The Netherlands accounted for the largest volume of green bonds being US$37 billion earmarked for GB. For example, 35 per cent of ING's US$3 billion green bond proceeds in 2018 were used to fund buildings with an energy performance certificate Label A and emissions performance in the top 15 per cent of the Dutch market. To finance the energy efficiency retrofits of 160 apartment buildings, Lithuania placed the first tranche of its sovereign green bond program, valued at US$24 million. Green bonds worth US$161 million were obtained by Industrial Development Corporation for low-carbon buildings in South Africa (IFC 2019). Similarly, in 2016, Bancolombia issued US$117 million green bond with IFC to grow its GB portfolio. Bancolombia provided green construction financing at an interest rate that was between 0.5 per cent and 2 per cent cheaper than that of conventional loans. The bank issued another green bond worth US$100 million in 2018. Since it attracted 72 domestic investors and was oversubscribed by 2.8 times, Bancolombia's cost of capital was reduced (IFC 2019). Berlin Hyp is the most active European commercial bank issuer of green bonds having accumulated €4 billion from eight green bonds. All proceeds have been earmarked for the development, renovation, or purchase of GBs (Winkler 2020). Additionally, the issuance of Emergence Plaza's first corporate green bond to re-finance the Cosmos Yopougon retail complex in Cote d'Ivoire at a reduced financing cost by 150 basis points (bps) reveals greater market access in emerging markets (UNEP 2021).

13.6.2 Green Loans for Green Building

Green building/construction loans that place more emphasis on the green credentials of the underlying assets are on the rise. As the need for GF rises, real estate lenders are increasingly basing their lending decisions on sustainability performance (JLL 2021a). Green loans are "any type of loan instrument made available exclusively to finance or re-finance, in whole or in part, new and/or existing green projects" (Loan Market Association 2021). Eligible green projects for green loans are similar to the descriptions under green bonds. However, green loans are usually smaller in volume and done privately (IFC 2021). The green loan market was estimated at US$70 billion in 2019 (Linklaters 2019) with real estate firms accounting for 32 per cent of the total (IFC 2019). Due to its greater accessibility, the green loans market is likely to outperform the green bonds market in future.

The EBRD remains an industry leader in green loans for GB. From 2006 to 2016, the EBRD's Green Energy Finance Facilities mobilised over €1148 billion in green loans to retrofit 60,832 buildings. For example, the EBRD signed a loan of up to €6.9 million with Al Ismaelia to finance the green retrofitting of several historic buildings in Egypt. Also, the bank extended a green loan of up to €50 million to the Lithuanian Public Investment Development Agency to finance demand-side energy efficiency investments in residential apartment buildings (EBRD 2017). Others, like Aviva Investors, plan to lend US$1.4 billion to GB within the next four years. To finance energy-efficient renovations, the Dutch bank ING is now providing GB incentive loans. For instance, ING gave Hines its first green loan in addition to a US$47 million financing for a prestigious office building in Frankfurt. An urban logistics park near London's Heathrow Airport will also be financed from the facility (JLL 2021a). Likewise, a consortium recently offered a loan of over US$635 million for a green mixed-use complex in Singapore, and Allianz Real Estate gave the Canary Wharf Group in London a £140 million green loan (JLL 2021a). IFC also provided a €73.5 million green loan to NE Property BV to support the retail property sector in Romania (World Bank 2021). KfW has invested €27 billion in loans and grants to retrofit 1 million homes, build 400,000 energy-efficient homes, and created 240,000 green jobs annually as a result (Sweatman 2012).

13.6.3 Green Mortgage

Many banks are taking advantage of the benefits associated with green mortgage to promote building sustainability and lower default risks of loans. Green mortgage offers preferential terms, such as lower interest rate or an increased loan amount, to either buy a new GB or to green-retrofit an existing building (WorldGBC 2022b). Green mortgage represents a lower investment risk for banks due to their low default rate and increased green value of GB thereby avoiding stranded assets in future (WorldGBC 2022b). To advance the use of green mortgages for GB, the IFC (2019) suggests that flexible mortgage pricing or underwriting that reflects the savings from resource efficiency is essential. For instance, due to utility savings redirected to the larger loan payment, banks may require lower down payments. Additionally, by considering savings as an increase in a customer's income, the debt-to-income ratio can be changed, broadening the pool of eligible customers. Lower interest rates for GF to end users can encourage the purchase of green homes and help create a portfolio of mortgages with higher value and lower risk (IFC 2019). Several global examples of green mortgages are provided below.

In 2016, the Banco Promerica in Costa Rica received a loan of US$30 million from the Dutch development bank for green mortgage construction financing. To encourage GB construction, the bank offered a discounted rate of 8.75 per cent (instead of 9.25 per cent), 0.75 per cent front-end fee (instead of 1 per cent), and a four-year term (instead of three years). Additionally, it provides 30-year fixed rates for green mortgages that are just a little bit higher than the 30-year variable market rate in Costa Rica. Similarly, Bancolombia provides green mortgages at a rate that is 65 bps lower than typical loans for the first seven years of the loan to eligible borrowers who purchase green-certified houses (IFC 2019). In addition, Banco Promerica intends to sell its portfolio of green mortgages as securities once it reaches US$20–30 million in value (IFC 2019). Likewise, the Energy Efficient Mortgages Initiative states that for a mortgage loan financing a renovation to be declared energy efficient, the renovation must result in a drop of at least 30 per cent in energy demand or consumption (Winkler 2020). Moreover, DNB Boligkreditt issued green bonds worth US$1.7 billion to finance or re-finance new and pre-existing mortgages for homes that adhere to the Norwegian building codes 2010 (TEK10) or 2017 (TEK17), corresponding to the top 7 per cent of Norwegian housing in terms of energy performance (IFC 2019).

13.6.4 Green Leases

In a green lease, the asset owner (the lessor) offers another party (the lessee) the sole right to use the asset for an agreed-upon term in exchange for an agreed-upon sum payable on a recurring basis (lease rents) during the agreed-upon lease duration. The agreement aims to guarantee that leased assets (e.g., lands and GBs) are used and managed sustainably and account for energy efficiency (Janda *et al.* 2016, CBI 2022a). Globally, almost 40 per cent of occupants and investors already have green lease provisions, and additional 40 per cent intend to do so by 2025 (JLL 2021b). It is estimated that green leases can save about 22 per cent in utility costs in US office buildings alone. The market could save over US$3 billion annually if all rented office buildings had green leases (The Institute for Market Transformation 2015).

13.6.5 Green Building Insurance

Green insurance refers to environmental pollution liability insurance (Zadek and Chenghui 2014). Insurance firms are essential to the advancement of GB. Overseeing US$31 trillion in assets and promoting GF products, institutional investors are key sources of financing for GB. Through existing GB certification, insurance companies can develop green insurance products that help developers to better identify and reduce risk. For new GB or green retrofits of existing building, GB insurance products offer premium discounts as well as coverage for replacing or rebuilding after loss using environmentally friendly materials in accordance with recognised green certification systems. Some insurers offer reduced rates and provide coverage for green power production equipment, alternative water systems, and vegetated roofs for buildings that have previously received green certification in the event of loss (IFC 2019).

13.6.6 Green Securitisation for Green Building

The process of turning a pool of illiquid assets (often hundreds of distinct assets) into marketable financial instruments (securities) is known as securitisation. The cash flows of the underlying assets, such as loans, leases, or receivables against other assets, form returns to investors on the securities. Low-carbon assets provide the cash flow for a green securitisation. Additionally, it describes any asset-backed instrument that raises money to pay for loans for green infrastructure (CBI 2018). Green securitisation is now required to access debt capital markets financing for small-scale, low-carbon, and climate-resilient assets, particularly GBs (CBI 2018). It increases capital access and decreases capital cost. For buildings, green securitisation has potential for mortgages for GBs, loans to green small-to-medium enterprises, and loans for energy efficiency upgrades. For instance, in June 2016, Dutch company Obvion released the first green Residential Mortgage-Backed Security in history, valued at US$557 million, and approved by the Climate Bonds Standard. Similar to this, FlexiGroup issued a US$39 million green securitisation backed by loans for residential rooftop solar power systems (CBI 2018).

13.6.7 Green Credits for Green Building

Green credit refers to the green deposit and loan industry, which includes mortgage loans and project loans for businesses and projects to support the transition and development to a more sustainable economy (Ren *et al.* 2020). It usually constraints finance to heavily polluting firms by increasing environmental protection investments leading to increasing green credits for GB (Hu *et al.* 2021, Debrah et al. 2022c). For instance, the Kozloduy International Decommissioning Support Fund, the Bulgarian government, and the EBRD all contributed to a credit line of €100 million. The green credit gave senior loans to Bulgarian banks so they could lend money to people, homeowners' associations, utility companies, and energy service providers. The loans were used to finance green retrofitting of residential buildings, installation of renewable energy resources, and promoting energy efficiency in buildings (EBRD 2017).

13.6.8 Green Building Investment Funds

Green investments are energy conservation and environmental protection investments from both private and public sectors (Ren *et al.* 2020). GB investment fund is therefore described as multi-stakeholder and publicly sponsored funds aimed at financing residential, commercial, and public facilities to meet emissions reduction targets of the built environment (Siew 2015, Akomea-Frimpong *et al.* 2022). Particularly, government fiscal investment in GB is very critical due to the uncertainties associated with GF instruments from commercial entities (He *et al.* 2022). As such, the financing arrangement leads to close partnerships between government, businesses, and industry towards mitigating climate change (Siew 2015). Hence, stakeholders such as the government, building owners, building occupiers, banks, renovation/building contractors, and energy service companies are key to forming partnerships for GB investment funds (Sweatman 2012). GB investment funds are seen as critical sources of GF-for-GB with some examples provided below.

In Hong Kong for instance, under the CLP Eco Building Fund, residential, commercial, and industrial buildings can receive up to HK$500,000 in subsidies for energy-saving improvement works. The fund awards up to HK$100 million annually and is intended to reduce energy costs, combat climate change, and create a greener Hong Kong (CLP Hong Kong 2022). As of 2021, 40 per cent of the total proceeds of US$3.5 billion sovereign green bonds in Hong Kong were allocated to GBs. In May 2021, the government launched a three-year Green and Sustainable Finance Grant Scheme (*formerly*, Green Bond Grant Scheme and the Pilot Bond Grant Scheme) to provide subsidy for eligible bond issuers and loan borrowers to cover their expenses and external review services (HKSAR 2021). Similarly, the Canadian government provided Canadians with 1 million free-energy audits and up to 700,000 grants, valued at up to C$5,000, to complete energy-efficient home improvements in 2020. It will also provide C$4.4 billion to the Canada Mortgage and Housing Corporation from 2021 to 2025 to help homeowners undertake deep home retrofits via interest-free loans worth up to C$40,000 (Government of Canada 2021). The UK plans to extend the Green Homes Grant by providing just about £1 billion to improve the energy efficiency of homes, schools, hospitals, and public buildings and replace fossil fuel heating. The UK's GB investment fund has the potential to deliver about £11 billion of private green investment in GB by 2030 (HM Government 2020). Under the Australian GB Fund, government grants of between A$50,000 and A$500,000 are available, for up to 50 per cent of the cost of energy-efficient retro-fitting and/or retro-commissioning of commercial office buildings, hotels, or shopping centres (Australian Government 2022). With US$213 billion, the US government is dedicated to GB and green retrofitting of more than 2 million sustainable affordable houses. An additional US$100 billion has been apportioned to constructing new schools and renovating existing ones to align with sustainability objectives. US$25 billion will be further used to develop and green-retrofit childcare facilities. Based on KfW's €30 billion energy-efficient construction and renovation programmes, Germany has doubled its energy-efficiency spending on buildings (UNEP 2021). Additional expenditure of €17 billion in energy-efficient buildings is expected by 2024. Japan's zero-energy housing represented about 16 per cent of private housing investment in 2021 (UNEP 2022).

Table 13.2 provides a summary of the unique characteristics and arrangements provided by the above GF instruments in terms of the procedures, provisions, and guidance for the development of GBs. Section 13.7 provides further details of the unique guidelines for GF-for-GB.

Table 13.2 Summary of the unique characteristics and special arrangements of GF-in-GB

Typology of GF	*Unique characteristics*	*Special arrangements: procedures, provisions, and guidance*	*References*
Green/climate bonds for GB	• Applied to finance or re-finance new/existing GB. • Bigger in volume • Higher transaction costs • Could be listed on an exchange or placed privately. • Lower interest rates	• Adherence to green bond frameworks: use of proceeds, process for project evaluation, management of proceeds and reporting • External reviews: consultant review (including second party opinions), verification, green certification, and rating by qualified third parties or rating agencies	ICMA (2021); IFC (2021); IFC (2019)
Green loans for GB	• Applied to finance or re-finance new/existing GB. • Smaller in volume • Higher transaction costs • Usually placed privately. • Lower interest rates	• Adherence to international standards or eligibility criteria, e.g., green loan framework • External review • Application is similar to green/climate bonds for GB above	ICMA (2021); IFC (2021); Loan Market Association (2021)
Green mortgage	• Lower interest rates • Higher loan-value ratio • Lower down payments	• Applicable to green-certified buildings or existing buildings with proof of green retrofits • Independent green certification/assessment may be required	WorldGBC (2022a–2022c); IFC (2019)
Green leases	• Landlord-tenant collaboration and agreement to implement green goals. • Landlord-tenant cost and profit sharing due to green features	• Tenants commit to or gain incentives by participating in building sustainable practices such as water/energy conservation and waste reduction. • Specifies who is responsible for larger projects such as energy efficiency renovations and who benefits from the resulting savings. • Penalties for non-compliance may apply	Green Building Alliance (2023); JLL (2021a, 2021b); The Institute for Market Transformation (2015)
GB insurance	• Identifies and reduces risk through GB certification. • Premium discounts for GB • Coverage for replacing/rebuilding after loss using GB technologies	• Green recertification costs following loss of or substantial damage to green property. • Extra expense coverage for GB features	Ochenkowski and Schinter (2008); IFC (2019)

(Continued)

Table 13.2 (Continued)

Typology of GF	*Unique characteristics*	*Special arrangements: procedures, provisions, and guidance*	*References*
Green securitisation for GB	• Financing GB through the issuance of asset-backed securities (ABS), which are backed by financial assets such as mortgages or lease receivables, i.e., securitisation. • Typically targeted at small-scale, low-carbon, and climate-resilient infrastructure. • Usually loans for small-to-medium GB firms and for energy efficiency upgrades • Sustainable securitisation • Liquidity in GF	• Loans and leases on existing assets, such as certified buildings, are issued, and new projects are financed. • These loans and leases are packaged together into different types of green ABS such as green mortgage-backed securities. • Securities are then sold to investors, who are currently signalling a high demand for green-labelled products. Investors can trade securities on the secondary market. • The proceeds and/or freed-up capital (depending on the type of deal), are used then to finance other green projects and assets. • Third-party verification or certification of GB assets is required	CBI (2018); Barmes (2019)
Green/carbon credits for GB	• Constraints investment to heavily polluting firms. • Incentives for environmental protection investments	• Earning green/carbon credits by meeting specific certification criteria in new GB or green retrofits of existing buildings, e.g., LEED, IFC EDGE • Trading green credits on secondary markets for additional liquidity • Some green credit programs may provide bonus credits or incentives for buildings in areas of high pollution or social vulnerability	Hu *et al.* (2021); Woo *et al.* (2021); Debrah et al. (2022c)
GB investment funds	• Interest free loans from government • Government grants for green retrofits • Green subsidies	• Multi-stakeholder and publicly sponsored funds for financing residential, commercial, and public facilities to meet emissions reduction targets. • Exposure to diversified GB portfolio	Siew (2015); Akomea-Frimpong *et al.* (2022)

13.7 Green Finance in Green Building: Standards, Certification, and Indices

Asset managers and their principals identify bonds and loans that improve the environment or the climate by using GF certification or standards. Particularly, external certification is a cost-effective way of proving that investments are green (Ehlers and Packer 2017). First, the "Green Bond Principles" developed by the ICMA are voluntary process guidelines that recommend transparency and disclosure of the green bond market by clarifying the approach for issuance of a green bond (ICMA 2021). Similarly, the "Green Loan Principles" build on and refer to the Green Bond Principles with a view of promoting consistency across financial markets (Loan Market Association 2021). Both the Green Bond Principles and Green Loan Principles are aimed at promoting environmental sustainability of global debt capital markets. They provide prospective issuers with guidance on the four key components of green bond/loan issuance: "use of proceeds", "process for project evaluation and selection", "management of proceeds", and "reporting". The ICMA (2021) and the Loan Market Association (2021) provide details of the above components of the Green Bond Principles and Green Loan Principles respectively. In addition, the Global Real Estate Sustainability Benchmark (GRESB) created the Green Bond Guidelines for the Real Estate Sector to give issuers and investors with additional sector-specific real estate guidelines. The guidelines expand on the Green Bond Principles by outlining environmental and social factors that could be used as green property bond eligibility criteria in the real estate industry (GRESB 2016). Since April 2021, European listed real estate companies align green project's eligibility to the technical screening criteria set by the "EU Climate Taxonomy for the construction of new buildings, renovation of existing buildings, acquisition and ownership of buildings, and individual energy efficiency and renewable energy intervention" (Duci *et al.* 2021). The EU Taxonomy is the first standardised and comprehensive classification system for sustainable economic activities. It provides a common definition for about 80 sustainable activities by setting performance thresholds (or technical screening criteria) for these activities. An objective of the taxonomy is to channel investments into low-carbon technologies, supporting the EU's transition towards carbon neutrality by 2050 (Sweatman and Hessenius 2020, Schütze and Stede 2021).

At the root of GF-for-GB are GB certifications and energy ratings. GB certifications or rating systems are used to assess and recognise buildings which meet certain green requirements or standards (WorldGBC 2022c). They use frameworks that entail project-based attributes and a common communication platform to share outcomes with various stakeholders. Green certifications also cover various environmental and social aspects, establish key terms and concepts that are simple to understand, and monitor outcomes throughout the project lifespan (GRESB 2016). They are usually voluntary and recognise and reward companies and organisations who build and operate greener buildings (WorldGBC 2022c). The Leadership in Energy and Environmental Design (LEED, US), Building Research Establishment Environmental Assessment Methodology (BREEAM,

UK), Green Star (Australia), Comprehensive Assessment System for Built Environment Efficiency (CASBEE, Japan), BEAM Plus (Hong Kong), High Quality Environmental Standard (HQE, France), Deutsche Gesellschaft für Nachhaltiges Bauen (DGNB, Germany), and Excellence in Design for Greater Efficiencies (EDGE, IFC), among others, are widely applied and known rigorous GB certification systems (Darko *et al.* 2025). Additionally, energy rating or labels such as the energy performance certificate (EU and the UK), ENERGY STAR (US), National Australian Built Environment Rating System (NABERS, Australia), among others, show how energy efficient a building is comparatively. They act as indicators for spending on energy efficiency and/or carbon emission reduction goals (GRESB 2016). These green certifications systems and energy ratings usually underpin GF-for-GB. For instance, Allianz's loan for solar power generation was based on an excellent BREEAM score and on the Green Loan Principles of the UK Loan Market Association. Likewise, the US Federal National Mortgage Association recently approved the BREEAM Residential Plus as an accepted standard for evaluating its US$13 billion green loans under its Multifamily Green Initiative (JLL 2021a). ING offers free on-site BREEAM and energy audit to GBs that indicate annual energy savings of more than €15,000 based on their app. Moreover, ING offers its clients free-energy performance certificate assessments, advice, and discounts on sustainable loans (IFC 2019). Similarly, Berlin Hyp requires reputable sustainability certifications such as LEED, BREEAM, DGNB, and HQE as prerequisite for GF-for-GB (Winkler 2020). According to the requirements for energy performance certificates, Dutch bank ABN AMRO has stipulated that all new buildings must have a minimum energy label of "A", and new commercial buildings must be LEED "platinum" or BREEAM "excellent" certified. Existing buildings must have a strategy for lowering "D", "E", "F", and "G" labels and an average label of "C" on gross floor area in square meters. The bank has provided €1 billion to finance the above improvements (IFC 2019). The tallest building in Côte d'Ivoire will be built using US$451 million investment capital from Société Générale France for Pierre Fakhoury Organization Africa and BESIX in Brussels. This EDGE certified green office building will have 64 floors (UNEP 2021). In Hong Kong, the GF Certification Scheme is a prerequisite for assessing the government's Green and Sustainable Finance Grant Scheme (HKSAR 2019).

Besides, several GF indices, especially for green bonds, are evolving. Green bond indices use a specified process to determine which bonds are green, and they enable investors to diversify their risks by purchasing a portfolio of green bonds (Ehlers and Packer 2017). At present, global green bond indices include Solactive Green Bond Index, S&P Green Bond Index and S&P Green Project Bond Index, Bank of America Merrill Lynch (BAML) Green Bond Index, Barclays MSCI green bond indices, ChinaBond China Green Bond Index, and ChinaBond China Green Bond Select Index. These indices portray market maturity and drive demand for green bonds among institutional investors (ICMA 2017). For instance, green bonds issued by Berlin Hyp are fully aligned with the Green Bond Principles and are included in the major green bond indices including Barclays MSCI, S&P, BAML, and Solactive. Only loans for environmentally friendly, energy-efficient commercial real estate qualify as eligible assets (Winkler 2020).

13.8 Green Finance for Green Building: the Role of Stakeholders

To bridge the GF-for-GB investment gap necessary to decarbonise buildings and construction, major stakeholders such as governments, industries, businesses, and consumers must act now. Together, these actors implement policies, regulatory measures, and incentives that promote priority-lending requirements, below-market-rate finance via interest-rate subsidies, or preferential central bank refinancing opportunities (Berensmann and Lindenberg 2016). This section discusses the role of public and private actors in accelerating GF-for-GB.

13.8.1 Government Policy and Regulation

Public policy is critical in spurring GB investment, producing energy and cost savings, lowering emissions, and creating millions of high-quality jobs (Kubba 2012). The Nationally Determined Contributions of countries reveal increasing commitments and investments in GB towards reducing global temperatures under the 1.5°C scenario by 2050 (Debrah et al. 2022d). Majority of governments require publicly owned and financed buildings to be green (IFC 2019). This is because public finance or governments' green fiscal investment is central to GB development and green retrofitting of existing buildings. In addition, government policy incentives and programmes are seen as major drivers of GF-for-GB (UNEP 2021) due to governments supervisory role of finance markets (He *et al.* 2022). Policymakers might encourage banks to provide direct capital to the sector by defining criteria for GF instruments and mandating climate risk reporting in investment portfolios (IFC 2019). Governments are progressively demanding that projects are built to GB standards. This is forcing the industry to prioritise sustainable projects for clients and communities (Kubba 2012). Others such as green labelling or performance certifications can assist investors in verifying, comparing, and managing their assets in order to diversify their green portfolios (IFC 2019).

Governments are also encouraging developers to build green through fiscal measures such as property tax incentives, technical support, grant provisions, and credit programs (EBRD 2017, IFC 2019). Italy, for example, announced a €500 million yearly subsidy for municipalities to invest in energy-efficient public buildings. In addition, the government provides a tax credit up to 110 per cent of the investment for energy-efficient retrofits. From 2021, the French government has invested about €7 billion on energy efficiency upgrades in residential homes, and office and public buildings such as schools and town halls (UNEP 2021). The Building Retrofit Energy Efficiency Financing Scheme in Singapore provides financing to assist the upfront costs of retrofits via energy performance contracts. The Singapore Building and Construction Authority shares 40 per cent of the risk of any loan default with the collaborating financial institutions. In addition, existing buildings must achieve and maintain the Green Mark certification during the loan period (UNEP 2021). Other non-fiscal incentives such as height bonus incentives, expedited permitting processes, and reduced permitting process promote GF-for-GB (IFC 2019). Further, regulators and central banks are developing polices and putting novel financial mechanisms in place to direct capital towards green real estate in order to diversify their incomes and limit climate-related financial risks (IFC 2019). It is expected that government investments in GF-for-GB will serve as precursor to unlocking the vast potentials from the private sector as discussed below.

13.8.2 Private-Sector Involvement

Private-sector involvement in GF-for-GB is growing. This is a reaction to investors' efforts to future-proof themselves against risks associated with climate transition, regulatory demands, and consumers' sentiment. As such, asset owners prefer certified GB portfolios. This change has increased interest in GF of both new GB and green retrofits of existing buildings (UNEP 2021). This opens opportunities for institutional investors and commercial banks to invest in greener assets. According to UNEP (2021), private and institutional investors are developing green taxonomies similar to the EU taxonomy on sustainable economic activities and sustainability disclosure requirements to scale up international investments. The institutional investors (i.e., pension funds, insurance companies, sovereign wealth funds, hedge funds, and mutual funds) participate in the GB market by injecting liquidity and giving primary lenders the opportunity to free up capital for new green lending products. In addition, institutional investors play a significant role in augmenting local currency finance for GB through direct lending or equity investments (IFC 2019).

Moreover, equity investors offering comparable green products include real estate funds and real estate investment trusts (REITs) (IFC 2019). For instance, Mexican REITs like Fibra Macquarie and Fibra Shop are working to certify all their portfolios as "green", through a combination of green-retrofit investments and setting higher quality-on-entry green standards for all new construction. Likewise, AshmireAVENIDA, a manager of a real estate fund, certified all its projects as green by 2021. It is also courting investors with a Latin American real estate fund with an all-green concept (UNEP 2021). In 2021, The IFC supported Business Partners Limited in South Africa with US$44 million GF to retrofit commercial properties. To provide homebuyers favourable rates for investing in an EDGE-certified green home, Balwin teamed up with Absa bank to create green mortgages. Arthaland is also utilising its US$60 million green bond offering to support a portfolio of forthcoming EDGE Zero Carbon projects in the Philippines (UNEP 2021).

To achieve sustainability goals in the built environment, significant green investments are necessary. A substantial part of these financial resources must flow from the private investors, institutional investors, banks, and real estate fund managers. This necessitates a redirection of private capital towards more sustainable investment (OECD 2018). For example, a coalition of 128 institutional investors managing assets worth US$43 trillion has committed to achieving net-zero emissions by 2050. These investors actively seek GF-for-GB finance, resulting in a higher demand than supply, ultimately leading to reduced capital costs for qualified green assets

(Net Zero Asset Managers Initiative 2021). Furthermore, commercial banks can introduce new GF products tailored for energy-efficient buildings. Building a portfolio of certified GBs presents opportunities for banks to increase their liquidity, develop new capital market products, expand access to lower cost capital, and allow institutional investors to invest in green assets (IFC 2019). Measures critical to stimulating private-sector green investments in buildings include credit enhancements for bonds, loan loss reserves, insurance products, performance guarantees, project completion coverage, political risk coverage, policy and regulatory risk coverage, currency risk protection, and securitisation. Additional strategies include promoting market transparency; standardised reporting; contract standardisation; co-investment, joint-ventures, and consortiums; and fostering cooperation, collaboration, and informal knowledge-sharing among institutional investors (Kaminker *et al.* 2015).

13.9 Benefits of Green Finance for Green Building

GF-for-GB is necessary to mitigate climate change and the impacts. While global building stock is likely to double by 2050, the resultant energy consumption and carbon emissions are expected to increase by 50 per cent (EBRD 2017). However, GB investments yield socioeconomic benefits such as green employment, increased worker productivity, improved occupant health, among others. For example, the building renovation market in the EU valued at €109 billion in 2015 created over 882,000 direct and indirect jobs. It is projected that for every €1 million energy efficiency investments in the European GB sector, 19 net jobs could be created (EBRD 2017). In the UK, GF-for-GB is expected to yield 50,000 jobs and new supply chains and factories by 2030 (HM Government 2020). Regarding health-related benefits of GB finance, the EU estimates an annual cost savings of €64–140 billion. Dublin (Ireland) invested over €23 million to upgrade over 1200 homes leading to improved respiratory system of occupants and 2tCO_2 annual emissions reduction per home (EBRD 2017). Hence, GF contributes to the built environment sector achieving its sustainability objectives.

Beyond environment and climate-related benefits, GF is becoming more mature and accessible to real estate companies with lower financial cost. While the weighted coupon rate for green bonds issued by European property markets in 2014 was 2.5 per cent, it dropped to 1 per cent in 2015, being a 150bps change in just a year. The same coupon rate had an average annual reduction of 25 bps between 2018 and 2021 (Duci *et al.* 2021). In emerging markets such as Cote d'Ivoire, Emergence Plaza sold a US$18.1 million eight-year private placement in August 2021 with a coupon of 7.5 per cent (EBRD 2017). Besides, development finance institutes usually offer green funds at lower rates compared to the debt markets (EBRD 2017). GF-for-GB also offers stable funding opportunities such as green bonds, green securitisations, and green credit facilities as discussed above. It enables banks and investors to increase their clientele and product offerings, create portfolios with better values and fewer risks, and gain access to new funding or capital sources (IFC 2019). This is due to the possibility of lower cost capital being made available by GF instruments to encourage the growth of financial solutions for GB. The challenge of high upfront cost of GB (Chan *et al.* 2018b) could be offset with GF benefits such as discounted rates, longer tenors, and longer grace periods, or lower front-end fees (IFC 2019). Some banks begin these loans with a conventional one until the preliminary green certification is obtained. The advantageous terms of the green loan then take effect, setting forth specific conditions and timelines to complete and certify the GB. Other banks cover the certification costs in part or in full as part of the GF benefits. A green mortgage product is usually pegged to the GB to offset the green investment. In effect, banks have two options for financing the same GB: green mortgages and GB finance. Without further appraisals, the building's green certification could satisfy the criteria needed for green mortgages (IFC 2019).

13.10 Summary and Future Directions

The Global Commission on the Economy and Climate estimates that about US$90 trillion of green investment in infrastructure is needed by 2030 (Mountford *et al.* 2018). This is critical to achieving the Paris Agreement Accord and the 2030 Sustainable Development Goals Agenda, particularly for buildings and construction. As a result, both the public and private sectors have adopted innovative financing mechanisms such as GF-for-GB. Enormous opportunities exist for GF in both new GB and green retrofitting of existing buildings. It is now anticipated that all new buildings will have net-zero emissions, and that renovations of comparable amount of existing buildings would reduce energy use by 50 per cent (Kubba 2012). Hence, understanding GF and the opportunities for GB are very essential.

In this chapter, an overview of GB development was provided. While GB still lacks adequate finance, it is explained that GF provides opportunities to bridge such gap. First, conventional finance was differentiated from GF. Further, the evolution, typologies, and examples of implementation successes of GF-for-GB were provided. It was explained that GF is largely underpinned by global green standards, green certifications, and emerging green indices. Examples of GB certifications and energy ratings pivotal to GF were discussed. Again, the roles of stakeholders such as governments, industries, businesses, and consumers towards the acceleration of GF-for-GB were expounded. It was demonstrated that while GF-for-GB leads to environmental and climate-related benefits, investors are becoming more interested in the market due to the increased accessibility to new and/or additional capital at lower financial cost.

In summary, GF-for-GB has the potential to drive industry transformation, reduce construction costs, attract investments, and stimulate the development of the GB sector. This approach is an effective means to mitigate the environmental impact of buildings and contribute to the overarching decarbonisation goals and promotes sustainability of the built environment. With continued government and market support, along with increased availability of GF, the GBs sector will flourish further, ultimately advancing the goal of achieving net-zero buildings by 2050. Furthermore, financial institutions must take leadership roles in facilitating the transition to a low-carbon economy by actively investing in GB projects, technologies, and consumer loans (IFC 2019). Given the current inadequacy of global investments in GBs, it is crucial to design more suitable sustainable financing models and tailor GF solutions specifically for GB projects. Notably, green bonds, green loans, green fiscal investments, GB insurance, and green credit support exhibit significant impact on GB development. Moreover, fostering green awareness is essential to ensure that stakeholders understand the imperative of transitioning to GF models for GB finance (Li *et al.* 2023). Lastly, more research is needed to comprehensively understand the evolving dynamics and economic feasibility of financing GBs. Future research should aim to develop more intelligent tools for assessing the opportunities of GF and to effectively evaluate and price the risks associated with GF-in-GB. Targeted international and national policies and incentives for GF-in-GB should be developed to stimulate increased green investments within the buildings and construction industry.

Acknowledgements

This book chapter is based on the first author's PhD thesis (Debrah, 2024) completed at The Hong Kong Polytechnic University entitled "Lifecycle economic feasibility assessment model for green finance in green building in Ghana", from which other papers have been produced with different objectives/scopes but sharing the same background and methodology. The research project described is a fully funded PhD Scholarship by the Department of Building and Real Estate at The Hong Kong Polytechnic University. The authors further express their appreciation for additional funding from the Hong Kong Polytechnic University under a Matching Fund Postdoctoral Fellowship Grant Scheme No. 1-W36V and Able Endowed Professorship in Construction Health Donation for UGC Matching Grant/ PJTD No. 847X.

References

Adabre, M.A., Chan, A.P.C., Darko, A., Osei-Kyei, R., Abidoye, R., and Adjei-Kumi, T., 2020. Critical barriers to sustainability attainment in affordable housing: International construction professionals' perspective. *Journal of Cleaner Production*, 253, 119995.

Agyekum, K., Opoku, A., Oppon, A.J., and Opoku, D.-G.J., 2020. Obstacles to green building project financing: An empirical study in Ghana. *International Journal of Construction Management*, 22(15) 2922–2930.

Akomea-Frimpong, I., Kukah, A.S., Jin, X., Osei-Kyei, R., and Pariafsai, F., 2022. Green finance for green buildings: A systematic review and conceptual foundation. *Journal of Cleaner Production*, 356, 131869.

Australian Government, 2022. Green Building Fund - Grants Spectrum [online]. *spectrumgrants.com.au*. Available from: https://grantsspectrum.com.au/page/green-building-fund [Accessed 19 Jul 2022].

Barmes, D., 2019. Securitisation is back, and green finance must stray far away. *Positive Money*.

Berensmann, K. and Lindenberg, N., 2016. *Green Finance: Actors, Challenges and Policy Recommendations*. Rochester, NY: Social Science Research Network, SSRN Scholarly Paper No. ID 2881922.

CBI, 2015. Buildings [online]. *Climate Bonds Initiative*. Available from: https://www.climatebonds.net/standard/buildings [Accessed 3 Aug 2021].

CBI, 2018. Green Securitisation: Unlocking finance for small-scale low carbon projects [online]. *Climate Bonds Initiative*. Available from: https://www.climatebonds.net/resources/reports/green-securitisation-unlocking-finance-small-scale-low-carbon-projects [Accessed 12 Jul 2022].

CBI, 2022a. Discussion Paper: Certification of Short-Term Debt [online]. *Climate Bonds Initiative*. Available from: https://www.climatebonds.net/resources/reports/discussion-paper-certification-short-term-debt [Accessed 22 Jul 2022].

CBI, 2022b. Market Data [online]. *Climate Bonds Initiative*. Available from: https://www.climatebonds.net/market/data [Accessed 15 Jul 2022].

CBI, 2022c. Certified Bonds - Low-carbon buildings [online]. *Climate Bonds Initiative*. Available from: https://www.climatebonds.net/certification/certified-bonds [Accessed 3 Jun 2022].

CBI, 2023. Certified Bonds [online]. *Climate Bonds Initiative*. Available from: https://www.climatebonds.net/certification/certified-bonds [Accessed 28 Feb 2023].

Chan, A.P.C., Darko, A., Ameyaw, E.E., and Owusu-Manu, D.-G., 2017. Barriers Affecting the Adoption of Green Building Technologies. *Journal of Management in Engineering*, 33 (3), 04016057.

Chan, A.P.C., Darko, A., Olanipekun, A.O., and Ameyaw, E.E., 2018a. Critical barriers to green building technologies adoption in developing countries: The case of Ghana. *Journal of cleaner production*, 172, 1067–1079.

Chan, A.P.C., Darko, A., Olanipekun, A.O., and Ameyaw, E.E., 2018b. Critical barriers to green building technologies adoption in developing countries: The case of Ghana. *Journal of Cleaner Production*, 172, 1067–1079.

Chegut, A., Eichholtz, P., and Kok, N., 2019. The price of innovation: An analysis of the marginal cost of green buildings. *Journal of Environmental Economics and Management*, 98, 102248.

Chiang, Y. and Cheng, E.W.L., 2010. Construction loans and industry development: the case of Hong Kong. *Construction Management and Economics*, 28 (9), 959–969.

Chien, F., Ngo, Q.-T., Hsu, C.-C., Chau, K.Y., and Iram, R., 2021. Assessing the mechanism of barriers towards green finance and public spending in small and medium enterprises from developed countries. *Environmental Science and Pollution Research*, 28 (43), 60495–60510.

CLP Hong Kong, 2022. Eco Building Fund [online]. Available from: https://www.clp.com.hk/en/business/low-carbon-solutions/funds-and-subsidies/eco-building-fund [Accessed 19 Jul 2022].

Darko, A. and Chan, A.P.C., 2017. Review of barriers to green building adoption. *Sustainable Development*, 25 (3), 167–179.

Darko, A., Debrah, C., and Chan, A. P., 2025. Introduction to Green Construction Projects. In Darko, A. and Chan, A.P. (Eds) *Developing a Body of Knowledge for Green Construction Project Management*, Singapore: World Scientific Publishing, 77–105.

Debrah, C., 2024. *Lifecycle economic feasibility assessment model for green finance in green building in Ghana* [Doctoral thesis, The Hong Kong Polytechnic University] Hong Kong. [online]. Available from: https://theses.lib.polyu.edu.hk/handle/200/13099 [Accessed 15 Jul 2022].

Debrah, C., Chan, A.P.C., and Darko, A., 2022a. Green finance gap in green buildings: A scoping review and future research needs. *Building and Environment*, 207, 108443.

Debrah, C., Chan, A.P.C., and Darko, A., 2022b. Artificial intelligence in green building. *Automation in Construction*, 137, 104192.

Debrah, C., Chan, A. P. C., Darko, A., Ries, R. J., Ohene, E., & Tetteh, M. O. (2024). Driving factors for the adoption of green finance in green building for sustainable development in developing countries: The case of Ghana. *Sustainable Development*, 1–22.

Debrah, C., Darko, A., and Chan, A.P.C., 2022c. A bibliometric-qualitative literature review of green finance gap and future research directions. *Climate and Development*, 15 (5), 432–455.

Debrah, C., Darko, A., Chan, A.P.C., Owusu-Manu, D.G., and Edwards, D.J., 2022d. Green finance in green building needs under the Paris Agreement. *In: IOP Conference Series: Earth and Environmental Science*. IOP Publishing, 012033.

Duci, G., Sabir, H., and Daliman, T., 2021. European Listed Real Estate - Green Bonds Monitor 2021. *European Public Real Estate Association*, 13.

Dwaikat, L.N. and Ali, K.N., 2016. Green buildings cost premium: A review of empirical evidence. *Energy and Buildings*, 110, 396–403.

EBRD, 2017. Green building investments.

EBRD, 2022. The EBRD's investment in green bonds passes €1 billion [online]. Available from: https://www.ebrd.com/news/2022/the-ebrds-investment-in-green-bonds-passes-1-billion-.html [Accessed 15 Jul 2022].

Ehlers, T. and Packer, F., 2017. Green bond finance and certification.

EIB and GFC, 2017. Joint White Paper by China Green Finance Committee and EIB set to strengthen international Green Bond market [online]. *European Investment Bank*. Available from: https://www.eib.org/en/press/all/2017-311-joint-white-paper-by-china-green-finance-committee-and-eib-set-to-strengthen-international-green-bond-market [Accessed 5 Jan 2022].

Gholipour, H.F., Arjomandi, A., and Yam, S., 2022. Green property finance and CO2 emissions in the building industry. *Global Finance Journal*, 51, 100696.

Government of Canada, 2021. Budget 2021 A Healthy Environment for a Healthy Economy [online]. Available from: https://www.canada.ca/en/department-finance/news/2021/04/budget-2021-a-healthy-environment-for-a-healthy-economy.html [Accessed 19 Jul 2022].

Green Building Alliance, 2023. Green Leases [online]. Available from: https://gba.org/resources/green-building-methods/processes/green-leasing/ [Accessed 22 Feb 2023].

GRESB, 2016. Green Bond Guidelines for the Real Estate Sector.

He, W., Liu, P., Lin, B., Zhou, H., and Chen, X., 2022. Green finance support for development of green buildings in China: Effect, mechanism, and policy implications. *Energy Policy*, 165, 112973.

HKSAR, 2019. Gov't backs green financing [online]. *Hong Kong's Information Services Department*. Available from: http://www.news.gov.hk/eng/2019/03/20190325/20190325_171653_647.html [Accessed 17 Apr 2022].

HKSAR, 2021. Green Bond Report 2021. *Hong Kong Special Administrative Region of the People's Republic of China*, 34.

HM Government, 2020. The ten point plan for a green industrial revolution [online]. *GOV. UK*. Available from: https://www.gov.uk/government/publications/the-ten-point-plan-for-a-green-industrial-revolution [Accessed 16 May 2022].

Hu, G., Wang, X., and Wang, Y., 2021. Can the green credit policy stimulate green innovation in heavily polluting enterprises? Evidence from a quasi-natural experiment in China. *Energy Economics*, 98, 105134.

ICMA, 2017. The GBP Databases and Indices Working Group – Summary of Green Fixed Income Indices Providers.

ICMA, 2020. Sustainable Finance High-level definitions.

ICMA, 2021. Green Bond Principles.

IFC, 2019. *Green Buildings: A Finance and Policy Blueprint for Emerging Markets*. Washington, D.C.:The World Bank.

IFC, 2021. *What You Need to Know about Green Loans* [online]. World Bank. Available from: https://www.worldbank.org/en/news/feature/2021/10/04/what-you-need-to-know-about-green-loans [Accessed 2 Jun 2022].

Janda, K.B., Bright, S., Patrick, J., Wilkinson, S., and Dixon, T.J., 2016. The evolution of green leases: Towards inter-organizational environmental governance. *Building Research & Information*, 44 (5–6), 660–674.

Ji, Q. and Zhang, D., 2019. How much does financial development contribute to renewable energy growth and upgrading of energy structure in China? *Energy Policy*, 128, 114–124.

JLL, 2021a. *How Real Estate Is Starting to Embrace Green Finance* [online]. Available from: https://www.jll.com.hk/en/trends-and-insights/investor/how-real-estate-is-starting-to-embrace-green-finance [Accessed 13 Jul 2022].

JLL, 2021b. *Decarbonizing the Built Environment: Ambitions, Commitments and Actions* [online]. Available from: https://www.jll.co.uk/en/trends-and-insights/research/decarbonizing-the-built-environment [Accessed 22 Jul 2022].

Kaminker, C., Eklin, K., Youngman, R., and Kawanishi, O., 2015. Mapping Channels to Mobilise Institutional Investment in Sustainable Energy: An OECD Report for G20 Finance Ministers and Central Bank Governors, Green Finance and Investment. *Green Finance and Investment.*

Kubba, S., 2012. *Handbook of Green Building Design and Construction: LEED, BREEAM, and Green Globes*. Oxford (UK): Butterworth-Heinemann.

Linklaters, 2019. *Sustainable Finance: The Rise of Green Loans and Sustainability Linked Lending* [online]. Available from: https://www.linklaters.com/en-hk/insights/thought-leadership/sustainable-finance/the-rise-of-green-loans-and-sustainability-linked-lending [Accessed 12 Jul 2022].

Li, S., Zheng, X., and Zeng, Q., 2023. Can Green Finance Drive the Development of the Green Building Industry?—Based on the Evolutionary Game Theory. *Sustainability*, 15 (17), 13134.

Loan Market Association, 2021. *Guidance on Green Loan Principles (GLP)*. LSTA.

Merna, T. and Njiru, C., 2002. *Financing Infrastructure Projects*. Thomas Telford.

Migliorelli, M. and Dessertine, P., 2019. The rise of green finance in Europe. *Opportunities and Challenges for Issuers, Investors and Marketplaces*. Cham: Palgrave Macmillan, 2, 2019.

Mountford, H., Corfee-Morlot, J., McGregor, M., Banaji, F., Bhattacharya, A., Brand, J., Colenbrander, S., and Stern, N., 2018. *Unlocking the Inclusive Growth Story of the 21st Century: Accelerating Climate Action in Urgent Times*. Global Commission on the Economy and Climate, Report.

Net Zero Asset Managers Initiative, 2021. Net Zero Asset Managers initiative announces 41 new signatories, with sector seeing 'net zero tipping point' – The Net Zero Asset Managers initiative [online]. Available from: https://www.netzeroassetmanagers.org/net-zero-asset-managers-initiative-announces-41-new-signatories-with-sector-seeing-net-zero-tipping-point/ [Accessed 6 Sep 2023].

Nguyen, H.-T., Skitmore, M., Gray, M., Zhang, X., and Olanipekun, A.O., 2017. Will green building development take off? An exploratory study of barriers to green building in Vietnam. *Resources, Conservation and Recycling*, 127, 8–20.

Noh, H.J., 2019. Financial strategies to accelerate green growth. *Handbook of Green Finance. Sustainable Development*. Singapore: Springer.

Ochenkowski, J. and Schinter, J., 2008. *Insurance for Green Buildings* [online]. Buildings. Available from: https://www.buildings.com/feature/article/10192678/insurance-for-green-buildings [Accessed 22 Feb 2023].

OECD, 2018. *Financing Climate Futures: Rethinking Infrastructure* [online]. Available from: https://www.oecd.org/environment/cc/climate-futures/ [Accessed 6 Sep 2023].

Pisani, F. and Russo, G., 2021. Sustainable finance and COVID-19: The reaction of ESG funds to the 2020 crisis. *Sustainability*, 13 (23), 13253.

Ren, X., Shao, Q., and Zhong, R., 2020. Nexus between green finance, non-fossil energy use, and carbon intensity: Empirical evidence from China based on a vector error correction model. *Journal of Cleaner Production*, 277, 122844.

Robinson, G., Leonard, J., and Whittington, T., 2021. *Future of Construction: A Global Forecast for Construction to 2030* | Oxford Economics.

Ross, E.J., Nichols, J.B., and Shibut, L., 2021. *Determinants of Losses on Construction Loans: Bad Loans, Bad Banks, or Bad Markets?*

Sachs, J.D., Woo, W.T., Yoshino, N., and Taghizadeh-Hesary, F., 2019. *Why Is Green Finance Important?* Rochester, NY: Social Science Research Network, SSRN Scholarly Paper No. 3327149.

.aker, D., 2017. *Investing for the Common Good: A Sustainable Finance Frame-.k.* Brussels: Bruegel, 80.

.iütze, F. and Stede, J., 2021. The EU sustainable finance taxonomy and its contribution to climate neutrality. *Journal of Sustainable Finance & Investment*, 14 (1), 128–160.

Siew, R.Y.J., 2015. Alternative framework for assessing sustainable building funds: Green building fund. *Building Research & Information*, 43 (2), 160–169.

Sweatman, P., 2012. Financing Mechanisms for Europe's buildings renovation. *Assessment and Structuring Recommendations for Funding European 2020 Retrofit Targets,* Climate Strategy & Partners and Climate & Company [online]. Available from: https://www.eurima.org/uploads/files/modules/articles/1577093388_Eurima-Financing_Mechanisms.pdf [Accessed 30 Sept 2024].

Sweatman, P. and Hessenius, M., 2020. "Applying the EU Taxonomy": Lessons from the Front Line. *Climate Strategy & Partners and Climate & Company. Accessed May*, 16, 2022.

The Institute for Market Transformation, 2015. *Announcing the 2015 Green Lease Leaders.* IMT.

Transparency Market Research, 2022. *Construction Waste Market – Global Industry Analysis, Size, Share, Growth, Trends, and Forecast 2017–2025* [online]. Available from: https://www.transparencymarketresearch.com/construction-waste-market.html [Accessed 20 Jan 2022].

UNEP, 2018. *Green Financing* [online]. UNEP - UN Environment Programme. Available from: http://www.unep.org/regions/asia-and-pacific/regional-initiatives/supporting-resource-efficiency/green-financing [Accessed 28 Jan 2022].

UNEP, 2021. 2021 *Global Status Report for Buildings and Construction* [online]. UNEP - UN Environment Programme. Available from: http://www.unep.org/resources/report/2021-global-status-report-buildings-and-construction [Accessed 25 Oct 2021].

UNEP, 2022. *2022 Global Status Report for Buildings and Construction.*

US Environmental Protection Agency, 2016. *Definition of Green Building.*

Volz, U., 2018. *Fostering Green Finance for Sustainable Development in Asia.*

Wang, W., Tian, Z., Xi, W., Tan, Y.R., and Deng, Y., 2021. The influencing factors of China's green building development: An analysis using RBF-WINGS method. *Building and Environment*, 188.

Winkler, B., 2020. *Green Buildings and Green Bonds – Environmental Finance* [online]. Available from: https://www.environmental-finance.com/content/the-green-bond-hub/green-buildings-and-green-bonds.html [Accessed 14 Jul 2022].

Woo, J., Fatima, R., Kibert, C.J., Newman, R.E., Tian, Y., and Srinivasan, R.S., 2021. Applying blockchain technology for building energy performance measurement, reporting, and verification (MRV) and the carbon credit market: A review of the literature. *Building and Environment*, 205, 108199.

World Bank, 2021. *Climate Explainer: What You Need to Know about Green Loans [online].* World Bank. Available from: https://www.worldbank.org/en/news/feature/2021/10/04/what-you-need-to-know-about-green-loans [Accessed 14 Jul 2022].

World Economic Forum, 2020. *What Is Green Finance and Why Is It Important?* [online]. World Economic Forum. Available from: https://www.weforum.org/agenda/2020/11/what-is-green-finance/ [Accessed 23 Feb 2022].

WorldGBC, 2021. *WorldGBC Annual Report – 2021* [online]. World Green Building Council. Available from: https://worldgbc.org/sites/default/files/WorldGBC%202021%20Annual%20Report.pdf [Accessed 18 Jan 2022].

WorldGBC, 2022a. *About Green Buildings.*

WorldGBC, 2022b. *What Are Green Mortgages & How Will They Revolutionise Home Energy Efficiency?* [online]. World Green Building Council. Available from: https://www.worldgbc.org/news-media/what-are-green-mortgages-how-will-they-revolutionise-home-energy-efficiency [Accessed 13 Jul 2022].

WorldGBC, 2022c. *Rating Tools* [online]. World Green Building Council. Available from: https://www.worldgbc.org/rating-tools [Accessed 21 May 2022].

Yi, X., Bai, C., Lyu, S., and Dai, L., 2021. The impacts of the COVID-19 pandemic on China's green bond market. *Finance Research Letters*, 42, 101948.

Zadek, S. and Chenghui, Z., 2014. *Greening China's Financial System: An Initial Exploration.*

Zeidan, R., 2020. Obstacles to sustainable finance and the covid19 crisis. *Journal of Sustainable Finance and Investment*, 12(2), 525–528.

Zhang, Y., Wang, Z., Deng, J., Gong, Z., Flood, I., and Wang, Y., 2021. Framework for a blockchain-based infrastructure project financing system. *IEEE Access*, 9, 141555–141570.

Zhang, L., Wu, J., and Liu, H., 2018. Turning green into gold: A review on the economics of green buildings. *Journal of Cleaner Production*, 172, 2234–2245.

Zhang, D., Zhang, Z., and Managi, S., 2019. A bibliometric analysis on green finance: Current status, development, and future directions. *Finance Research Letters*, 29, 425–430.

14 An Evaluation of the Implementation of Sustainable Construction Practices in the UK

An Analysis of the UK Government's Role

Harry Deane and Serik Tokbolat

14.1 Introduction

14.1.1 Background

The construction industry is a significant contributor to the UK's economy. It is an industry that generates more than £110 billion a year and accounts for 7% of the UK's GDP (Construction 2025, 2013). However, it's also largely responsible for the negative impacts that it has on our environment. Due to its vast size, the construction industry is a major consumer of electricity, materials, and water and is a formidable producer of carbon emissions (Akadiri et al., 2012). The term 'Sustainable Construction' (SC) generally refers to the action and the accountability of the construction industry in achieving the concepts of sustainable development (SD) (Gunatilake, 2013). SD has been a common policy topic with the UK Government and voluntary organisations for many years due to the impact that the construction industry has on our environment.

Roughly 40% of the UK's carbon emissions are associated with the built environment and the construction industry is responsible for 18% of large particle pollution in the UK (Smith, 2022). As a result, we are therefore seeing the introduction of legislation and initiatives such as The Planning and Energy Act 2008, The Environment Act 2021, The Paris Agreement, and the UK Climate Change Act 2016 all intended to tackle the growing impact the construction industry has on our environment. These acts hope to promote the use of renewable and recyclable materials, reduce embodied energy in building materials, reduce on-site waste, and reduce energy consumption within the built environment (Gov UK, 2022). However, despite their intentions, there still seems to be a lack of trust in the Government's ability to successfully meet its own targets.

Comments from climate change secretary 'Amber Rudd' showcase the need for the construction industry, consumer groups, and climate change professionals to collaborate in order to successfully implement new policy and create '*a system that works*' (Carrington, 2015). Despite positive contributions from government, voluntary agencies, and the construction industry, there still seem to be a number of barriers hindering the UK's process towards the sustainability agenda. Research has found that the perception of high costs, client understanding, conflicting targets,

DOI: 10.1201/9781003317890-19

and the size of the company have all been identified as barriers within the construction industry to implementing SC practices (Opoku et al., 2015).

Although the industry is inextricably linked to efforts to reach SD goals, the requisite changes and improvements in industry practices are not occurring quickly enough (Kemp, 2017). In 1997 the Government signed the Kyoto Protocol and committed to the Paris Climate Agreement in 2016 (GOV, 2013). Recently, the Government announced £440m funding towards green construction projects showcasing further commitment to curbing the effects of climate change (Lowe, 2021). Recent publications from the UK Government have reiterated their obligation to sustainability by making it a key aspiration for the future. The publication of 'Construction 2025, Net Zero Strategy: *Build Back Greener*' expresses the desire to have an industry that leads the world in low-carbon and green construction exports by 2025 and to be Net Zero by 2050 (GOV.UK, 2013).

Construction 2025 lists various targets, one of which is a 50% reduction in the trade gap between imports and exports for building materials and components (GOV, 2013). However, there is still a trade deficit that has rather grown by over 50% in 2019, thus evidencing a target clearly not being met by the Government (GOV.UK, 2020). Contribution from the Government has significantly grown in recent years which quickly adds to the mounting pressure on the construction industry to facilitate the move towards a SC industry. The UK Government expects to reduce carbon emissions by 78% by 2035 (compared to 1990 levels), become 'net zero' by 2050, and has more recently committed the UK to achieving a 68% reduction in the UK's carbon emissions by 2030 following COP26 (Environmental Audit Committee, 2022).

This study looked at evaluating the approaches used by the government and assessed how successful they are in promoting the implementation of SC practices. In view of this, Sections 14.2–14.5 of this study address the four key topics; Sustainability, The UK Construction Industry, The UK Government, and The Drivers & Barriers. Section 14.6 outlines the Research Methods that were chosen for the study. The Findings & Analysis and Conclusion sections subsequently followed on from the research methods to create Sections 14.7 and 14.8 of the study. The main aim of this chapter is to provide an insight into the role of the UK Government and its approach to delivering a more SC industry. In addition to this, the study hopes to outline the forces both driving and hindering the adoption of SC practices to produce further knowledge that will help guide both the Government and the construction industry to a more effective approach in delivering the sustainability agenda.

14.1.2 Research Gaps

Sustainability within the construction industry has been studied for a long time. Studies have shown the negative impacts that the industry has on the environment and further highlighted the key relationship between the construction industry and SD (Sev, 2009). Sustainability within the construction industry has been a common subject for researchers due to the contributions the industry makes both to the economy but also to the production of carbon emissions (Akadiri et al., 2012). Research has outlined many SC methods; however, there is little research outlining the role of the UK

Government in successfully achieving them. Whilst the Government set out policies and regulations for the construction industry to abide by, there is little research that evaluates how effective the Government is in delivering its requirements. Hence, this study looks to research that gap to conclude effective approaches that are currently being used, but also to initiate further research into the subject.

14.1.3 Research Rationale and Key Questions

Organisations and government bodies are focussing on implementing a healthy and robust strategy for sustainable procurement within the UK. The construction industry has been highlighted as being heavily responsible for the impact it has on the environment and its occupants. As a result, the construction industry is at the forefront of change needed to mitigate the negative impacts it creates. The Government continues to update and implement relevant strategies in the form of policy, legislation, and initiatives to help promote the construction industry to become more sustainable. However, considering the ambitious targets that are being asked by the Government, and recent failures that have looked at achieving certain targets, it poses several key questions that have been used to guide the research approach:

- How effective is the UK Government in driving SD within the Industry?
- What are the drivers and barriers currently impacting the adoption of SC practices within the UK construction industry?

14.2 What Is Sustainability?

To understand the growing role that sustainability has worldwide, it is important first to provide clarity on the meaning of sustainability and how it is perceived. According to Spence and Mulligan (2005) there are two common aims when accounting for sustainability:

1 'To accelerate human growth, especially in the poorest countries'.
2 'To eliminate the current world's inequities. All whilst avoiding the depletion of the earth's natural resources to establish no risk on future generations becoming destitute'.

14.2.1 The Brundtland Report

The concept of 'Sustainable Development' dates to 1987 when it first appeared in the 'Brundtland Report'. The report was a document produced by the UN (United Nations) under the 'World Commission on Environment and Development' (Acciona, 2019) The report defines SD as such:

> Sustainable Development is development that meets the needs of the present without compromising the ability of future generations to meet their own needs.
> (WCED, 1987)

Known worldwide, this report is considered a significant landmark in the promotion of sustainability and has in turn encouraged both industrial and organisational change across the globe. Initiatives such as the UN Earth Summit, UN Framework Convention on Climate Change, and Agenda 21 have all materialised because of the report (Holden et al., 2014). It emphasises the importance of global awareness in the face of environmental disasters that threaten the global ecosystem. Climate change, ozone depletion, industrial pollution, species extinction, and nuclear destruction are all described as 'new realities' in the report (WCED, 1987). However, despite the recognition of environmental threats, the report also identifies the role of social and economic factors. The definition portrayed above contains two key elements towards SD:

> The concept of needs, particularly the vital needs of the world's poor in which overriding priority should be given.
>
> (WCED, 1987)

> The idea of limitations which is imposed by the state of technology and social organization on the environments' ability to meet both present and future needs.
>
> (WCED, 1987)

The report therefore creates a much broader approach to the often wrongly perceived subject of sustainability being exclusive to the environment. The report refers to three elements: social, economic, and environmental in which an integrated balance is required to achieve long-term SD goals (Mcchesney, 1991). This report has acted as an eye-opener for individuals and organisations across the globe with a clear message portrayed through its signified definition (Anand and Kumar, 2020). It is therefore essential that it is well understood and clear in its objectives.

Although clear in its meaning, the definition is difficult to measure. Holden et al. (2014) criticise the approach by explaining that the SD concept is overly complex and therefore ineffective in guiding policymaking. Likewise, in the opinion of Combet (2020), the meaning of SD says nothing specific about organisations and their ability to achieve sustainable goals. Combet (2020) also states that it can therefore '*mean anything to anybody*'. Though Combet criticises the report, there is an understanding of the importance of '*managing the relationship between the needs of humans and their environment (biophysical and social) in such a way that critical environmental limits are not exceeded and modern ideals of social equity and basic human rights (including the 'right to development') are not obstructed*' (Du Plessis, 2007).

14.2.2 The Triple Bottom Line Theory (TBL)

A similar approach to the report is one created by John Elkington in 1994 who introduced the Triple Bottom Line Theory (TBL) theory to create the idea that businesses should focus more on people and the planet rather than just profits (Kenton,

2020). The key aspect of the TBL theory promotes the need for businesses to integrate the three bottom lines:

- Profit
- People
- Planet

The aim of John Elkington was to create a different mindset for businesses to adopt, a mindset that steered away from the general aim to achieve financial success and look more into the consideration of the other two factors (Elkington, 1998).

Again, criticism of this theory can be shown in its inability to measure. Where profits can be quantified it is difficult to do the same for the latter two aspects of the approach (Sridhar and Jones, 2012). Criticisms have outlined complying as proving a difficulty as organisations find it an issue to translate the concept into a strategic plan (Srivastava et al., 2021). Both the theories created by Elkington and Brundtland seem to share the same characteristics by focusing on a balance of Economic (Profit), Social (People), and Environmental (Planet) factors when accounting for sustainability. It is quite clear that there is a common understanding and need for all three of these subjects to be integrated and maintained. With that, the focus on sustainability should not be subject to only improving the environment. The 'Three-Pillar Approach' (TPA) is commonly represented as three overlapping circles and is what shapes the Brundtland report. In its entirety, it is the simplest way to break down something as complex as SD (Gunatilake, 2013). The concept has been favoured due to its flexibility which has allowed it to provide an open and adaptive approach that can be applied to a variety of situations and organisations (Grossman, 2012). Despite criticisms, the above concepts have all acted as significant milestones in the evolution of SD with the TPA being one of the most used ways of representing it today (Gunatilake, 2013).

14.3 The Construction Industry

Worldwide, the construction industry has a huge role to play in achieving SD. It is an industry targeted often by governments due to its large economic output and its ability to consume and produce mass quantities of unsustainable products. The construction industry is responsible for more than a third of the world's energy usage (Cheng et al., 2008). The industry is a mass consumer of concrete, plastics, and fossil fuels and is an industry that is expected to reach £16.6 trillion by 2025 with a compound annual growth rate of 7% (Business Wire, 2021). The UK Government have estimated the private residential sector will need 1.7–2.5 million new homes by the year 2025 to meet rising housing demand (Committee on Climate Change, 2019). Boris Johnson back in June 2020 stated that we need to '*Build, Build, Build*'. He implemented a 'New Deal' which was released in the wake of the COVID-19 pandemic promising huge investment from the UK Government to help develop new infrastructure and promote economic recovery (GOV.UK, 2020).

The UK Government is ambitious in developing and delivering the built environment, it is also ambitious in its sustainability targets. The Government wishes to reduce greenhouse gas emissions in the built environment by 50% and become 'Net Zero Carbon' by 2050 (GOV.UK, 2020). Considering these objectives with the recent need to '*Build, Build, Build*' it adds to the already immense pressure on the construction industry to deliver a sustainable built environment. The UK construction industry already exploits largely water resources, carbon emissions, and land usage and continues to impact communities across the UK (Adetunji et al., 2003). It is also an industry that is a formidable consumer of non-renewable energy sources. Setting aside its consumption of fossil fuels, the construction industry is a major user of heavy metals such as copper, zinc, and lead, all of which are becoming limited in their existence (Spence and Mulligan, 2005). As construction has significant economic importance and has large environmental and social impacts, the relationship between the construction industry and SD is becoming more and more significant (Sev, 2009).

One of the major issues the construction industry faces is its ability to produce significant amounts of waste. In the UK alone the industry consumes more than 400 million tonnes a year of natural materials (Gould, 2019). A total of 13% of this acts as waste which is sent to landfill sites directly without even being used, 32% of the overall waste within landfill comes from the construction industry (Gould, 2019). With statistics such as these, it comes as no surprise to see the industry targeted for much of the UK's sustainable strategies. The role the construction industry must play in reducing its impact on the environment is apparent. But how effective are current sustainable practices? It is shown that typically buildings contribute to around 50% of all energy produced (UKGBC, 2017). However, findings have shown that the implementation of smart designs, low-energy appliances, improved insulations, etc. can help reduce this energy usage by around 30–50% without having any substantial detriment to the cost of investment (Cheng et al., 2008).

The construction industry continues to adopt SC practices with some being more effective than others. The main approach implemented by the Government is to focus on planning and construction practices (Akadiri and Fadiya, 2013). The UK Government updates and utilises 'UK Building Regulations Document L2A, Part L' to guide the industry as well as using sustainable assessment methods such as 'BREEAM' (Building Research Establishment Environmental Assessment Method) to assess the implementation of sustainable practices on buildings. These methods aim to give the construction industry the knowledge to operate within a sustainable means by reducing energy usage, water, waste, and materials (Williams, 2023). Literature has shown that businesses that meet regulatory requirements and have strategically shifted to adopt them can avoid severe costs later down the line (Khanna and Brouhle, 2009). Likewise, O'Malley et al. (2014) state that sustainable building assessments provide construction organisations with the right guidance for developers to measure, evaluate and implement sustainable goals through all means of the building's life cycle. However, criticism can be found within the writings of Ding (2008) who found that the interaction between the environment and the construction industry is largely still unknown which limits the effectiveness of the assessment methods used. Furthermore, Janna Laan (a natural materials expert) has stated that when it comes to SC '*building regulations are nowhere near good enough*' (Harby, 2021).

Collaboration has been outlined as a key element towards the successful implementation of SC practices within the UK. To successfully adopt them, a true collaboration between government and industry stakeholders is needed (Rahmawati et al., 2014). It is therefore essential that an efficient and well-collaborated framework is applied between the construction industry and the UK Government. The nature of the industry is large, diverse, and fragmented which currently poses a threat to governmental targets within the UK (Construction Excellence, 2008). There is certainly pressure on the construction industry to deliver. The Government is expecting a contribution from the construction industry to account for an overall reduction in 'Greenhouse Gas Emissions' (GGE) by 80% by the year 2050 (GOV.UK, 2013).

Despite attempts, it seems not much has been achieved in counteracting the impacts the construction industry has and continues to have on the environment (Kibert, 2007). Research conducted by Opoku et al., (2015) showed the fragmented nature of the construction industry as a large reason for the lack of progression towards SD. The UKGBC (UK Green Building Council, 2018) explains that to create a more sustainable built environment '*true collaboration and compliance*' is needed. Similarly, Lendlease (named the UK's most sustainable housebuilder) has also outlined the necessity of collaboration within the construction industry when improving sustainability (Hazlegreaves, 2019).

14.4 The UK Government

For many years, the UK Government has developed several policies and guidance documents intended to improve the implementation of SC practices. The UK was the first country to produce its 'national strategy' towards SD in 1994 following on from the 1992 'Rio Earth Summit' (HM Government, 2005). The strategy first analysed the common principles of SD starting with the assumption that most societies seek to achieve economic development to obtain a 'better quality of life'. Second, it outlines the 'need to protect and enhance the environment' (HM Government, 2005). The 1994 strategy provided a foundation for the UK's response to the global issue of climate change.

Since 1994, the Government has continued to develop their strategies. In 2008 the Government created the 'Strategy for Sustainable Construction' which outlines drastic changes needed to achieve SD (HM Government, 2008). The strategy was developed to promote leadership and common change in SC methods, as well as bring benefits to both the economy and the industry. As a result, various government bodies and several industry-related institutions have produced various policies, guidance, reports, etc. outlining SC objectives. However, the Government has not always stayed true to these, David Cameron was elected as Prime Minister in 2010 and during his term, the Government cut back on several green policies, reduced subsidies to renewables, and scraped a project intended to capture carbon from the atmosphere (Herbert, 2019). In addition to this, the Climate Change Committee (CCC) warned that the Fourth and Fifth Carbon Budget set will not be achieved based on current strategies and policies.

British and International standards are the main documents referred to in guiding and dictating common building requirements. As a result, it is apparent that these standards refer to appropriate and relevant guidance to implement necessary sustainable practices. Part L of the Building Regulations (BR) has been criticised due to the volume of regularity changes (Bell and Lowe, 2000). Further literature has criticised BR for failing to drive the industry to its full potential in terms of reducing CO_2 emissions (Gunatilake, 2013). Similarly, Halliday (2018) states that the existing SC requirements appear to be substantially lower than what is expected of, and achievable by the industry. These regulations seem to promote minimum requirements with best practices not being encouraged. Therefore, it seems most organisations within the industry that meet the minimum requirements are failing to address certain basic SC issues (Rydin et al., 2007).

In 2013, the UK Government published the 'Construction 2025' strategy in which sustainability was a key element. The strategy looks at reaching ambitious targets: a 33% reduction in the cost of construction; a 50% reduction in time taken to deliver projects; a 50% reduction in greenhouse gas emissions; and a 50% reduction in the trade gap of construction products and materials (Construction 2025, 2013). Whilst these targets are in line with the concept of SD by accounting for people, the environment, and the economy there is scepticism surrounding the ability to meet them. The 'Construction Leadership Council' (CLC) was responsible for the 2025 strategy and admitted that they themselves were not measuring progress on its targets (The Construction Index, 2017). Although the strategy promises to be hugely beneficial to the economy, environment, and the public, Smith (2014) argues that the strategy fails to identify how the ambitious targets are expected to be achieved. Smith (2014) also highlights that without a strong plan, it could have serious repercussions by causing overruns on cost (upwards of £15–£20 billion per year) and incurring delays in overall project delivery.

One of the major UK targets associated with the environment is 'Net Carbon 2050'. In June 2019, legislation passed which obliged the Government to reduce the UK's net emission of greenhouse gases by 100% before 2050 (GOV.UK, 2020). Net zero entails achieving a balance between the production of greenhouse gases and the amount taken out of the atmosphere (Perry, 2020). The UK intend to '*lead*' the world in becoming the first major economy to pass a law to reduce emissions to net zero by 2050 (GOV.UK, 2020). However, this is not the first time the UK has been introduced to a target that has urged to meet zero carbon. In 2015, the Government axed the 'Zero Carbon Homes by 2016' policy after again stating the UK to be the '*first country to make such a commitment*' (Oldfield, 2015). The policy aimed to significantly reduce CO_2 emissions from housing which is currently the fourth most significant contributor to the UK's greenhouse gas emissions (Shepheard, 2020).

The Government declared in 2021 that support for the 'Green Homes Grant' (GHG) will cease funding in 2022 (Shankleman and Moralws, 2021). The Government launched the GHG to help minimise energy consumption in UK households, but it was terminated due to the '*lack of interest from consumers*' (Johnson, 2021). In 2015, ministers discarded the 'Green Deal' which was a policy created in 2013 that gave homeowners the opportunity to save on energy bills and therefore reduce energy usage (Vaughan, 2015). This is another clear example of an initiative produced by the Government that has failed to achieve its goal. The Government's growing inability to stick to policy is more and more evident within the built environment and the constant scrapping of decarbonisation goals gives the impression to the public that our goals are too difficult to meet (Mace, 2022).

14.5 Drivers and Barriers

14.5.1 Drivers

There are several drivers currently pushing the construction industry to adopt more SC methods. The association between SC and the economic reward of adopting such practices has been identified as a factor that encourages implementation within the construction industry's supply chain (Opoku et al., 2015). Financial incentives, legislation/regulations, and client demand have also been found to be some of the main factors in driving the adoption of SC methods (Rademaekers et al., 2012). However, the driving force behind the approach towards the sustainability agenda is inevitably the Government. They enforce the need for change through incentives, policy, and legislation, the likes of which empower drivers and reduce barriers (Construction Products Association, 2007).

Cost savings, better quality products, increased market share, enhanced health and safety, and corporate image have all been associated with encouraging the implementation of SC practices (World Green Building Council, 2009). For businesses, sustainability can help to build a better image, provide a competitive advantage within the market, attract new work as well as employees, and promote repeat purchases (Liu et al., 2020). SC practices are considered one of the main success factors in growing efficiency and competitiveness within business organisations (Tan et al., 2011). Findings from Alex Opoku et al. (2015) illustrate that the most influential driver behind construction organisations adopting sustainable practices is the commercial benefit which superseded drivers such as energy efficiencies and green reputation. Winning jobs, gaining reputation, attracting employees, boosting competitive advantage and stakeholder influence have all been identified as significant drivers.

Despite these being drivers, they suggest that construction organisations are prioritising the business element of sustainability by focusing on the potential profit gain. This would contradict the SD concept where it has been established that a balance is required and that less focus should be on profits. The list of drivers towards the implementation of sustainable practices extends beyond the Government and delves more into what is effective for the business. The advantages sustainability can bring to construction organisations through reputation and competitive advantage are becoming apparent (Holton et al., 2008). The construction industry is very much driven by client demand and as a result plays a major role in the adoption of SC within organisations (CIOB, 2016).

14.5.2 Barriers

The UKGBC (2009) has outlined that the number of varying guidance documents is currently acting as a barrier to implementing SC practices within the industry. Despite there being positive contributions from the UK Government, there is still a growing concern regarding the lack of investment, research, development, and skills within the field of SC (Shan et al., 2017). More recently, literature has highlighted that lack of collaboration between Government stakeholders is posing as a barrier. Fiona Harvey (2021) states that MPs are failing to guide their departments to adopt the net zero goal and that minimal coordination between central and local governments is occurring. Meg Hiller (chair of the public accounts committee) comments that '*Government has set itself a huge test in committing the UK to a net zero economy by 2050, but there is little sign that it understands how to get there*' (Meg, 2021). A recent report published by the Environmental Audit Committee (EAC) comments that the industry had received '*little government guidance*' when looking to achieve UK environmental targets (Haynes, 2022). Considering recent literature paired with the Government's inability to deliver past policies such as the Zero Carbon 2016 scheme, it does test the faith one can put in the Government and its ability to achieve such targets.

Furthermore, findings from Alex Opoku et al. (2015) show that a lack of understanding of sustainability and the perception of high costs related to sustainability are acting as considerable barriers to implementing SC practices. Furthermore, the challenge of managing the risk associated with SC practices has also been highlighted as a key challenge for businesses when seeking to implement SC practices (Opoku et al., 2015). This is reinforced by the findings of Apine and Valdés (2017) who found that more experience and expertise are needed to control and manage the risk associated with the adoption of new practices and technologies, which could have unforeseen impacts on project deadline and cost. Research from Pitt et al. (2009) states that the lack of general awareness and demand for sustainable buildings is proving a challenge for businesses to implement sustainable solutions.

In general, SD aids in the reduction of energy use and operating costs whilst preserving natural and social habitats (Liu et al., 2020). It also supports the provision of a safe and pleasant living climate as well as boosting the financial performance of both developers and occupiers (Liu et al., 2020). The perception that sustainable buildings '*cost more*' is in fact misinterpreted as it has been found that the whole life cost of sustainable buildings is generally cheaper (Opoku et al., 2015). Sustainability should not be considered an 'add-on' element but rather a necessary aspect to be considered throughout the whole project life cycle (Eley, 2011).

14.6 Research Methods

14.6.1 Aims and Objectives

The purpose of this study is to establish how effective the UK construction industry is in achieving SD. Reviewing both the role of the Government as well as the construction industry to suggest where improvement is needed. The study was guided by several objectives which helped feed the overall aim, these were (1) to evaluate the current meaning of sustainability and review how sustainability is perceived, (2) to analyse the economic impact that sustainable measures have on UK construction organisations, (3) to analyse the drivers and barriers impacting the adoption of SC practices within the UK, and (4) to evaluate the UK Government's approach to achieving SD.

14.6.2 The Mixed Method Approach

Based on the subjective and broad meaning of the definition of sustainability, the study aimed at using a mixed method approach. Research has described the mixed method approach as a '*type of research where the researcher combines elements of quantitative and qualitative research approaches for the broad purposes of breadth and depth of understanding and corroboration*' (Johnson et al., 2007). According to Creswell (2009), the use of mixed methods creates the opportunity to overcome biases obtained from one method by using another. Due to the complex and fragmented nature of the construction industry, the study utilised the mixed method approach to gain relevant data through both primary and secondary forms of research, providing the study with solid data in which to analyse.

14.6.3 The Qualitative Approach

The 'Qualitative' approach of this study utilised semi-structured interviews with a mix of industry professionals. Fellow and Liu (2008) argue that interviewing helps the study to harness detailed information on an individual's perceptions, opinions, and feelings. Interviews allowed the study to adapt certain questions with regard to the individuals' subjective responses, helping to obtain relevant information based on the opinionated topic of sustainability. Furthermore, research has shown qualitative methods to be effective approaches when addressing real-life situations (Dainty et al., 2000).

The subject of sustainability can be perceived in many forms; hence interviews were chosen as the most effective method of harnessing thoughts and feelings on the subject. Interviews were targeted at a minimum of 15 well-experienced construction professionals sampled through formal emails which were issued to the professionals requesting them to participate in the study. The professionals were targeted randomly through the social media platform LinkedIn. The study targeted participants who had industry experience of over 30 years, helping to establish reliable and valid data for the research project. One individual was chosen to have less than ten years' experience to see how much of a factor experience, and age were in the answers given. All interviews were performed on Microsoft Teams and recorded (with consent). The research sample was carefully chosen to help create a pool of professionals that ranged in background and experience (see Table 14.1). The subject of sustainability is broad and often wrongly perceived, the sample was chosen to obtain a strong range of opinions from a number of different sectors and occupations needed to effectively analyse the results to help establish the study's aim and objectives.

Table 14.1 Interviewee Industry Background

Interviewee	*Company*	*Background*	*Years of experience*
1	Company A	Contractor	35 Years
2	Company B	Sustainability Consultant	6 Years
3	Company C	Contractor	30 Years
4	Company D	Sustainability Consultant	35 Years
5	Company E	Property Developer	35 Years

14.6.4 The Quantitative Approach

The 'Quantitative' research approach utilised online questionnaires issued to construction professionals with vast experience within the industry. The questionnaires helped to establish the objectives of the study that were being explored. This brought clarity on the industry's perception towards SD as well as the UK Government's role. The questionnaire consisted of open and closed questions which were statistically analysed, and rating scale questions were also used to enhance the generation of numerical data. Although most questions remained closed, the involvement of open questions has been shown to be an effective tool in harnessing key data when undertaking questionnaire research (Krosnick et al., 2009). The questionnaires were issued through an e-mail using an anonymous 'Qualtrics' survey link. The study targeted a sample of 100 professionals to participate in the questionnaire. Social media was a significant tool used to promote engagement for the questionnaires, an anonymous link was broadcast throughout several social media platforms such as Twitter, Facebook, and LinkedIn aimed at random construction professionals who were open to participating in the study.

14.6.5 Method of Analysis

A total of 100 participants were targeted for the questionnaires and a total of 15 participants were invited for the semi-structured interviews. All participants remained anonymous which helped to ensure responses that were given were true to the social style and personality of that participant, thus reducing bias through the 'Interviewer Effect' (Marsden and Wright, 2010). The 'Interviewer Effect' is where the interviewee responds with answers that they believe to be the desired response from the question, rather than their own opinion (West and Blom, 2016). The results of the questionnaire were statistically analysed to showcase the relevant findings through a visual representation thus highlighting the key findings of the study. DePoy and Gitlin (2015) explain that numerical data provides an accurate and standardised language in which phenomena can be described.

The semi-structured interviews were crucial to the research by establishing relevant detailed opinions with regard to sustainability in the construction industry. The interviews were first transcribed to allow the study to obtain a clear and visual account of the interview. Next, the transcript was coded and categorised into themes. Coding qualitative data makes interpreting interview responses easier; it is done by assigning 'codes' to words and phrases to help better the analysis of the results

(Thematic, 2019). The themes aid the analysis of data and show the patterns or trends obtained. Coding the interviews ensured there was a concise evaluation of the data within key areas that were being explored. It is one of the most crucial steps in the analysis process and is used to arrange and make sense of the textual data obtained (Basit, 2003). The overall results obtained have allowed the study to compare the findings from the literature review and relate that to the results obtained from the interviews/questionnaires to generate areas of strength and limitation.

14.7 Findings and Analysis

The questionnaire results were tabled, put into charts, and statistically analysed. The interview results were coded manually through a 'Vivo Code' system. Each piece of data obtained is directly analysed and interpreted, thus allowing the study to outline the findings and draw several conclusions.

14.7.1 Questionnaire Findings

The sample obtained for the questionnaire can be seen above in Table 14.1. Despite having a range of job professions involved in the study, most of the answers generated are coming from professionals working within consultancy firms (30/50).

14.7.1.1 Question 1. What Type of Company Do You Work for?

Table 14.2 Participants' Place of Work

Type of company	*Number of participants*	*Percentage of participants (%)*
Consultant	30	60
Main Contractor	8	16
Designer	3	6
Developer	6	12
Self-Employed	3	6
Total	**50**	**100**

14.7.1.2 Question 2. How Many Years of Industry Experience Do You Have?

The results obtained are displayed above in Table 14.2 which showed that significantly 50% (25/50) of respondents have an industry experience of more than 20 years. The second highest (20%) shows participants having an experience between 11 and 15 years. Only three participants were recorded as having an industry experience between one and five years, with no participants registering less than one year.

14.7.1.3 Question 3. Using a Scale of 1–10, Please Rate Your Knowledge and Understanding of the Three Pillars of SD, According to How Important Each Is to the Construction Industry

Question 3 was included to gauge the participant's knowledge surrounding the popular 'Three Pillars' concept towards SD. Only nine participants were not aware of the concept and so did not score a significance rating. The 'Social' element scored the highest average at 7.36/10 but this was closely followed by 'Economical' and 'Environmental' which both scored above 7.10. These results display that most of the participants are aware of the concept and consider both Economical and Environmental factors as being the most important. However, as outlined by both Elkington and Brundtland, a balance of each three of the pillars is needed to achieve SD (WCED, 1987).

14.7.1.4 Question 4. The UK Government Is Providing the UK Construction Industry Adequate Support to Implement SC Practices, Do You Agree?

Question 4 looked at gaining an opinion on the UK Government and its role in promoting SC. Most participants (21/50) 'Disagree' with the statement and do not believe the government is providing sufficient support needed to implement SC. A further 9 respondents out of 50 'Strongly Disagree' with the statement, with only 6 participants stating that they 'Agree' and 4 stating they 'Neither Agree nor Disagree'. None stated that they strongly agreed.

14.7.1.5 Question 5. Do You Know Any of the Following Policies/Legislations?

Table 14.3 is the list of recent government policies, legislations, and initiatives. Question 5 looked to establish how aware the respondents are of these documents put forward by the Government. The most well-known is the 'Net Zero 2050' policy of which 84% of respondents were aware of; this was closely followed by the following:

- Climate Change Act (80%)
- Building Regulations (60%)
- The Construction 2025 (50%)

4% of the participants were not aware of any of these documents.

Table 14.3 A Table of Recent Government Policies, Legislations, and Initiatives

Government policies/legislations/initiatives
Construction 2025
Net Zero 2050
Climate Change Act
Energy Performance and Building Regulations
Clean Growth Strategy
Industrial Strategy
Government Construction Strategy 2016–2020
Construction Sector Deal

14.7.1.6 Question 6. How Has Been Your Past Experience When Dealing with Projects Considering SC Practices? (Very Easy – Very Difficult)

Question 6 was included to help develop an understanding of the experience professionals had when working on projects that adopted SC practices. Most respondents found working on projects adopting SC practices to be a 'Difficult' experience as 28/50 participants had this opinion. Only one respondent found the experience to be 'Very Difficult' with no participants registering a 'Very Easy' experience. The majority of 'Difficult/Very Difficult' answers were backed up with comments stating that there was 'Difficulty complying with governmental regulations/guidance'.

14.7.1.7 Question 7. On A Scale of 1–10, How Relevant Do You Consider the Following 'Drivers' Impacting the Implementation of SC in UK Construction Organisations?

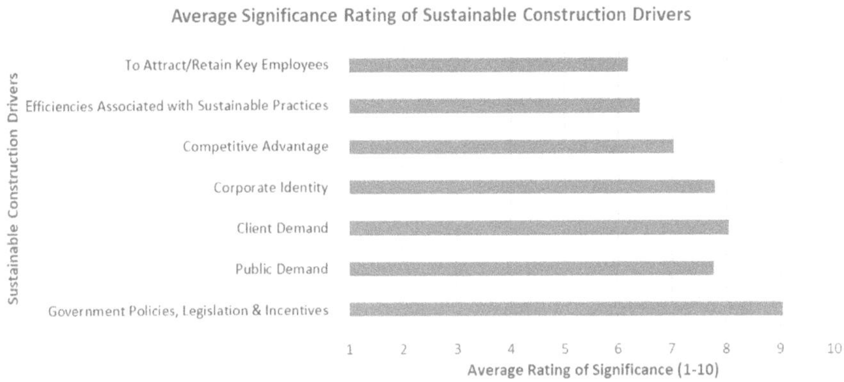

Figure 14.1 On a scale of 1–10, how relevant do you consider the following 'drivers' impacting the implementation of SC in UK construction organisations?

The driver found to have the highest average rating of significance was 'Government Policies, Legislation & Initiatives' (9/10). This result seems to compare with that of Rademaekers et al. (2012) who found that 'Financial incentives, legislation/regulations and client demand' are some of the main factors in driving the adoption of SC methods. Significantly, the second highest driver found within Figure 14.1 was client demand with an average score of 8.1. The lowest driver was found to be 'Attracting/ Retaining Key Employees' with a rating of only 6.2. Both 'Competitive Advantage' and 'Corporate Identity' scored relatively high but not as high as 'Client Demand'.

14.7.1.8 Question 8. On a Scale of 1–10, How Relevant Do You Consider the Following 'Barriers' Impacting the Implementation of SC in UK Construction Organisations?

Most participants found that the 'Perception of cost associated with sustainable practices' was the highest barrier currently hindering the adoption of SC (8.7/10).

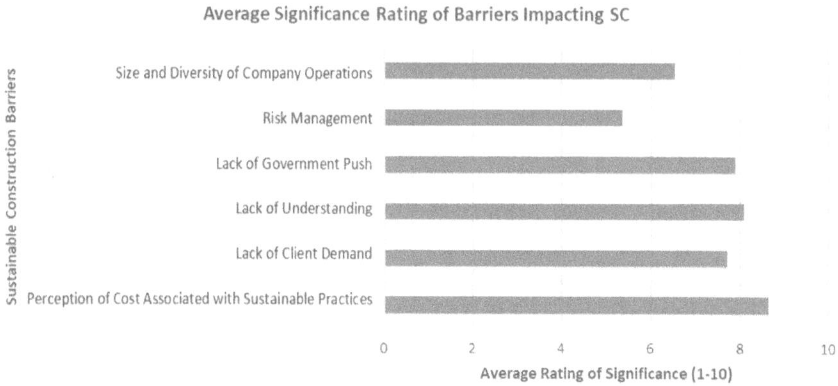

Figure 14.2 On a scale of 1–10, how relevant do you consider the following 'barriers' impacting the implementation of SC in UK construction organisations?

Although the literature review had found this perception to be 'wrong' (Cheng, 2008), most responses that were given back in question 12 had been related to 'unknown costs' as a difficulty when adopting SC. Not shortly behind were: Lack of Understanding (8.1), Lack of Government Push (7.9), and Lack of Client Demand (7.6)

14.7.1.9 Question 9. Would You Agree that Current Set Targets by Government on Sustainability within the Construction Industry Are Achievable? (Strongly Agree – Strongly Disagree)

Question 9 was crucial in understanding whether professionals within the construction industry believe the UK Government can deliver their own set targets. From the results, it can be determined that most professionals DO NOT believe that government targets towards sustainability are achievable. A total of 20/50 (40%) participants registered 'Disagree' with 6/50 (12%) stating that they 'Strongly Disagree'. Only 14/50 stated that they agreed and 10/50 neither agreed nor disagreed.

14.7.2 Questionnaire Analysis

The questionnaires were an integral part of the study as they provided a set of standardised results which could be compared and analysed against the literature and interview results. Questions 1, 2, 3, and 5 were structured to give the validity and representativeness of the results by helping to build a profile of the participants' experience and knowledge of sustainability in the construction industry. Despite there being a ranged sample, only 50 professionals were obtained for the questionnaire with most professionals coming from a consultancy background. The sample obtained was not significantly diverse and therefore results cannot be representative of the whole construction industry but can still be analysed and interpreted as valuable to the study.

The results obtained from Question 4 showed that 60% of participants believe the UK Government is not providing adequate support for the UK construction industry to implement SC practices. This would suggest that the Government is not successfully working with the industry and would contradict the current Construction 2025 agenda which aims to seek a collaborative approach (Gov, 2013). It could be said that existing government policies and regulations are not aligned with the needs and goals of sustainability within the industry, suggesting a re-evaluation may be necessary. However, the results could indicate a diverse range of opinions, although most disagreed with the statement, there is little to suggest why. Some respondents may be satisfied with the government's initiatives, while others may see room for improvement.

The results from Question 6 indicate that most professionals who had worked on projects that adopted SC practices found the experience to be somewhat difficult. Despite most participants experiencing some level of difficulty it may not necessarily mean this is representative of all SC practices nor does it indicate that this had extreme or serious consequences on the project. Whilst the results can determine that adopting SC practices comes at a level of difficulty, the results are not significant enough to assume these are detrimental to projects and/or their ability to adopt SC practices. The responses given here can suggest there is a lack of collaboration, as professionals are unable to comply with government guidelines thus creating a difficult experience for industry stakeholders. However, those who answered 'Easy' generally commented that working with sustainable consultants made it easier, showcasing the effectiveness of specialist assistance in delivering SC practices. While the results can indicate there is a level of difficulty, the limited sample size and the subjective nature behind people's experiences and opinions mean it can be difficult to generalise the findings.

Questions 7 and 8 were essential in finding out what professionals believe to be the most significant factors currently driving/hindering the implementation of SC practices. Several drivers were outlined as significant. Government policies, legislation, and incentives were scored as the most significant, no doubt due to the important role each has in guiding the use of SC practices. However, the literature review identified that government policies, legislations, and incentives can be subject to change and even ceased (Shankleman and Moralws, 2021). The identified drivers are interconnected and can vary in significance depending on the context and individual circumstances of construction firms. While government policies, client demand, and corporate identity are recognised as important divers, the lower ranking attracting key employees and competitive advantage may indicate a potential blind spot in the industry's understanding.

Results showed the perception of cost to be the most significant barrier to implementing SC practices within the construction industry, shortly followed by lack of client demand, lack of understanding, and lack of government push. It is not surprising to see that cost is considered the most significant barrier as the findings from the literature review had already listed this as a barrier (Opoku et al., 2015). However, despite being outlined here as a 'barrier', the cost has also been identified as a driver due to the long-term savings that can come from implementing

SC practices (World Green Building Council, 2009). This could suggest that the perception of high costs associated with the implementation of SC practices is just a perception and not a reality and that potentially there is a gap in knowledge and understanding around the subject of cost. Lack of understanding was also found to be a significant barrier. Considering government legislation and policies were found to be the most significant driver it is interesting to see a lack of understanding as an equally significant barrier. This could suggest that there is little understanding within the construction industry as to how the government is effectively implementing policy and legislation towards the adoption of SC practices.

The final question was included to understand whether the construction industry is generally trusting in the UK Government and its ability to meet set targets. The results show that most professionals do not believe that current government targets towards sustainability are achievable. The ten respondents who neither agree nor disagree indicate that there is some uncertainty or lack of information among a portion of the sample. Most professionals disagreed which shows there is a level of scepticism within the construction industry. This scepticism towards the government could be due to various reasons, such as a perceived lack of resources, insufficient strategies, or past failures in meeting goals. It is therefore important to understand what exactly the professionals believe is not achievable and why. That said, the results clearly show a level of scepticism that regardless of the reason, is acting as a barrier towards SD within the construction industry.

14.7.3 Interview Analysis

Five semi-structured interviews were conducted as part of the primary research. Each interview was transcribed and coded into relevant themes drawn from the data obtained. 'In Vivo Coding' was used in order to help create the themes. For ethical considerations, the participants' names were left out and each is referred to as 'Interviewee 1, 2, 3, 4 and 5' (see Table 14.1).

14.7.3.1 Thematic Analysis (In Vivo Code)

The following themes were chosen due to the frequency in which they occurred throughout the interview transcripts: Meeting Generational Needs (A), Cost Implications (B), Competitive Advantage (C), Legal Requirement (D), and Compliance (E). These themes are related back to the studies' objectives and are compared against the literature review and the findings of the questionnaire.

14.7.3.2 Theme A: Meeting Generational Needs

Theme A was attributed to the responses of all interviewees regarding how they interpreted the meaning of sustainability within the construction industry. All five interviewees identified that SD relates to the ability to meet current generational needs, whilst ensuring future generations to meet their own. The definition provided by each interviewee shares the same definition within the 'Brundtland Report' (WCED, 1987). Collectively, all interviewees provided reassuring responses with regard to this question which shows the interviewees have a good awareness of the definition of SD.

Interviewee 5 comments '*myself and the company share the same understanding to ensure that future generations are able to meet their own needs without our generation diminishing their ability to do so*'. Similarly, comments from Interviewee 4 state '*this is the definition our company roles out too, and I think this is the general understanding of sustainability*'. Responses from both interviewees suggest that large property companies within the industry are aware of sustainability and its definition in relation to the Brundtland report. It can be said that the construction industry is aware of sustainability at the top level right through to the employees. Questions should be directed as to why there is a slow progression into adopting SC practices if there is a good understanding of the subject. Understanding the definition is key as research gathered from the literature review has stated that a 'lack of understanding towards sustainability' is threatening the successful adoption of SC practices (Holden et al., 2014).

14.7.3.3 Theme B: Cost Implications

Cost Implications have proven to have a strong relationship with regard to SC. Throughout the literature review and questionnaire findings, the 'Cost' of sustainable practices has been found to deter investors. Four of the five interviewees relate to cost as a barrier. It seems that the cost benefits of adopting SC practices are not being seen by clients and therefore there is a lack of desire to push for them. Interviewee 3 states that '*clients are intimidated by the cost, whether it's the cost of adopting sustainable methods or the cost involved with the research and development, it isn't worth risking high costs associated with sustainability where familiar, cheaper and just as effective solutions can be utilised*'. Similarly, Interviewee 5 refers to the concern of cost implications surrounding the adoption of sustainable methods by stating that '*value must be seen, but there are still some companies who do not see the benefits of sustainability. One is because they do not see the value. Two, because of the lack of productivity and quality of some elements*'.

These responses can be referred back to opinions given by participants in the questionnaire. Many explain that their 'difficult' experience was due to 'cost being a concern to the client'. Similarly, 'In Vivo Coding' shows that participants who were asked about the barriers impacting SC answered '*costs related to SC practices*' to be the most significant barrier. The findings from A. Opoku et al. (2015) state that this perception is in fact misinterpreted as they found that the whole life cost of sustainable buildings is generally cheaper; however, the findings obtained within this study show that the perception of high costs is still very much present within the industry, contradicting the findings of Opoku et al. (2015).

14.7.3.4 Theme C: Competitive Advantage

Mentioned throughout the interviews was the competitive advantage that sustainability can bring to a company. When asked the main reason behind considering sustainability in past projects, Interviewee 5 mentioned, '*it is our value, we look at driving sustainability to use it as a competitive advantage as well as a core business value that looks to improve living*'. Likewise, Interviewee 2 states that '*sustainability is considered due to the value and competitive advantage that it gives our business, we look at construction methodology and material components to achieve this*'. Also, when asked if sustainable approaches give companies a competitive advantage over companies that do not 4/5 interviewees answered yes. Interviewee 4 comments that '*companies who adopt further sustainability do have a competitive advantage yes, but the client's budget does limit how far they can go*'.

The findings obtained can be compared to that of Tan et al. (2011), who found SC practices to be one of the main factors in boosting competitiveness within construction businesses. However, the results found from the interviews seem to contradict that of the questionnaires. Competitive Advantage was discovered to be the third lowest driver in the adoption of SC practices whereas the interview responses highlighted competitive advantage as a key factor towards the implementation of SC practices.

14.7.3.5 Theme D: Legal Requirement

Legal Requirement was one of the key themes relating to this study. Given the objective and overall aim of the study, this theme was critical in gauging the interviewees' opinions on governmental involvement in implementing SC. When asked 'what were the main reasons for considering sustainability within past projects?' most interviewees (3/5) responded that it was due to '*legal requirement*'. Interviewee 5 said, '*it is such a legal requirement now, you won't get your planning without meeting certain sustainable requirements*'. Similarly, Interviewee 4 (2021) stated that '*the main reasoning behind considering sustainable practices was down to the legality of it, we are required to meet certain targets i.e., energy usage, water usage and production of carbon*'.

Both the literature review and questionnaire results highlighted government legislations, policies, and initiatives being the main drivers towards the sustainability agenda within the UK construction industry (Figure 14.2). Interviewees outlined that the main documents referred to were BR and Sustainable Assessment Methods. However, Interviewee 5 mentioned, '*we have our own sustainability brief and framework that we drive internally, more often than not, this looks at complying with BREEAM standards*'. Despite the Government's legal requirements, it seems companies adopt their own internal framework that they use in accordance with the legal guidelines and regulations. Interviewee 3 states, '*we have our own Corporate Social Responsibility (CSR) policy and sustainability programme that works in line with government standards*'.

14.7.3.6 Theme E: Compliance

Theme E highlights the interviewees' opinions on complying with government documents and guidelines. In general, the interviewees found complying with documents issued by the government to be a mixed experience. Interviewee 1 comments that '*most documents are easy to understand, but I must admit the policies and targets can get complicated to comply with*'. Similarly, Interviewee 4 stated, '*for one project the Government wrote the national policy statement, to which the client needs to comply. However, it was found that the government did not comply with the Paris Climate Agreement within the statement, and this incurred delays in the project as the Government had to correct this issue*'. Despite interviewee 4 having a bad experience in complying with government documents, it seems that most considered complying an easy experience as three out of five of the interviewees mentioned having no difficulty in complying. In contrast to this, Interviewee 3 stated that '*The Government makes certain requirements but don't consider the nature of the business in achieving them, they have requirements, but these sometimes create disputes in attempting to comply*'.

14.8 Discussions and Conclusion

14.8.1 The Definition of 'Sustainability'

Findings from the literature review, coupled with results from the primary research led this study to believe that there is one common definition of sustainability. It was largely evident throughout the literature review that to achieve sustainability an integrated balance of economic, environmental, and social elements needs to be established. From the questionnaire findings, it can be determined that most professionals share the same understanding and importance between the balance of these three elements. Further to this, Theme A concluded that all five interviewees and their companies share the same understanding of sustainability. The results obtained from the primary and secondary research allowed the study to determine that the most relevant and referred to the definition of sustainability is as such; 'meeting the needs of current generations without threatening future generations to meet their own'. Furthermore, this definition is clearly reflected within the industry and the overall perception towards sustainability is strongly understood.

14.8.2 The Cost of Sustainability

The literature review allowed the study to outline that the cost of implementing SC practices is wrongly perceived (Opoku et al., 2015). From the study, it can be assumed that there is a strong perception within the industry that implementing sustainable practices can come at a substantial cost. Findings from the questionnaire showed that 'perception of cost' is the most significant barrier currently hindering the implementation of sustainable practices. Likewise, Theme B also shows this perception by proving there is a lack of awareness around the cost benefits of adopting such practices, both at the client and business level.

However, Theme C showcased that SC practices can have a positive impact on businesses by giving them a 'Competitive Advantage', likewise the literature review showed that sustainability can promote new work and repeat purchases (Liu et al., 2020). It can be assumed that businesses that meet sustainable requirements and have developed to do so, can avoid severe costs later down the line for not only their clients but themselves. This is also apparent in the interview findings. Although there is a negative perception that implementing SC practices comes at a substantial cost, doing so earlier can save costs later down the line by reducing energy, electricity, and water usage. This is reinforced by the findings of the World Green Building Trends (2016) in which their 'Smart Market Report' showcased that on average, green buildings are 14% less costly to operate than traditional buildings. The Government will continue to strengthen and develop its sustainability strategies which will impact businesses eventually. Therefore, the cost of adopting such practices now can help reduce the risk of higher costs in the future.

14.8.3 The Drivers vs Barriers

In terms of drivers, there were several different factors impacting the adoption of SC practices. Corporate identity and client demand were highlighted as key factors in driving the construction business to implement sustainable practices. However, findings from the questionnaires show government requirements to be the main driver towards the implementation of SC practices. This was also the case with the interviews where many responses refer to 'legal requirements and compliance', thus highlighting the importance of the Government's role. As a result of the study, it can be assumed that the Government is the driving force towards businesses implementing SC methods.

On the other hand, several barriers were clearly identified through the literature review and reinforced as a result of the research findings. It was established that the main barrier hindering the implementation of sustainable practices was the perception that they incur a high upfront cost. Although the literature review had uncovered that this perception is in fact misinterpreted, findings from both the questionnaires and interviews showcase this to be the most influential barrier. The interview findings suggest that this is because businesses and developers are not seeing the cost benefits of adopting such methods and that it is too intimidating. Whilst there is research to show the cost benefits behind adopting SC practices, many chose to avoid the upfront cost, suggesting there is a strong misinterpretation within the industry.

14.8.4 The UK Government and Sustainability

Evaluating how effective the Government is in the process towards SD is critical considering the role they have. This study has outlined that the relationship between the Government and the construction industry is critical and good collaboration is needed for both to deliver a SC industry. The Government sets the regulations and requirements in which we construct buildings within the UK and therefore responsibility is on the construction industry to adhere to these requirements. However, most participants in the questionnaires and interviews criticised their experience when attempting to meet such requirements and adopt SC practices. Many participants outlined that the vast number of documents makes it difficult to comply. As shown from Theme E, it seems the Government expects requirements and targets to be met but does not consider the business's ability to achieve them. It can be suggested that the approach taken by the Government is not so effective and that the 'Collaborative' theme which is so apparent within the Construction 2025 publication by the Government, is yet to be proven.

Two common elements outlined from the research findings highlight 'lack of collaboration' and difficulty in 'complying' as barriers for businesses in implementing SC practices. The literature review also highlighted 'collaboration' as being a key element in the success of a sustainable built environment (Rahmawati et al., 2014). However, judging from the results obtained, the study can suggest that the collaborative approach needed is not being undertaken and there is a fragmented approach towards the government and construction industry's approach.

14.8.5 Conclusion

The utilisation of the mixed method approach proved to be an effective research strategy when harnessing data for this study. Communicating with industry professionals enabled the study to harness live data that was personal to the industry. The 'Thematic Analysis' was very beneficial to the study as this helped to identify themes within the interview results that were relevant to the study's objectives. The questionnaire results were critical in allowing the study to quantify participant's responses and make generalisations of the findings, whilst the literature review set the framework for the primary research of the study. From the research findings, the study can suggest that the Government must ensure that the targets they set are realistic and can be achievable by the construction industry if they are expected to see results. There is currently a need to ensure that construction industry stakeholders are aware of the progression towards key targets and whilst these can be ambitious, the UK Government needs to make sure these are achievable. By axing policies, the Government is slowing the progression towards the sustainability agenda within the UK.

The Government needs to improve its ability to collaborate with the construction industry in a way that is beneficial to both. The construction industry is heavily governed by client demand and is a highly competitive industry, if the Government is setting out drastic targets to promote change within the industry, it must address the business's ability to generate profit through SC practices. The study has unveiled that the industry struggles to comply with Government requirements, findings have shown that the number of documents is too vast, and the Government isn't accounting for businesses' abilities to meet requirements. It can therefore be concluded that there is a need to collaborate effectively to create requirements (whilst ambitious) that can be adopted by developers. Setting these high standards will govern the industry to perform and deliver sustainably to meet the client demand.

Effective collaboration is vital between the construction industry and the Government. Requirements must be achievable for the construction industry and SC practices must be seen as a good investment opportunity for the businesses. Therefore, the UK Government should provide additional resources, better incentives, investment opportunities, and clearer goals for businesses to achieve. There is also a need to streamline the number of documents and regulations. The findings within the study show that the vast number alone is currently acting as a barrier towards the agenda. Whilst the Government publicise many documents/strategies on how to approach the sustainability agenda, they must further consider the construction industry's ability to comply. Suggestions can be made to develop a shared platform that can be used by various bodies to determine what is required and in detail. Allowing an open forum for discussion on how to improve certain approaches for the benefit of both the Government and the construction industry.

This study has shown that client demand is a key driver. The supply chain within the industry is fundamental to its own economy and responsibility needs to be put on larger construction organisations to operate sustainably. They have the financial backing to implement change and have a huge influence on the rest of the industry and its supply chain in adopting more SC methods. The interview results show several jobs are won based on how sustainably the business operates. As the supply chain competes for work, sustainability will be more of a factor in winning new work, larger organisations that operate as developers and contractors should therefore be targeted to implement these sooner as they can act as a role model for the rest of the supply chain.

The study suggests that working with businesses that share similar values in sustainability will help to promote SD throughout the supply chain and in general. Companies should utilise and focus on their own internal policies towards sustainability and how this relates to the government's strategies. This will help for a more fitting adaption of sustainable practices, one that is more collaborative. This study therefore adds to the current body of knowledge on sustainability within the construction industry. It analysed the approach from the government's perspective and focused on how effective the construction industry is in adopting SC practices in accordance with government guidelines. This study has expanded on current knowledge by outlining further the drivers and barriers impacting the implementation of SC practices. The study challenges the government's current approach towards sustainability within the construction industry and recommends areas where improvement can be made. Whilst making suggestions and recommendations of its own, this study has explored the very lacking topic of the government's role on the sustainability agenda within the construction industry. This study hopes to add to the field of knowledge and pioneer further academics to generate their own research and findings.

14.8.6 Limitations

Whilst recommendations can be made this study has come with its limitations. Primarily, due to the size and diversification of the industry, the results obtained cannot be reflective of the overall construction industry. Naturally, bias may have occurred due to the small sample size obtained. The sample (whilst randomly allocated) was based on construction professionals available at the time and was largely based on participants working within London. Again, this therefore suggests the results can't be reflective of the whole of the UK.

References

Acciona (2019). *Do You Know When Sustainability First Appeared?* [online] Available at: https://www.activesustainability.com/sustainable-development/do-you-know-when-sustainability-first appeared/#:~:text=It%20seems%20incredible%20but%20the.

Adetunji, I., Price, A., Fleming, P. and Kemp, P. (2003). Sustainability and the UK Construction Industry—A Review. *Proceedings of the Institution of Civil Engineers – Engineering Sustainability*, 156(4), pp. 185–199.

Akadiri, O.P., et al. (2012). *Design of a Sustainable Building: A Conceptual Framework for Implementing Sustainability in the Building Sector*. [online] Available at: file:///C:/Users/User/Downloads/buildings-02-00126.pdf.

Akadiri, P. and Fadiya, O. (2013). *Empirical Analysis of the Determinants of Environmentally Sustainable Practices in the UK Construction Industry*. [online] Available at: https://www.emerald.com/insight/content/doi/10.1108/construction industry-05-2012-0025/full/pdf?title=empirical-analysis-of-the-determinants-of-environmentally-sustainable-practices-in-the-uk-construction-industry

Anand, A. and Kumar, R. (2020). Importance of Brundtland Report in the Protection of Environment: A Legal Analysis. *South-Asian Journal of Multidisciplinary Studies (SAJMS)*, [online] 3(3). Available at: http://sajms.com/wp-content/uploads/2016/04/Sustainable-Development-Paper.pdf.

Apine, A. and Valdés, F.J.E. (2017). *Risk Management in Sustainable Projects in the Construction Industry Cases of Swedish Companies Supervisor: Nils Wåhlin*. [online] Available at: https://www.diva-portal.org/smash/get/diva2:1078225/FULLTEXT01.pdf.

Basit, T.N. (2003). Manual or Electronic? The Role of Coding in Qualitative Data Analysis. *Educational Research*, 45(2): 143–54] Available at: https://citeseerx.ist.psu.edu/viewdoc/download?doi=10.1.1.541.4943&rep=rep1&type=pdf.

Bell, M. and Lowe, R. (2000) Building Regulations in Energy Efficiency: Compliance in England and Wales | Request PDF, Building regulation and sustainable housing. Part 1: A critique of Part L of the Building Regulations 1995 for England and Wales. Available at: https://www.researchgate.net/publication/254409071_Building_regulations_in_energy_efficiency_Compliance_in_England_and_Wales

www.businesswire.com (2021). *Global Construction Market Expected to Reach $16.6 Trillion by 2025, Growing at a CAGR of 7%*. ResearchAndMarkets.com. [online] Available at: https://www.businesswire.com/news/home/20210309005459/en/Global-Construction-Market-Expected-to-Reach-16.6-Trillion-by-2025-Growing-at-a-CAGR-of-7—ResearchAndMarkets.com.

Carrington, D. (2015) Green groups welcome Amber Rudd's appointment as Climate secretary, The Guardian. Available at: https://www.theguardian.com/environment/2015/may/11/green-groups-welcome-amber-rudd-climate-secretary.

Cheng, C.-C., Pouffary, S., Svenningsen, N. and Callaway, J.M. (2008). *The Kyoto Protocol, The Clean Development Mechanism and the Building and Construction Sector: A Report for the UNEP Sustainable Buildings and Construction Initiative*. Paris: United Nations Environment Programme.

CIOB (2016). *Search Results. Construction IndustryOB*. [online] Available at: https://www.ciob.org/sites/default/files/View%20Inga%27s%20full%20report%20here.pdf.

Combet, E. (2020). Planning and Sustainable Development in the Twenty-First Century. *Œconomia. History, Methodology, Philosophy*, [online] (10–3), pp. 473–506. Available at: https://journals.openedition.org/oeconomia/9558 [Accessed 13 Apr. 2021].

Committee on Climate Change (2019). *UK Housing: Fit for the Future? Committee on Climate Change*. [online] Available at: https://www.theccc.org.uk/wp-content/uploads/2019/02/UK-housing-Fit-for-the-future-CCC-2019.pdf.

Construction 2025 (2013). [online]. Available at: https://assets.publishing.service.gov.uk/government/uploads/system/uploads/attachment_data/file/210099/bis-13-955-construction-2025-industrial-strategy.pdf.

Construction Excellence (2008). *Key Issues in Sustainable Construction – Constructing Excellence*. [online] Available at: https://constructingexcellence.org.uk/key-issues-in-sustainable-construction/.
Construction Products Association EU Regulations and Directives Guidance Document 2 Contents (2007). [online] Available at: https://www.constructionproducts.org.uk/media/261715/cpa-guide-to-eu-regulations-and-directives.pdf.
Creswell, J. (2009). *Research Design Qualitative, Quantitative, and Mixed Methods Approaches*. [online]. Available at: http://www.drbrambedkarcollege.ac.in/sites/default/files/research-design-ceil.pdf.
Dainty, A.R.J., Neale, R.H. and Bagilhole, B.M. (2000). Comparison of Men's and Women's Careers in U.K. Construction Industry. *Journal of Professional Issues in Engineering Education and Practice*, 126(3), pp. 110–115. doi:https://doi.org/10.1061/(asce)1052-3928(2000)126:3(110).
DePoy, E. and Gitlin, L. (2015). *Introduction to Research – 5th Edition*. [online] Available at: https://www.elsevier.com/books/introduction-to-research/depoy/978-0-323-26171-5 [Accessed 23 Mar. 2021].
Ding, G.K.C. (2008). Sustainable Construction—The Role of Environmental Assessment Tools. *Journal of Environmental Management*, 86(3), pp. 451–464.
Du Plessis, C. (2007). A Strategic Framework for Sustainable Construction in Developing Countries. *Construction Management and Economics*, 25(1), pp. 67–76. doi:https://doi.org/10.1080/01446190600601313.
Eley, J. (2011). *Sustainable Buildings: The Client's Role*. RIBA Publishing – Publication Index | NBS. [online] Available at: https://www.thenbs.com/PublicationIndex/documents/details?Pub=RIBAPUBS&DocID=297404.
Elkington, J. (1998). Accounting for the Triple Bottom Line. *Measuring Business Excellence*, 2(3), pp. 18–22.
Environmental Audit Committee (2022). *Building to Net Zero: Costing Carbon in Construction*. [online] Available at: https://committees.parliament.uk/publications/22427/documents/165446/default/.
Fellows, R. and Liu, A. (2008) *Research methods for construction*. Oxford: Wiley-Blackwell.
Gould, C. (2019). *Can We Tackle Construction's Waste Problem?* [online] WP Group. Available at: https://thewp-group.co.uk/blog/can-we-tackle-constructions-waste-problem/?cn-reloaded=1 [Accessed 18 Apr. 2021].
GOV.UK (2013). *Construction 2025: Strategy*. [online] GOV.UK. Available at: https://www.gov.uk/government/publications/construction-2025-strategy.
GOV.UK (2020). *"Build Build Build": Prime Minister Announces New Deal for Britain*. [online] Available at: https://www.gov.uk/government/news/build-build-build-prime-minister-announces-new-deal-for-britain#:~:text=In%20a%20speech%20in%20the [Accessed 19 Apr. 2021].
GOV.UK (2022). Promoting Net Zero Carbon and Sustainability in Construction. Available at: https://assets.publishing.service.gov.uk/media/631222898fa8f54234c6a508/20220901-Carbon-Net-Zero-Guidance-Note.pdf
Gunatilake, S. (2013). *The Uptake and Implementation of Sustainable Construction: Transforming Policy into Practice*. [online]. Available at: https://core.ac.uk/download/pdf/19084173.pdf.
Grossman, D.L. (2012) (PDF) The Three Pillars of sustainable development: Critical issues and perspectives, The Three Pillars of Sustainable Development: Critical Issues and Perspectives. Available at: https://www.researchgate.net/publication/272508647_The_Three_Pillars_of_Sustainable_Development_Critical_Issues_and_Perspectives
Halliday, S. (2018). *Sustainable Construction* – NTU Single sign-on – CAS – Central Authentication Service. [online] Available at: https://ebookcentral.proquest.com/lib/ntuuk/reader.action?docID=5606180 [Accessed 26 Mar. 2021].

Harby, A. (2021). *What Are the Challenges to Sustainable Construction.* [online] City Changers.org – Home Base for Urban Shapers. Available at: https://citychangers.org/challenges-sustainable-construction/.

Harvey, F. (2021). *Boris Johnson Failing on UK Plan to Reach Net Zero, Say MPs.* [online] Available at: https://www.theguardian.com/environment/2021/mar/05/boris-johnson-failing-on-uk-plan-to-reach-net-zero-say-mps.

Haynes, L. (2022). *Government 'Dragging Feet' Over Efforts to Cut Construction Carbon Footprint.* [online] Construction News. Available at: https://www.constructionnews.co.uk/sustainability/government-dragging-feet-over-efforts-to-cut-construction-carbon-footprint-27-05-2022/.

Hazlegreaves, S. (2019). *Is Industry-Wide Collaboration the Key to Sustainability in Construction?* [online] Planning, BIM & Construction Today. Available at: https://www.pbctoday.co.uk/news/planning-construction-news/collaboration-in-construction-2/68721/ [Accessed 11 Mar. 2021].

Herbert, R. (2019). *This Government Has Failed the Environment.* [online] Conscious Creatives. Available at: https://www.consciouscreatives.co.uk/environment/this-government-has-failed-the-environment/ [Accessed 14 Feb. 2023].

HM Government (2005). *Securing the Future Delivering UK Sustainable Development Strategy* [online]. Available at: https://assets.publishing.service.gov.uk/government/uploads/system/uploads/attachment_data/file/69412/pb10589-securing-the-future-050307.pdf.

HM Government (2008). *Strategy for Sustainable Construction* [online]. Available at:https://webarchive.nationalarchives.gov.uk/20121205201814/http://www.bis.gov.uk/files/file46535.pdf.

Holden, E., Linnerud, K. and Banister, D. (2014). Sustainable Development: Our Common Future Revisited. *Global Environmental Change*, [online] 26, pp. 130–139. Available at: https://www.sciencedirect.com/science/article/pii/S0959378014000727.

Holton, I., Glass, J. and Price, A. (2008). Developing a Successful Sector Sustainability Strategy: Six Lessons from the UK Construction Products Industry. *Corporate Social Responsibility and Environmental Management*, 15(1), pp. 29–42.

Johnson, R.B. et al. (2007). *Mixed Methods – An Overview.* ScienceDirect Topics. [online] Available at: https://www.sciencedirect.com/topics/psychology/mixed-methods#:~:text=%E2%80%9CMixed%20methods%20research%20is%20the.

Johnson, B. (2021). *"A Shambles": Is the Green Homes Grant Scheme Budget Being Cut?* [online] Available at: https://www.businessgreen.com/news/4027031/shambles-green-homes-grant-scheme-budget-cut [Accessed 6 Mar. 2021].

Kemp, D. (2017). *State of Sustainability: Where Is UK Construction Now?* [online] Construction News. Available at: https://www.constructionnews.co.uk/sustainability/state-of-sustainability-where-is-uk-construction-now-04-12-2017/.

Kenton, W. (2020). *Triple Bottom Line (TBL).* [online] Investopedia. Available at: https://www.investopedia.com/terms/t/triple-bottom-line.asp.

Khanna, M. and Brouhle, K. (2009) *The effectiveness of voluntary environmental initiatives.* Available at: https://www.researchgate.net/publication/292282316_The_effectiveness_of_voluntary_environmental_initiatives

Kibert, C.J. (2007). The Next Generation of Sustainable Construction. *Building Research & Information*, 35(6), pp. 595–601.

Krosnick, J. et al. (2009) Question and questionnaire design | request PDF, Researchgate. Available at: https://www.researchgate.net/publication/228383338_Question_and_Questionnaire_Design

Liu, Z. et al. (2020). *Sustainable Construction as a Competitive Advantage* [online] Available at: file:///C:/Users/User/Downloads/sustainability-12-05946%20(1).pdf.

Lowe, T. (2021). *Government Announces £440m Funding for New Green Construction Projects.* [online] Building. Available at: https://www.building.co.uk/news/government-announces-440m-funding-for-new-green-construction-projects/5114266.article.

Mace, M. (2022) The good, the bad and the unlawful: Where did it all go wrong for the UK's approach to net-zero?, edie. Available at: https://www.edie.net/the-good-the-bad-and-the-unlawful-where-did-it-all-go-wrong-for-the-uks-approach-to-net-zero/

Mcchesney, I. (1991). *The Brundtland Report and Sustainable Development in New Zealand.* [online]. Available at: https://core.ac.uk/download/pdf/35461353.pdf.

Meg, H. (2021). *Boris Johnson failing on UK Plan to Reach Net Zero, Say MPs.* [online] Available at: https://www.theguardian.com/environment/2021/mar/05/boris-johnson-failing-on-uk-plan-to-reach-net-zero-say-mps.

O'Malley, C., Piroozfar, P.A.E., Farr, E.R.P. and Gates, J. (2014). Evaluating the Efficacy of BREEAM Code for Sustainable Homes (CSH): A Cross-Sectional Study. *Energy Procedia*, 62, pp. 210–219.

Oldfield, P. (2015). *UK Scraps Zero Carbon Homes Plan.* [online] The Guardian. Available at: https://www.theguardian.com/environment/2015/jul/10/uk-scraps-zero-carbon-home-target.

Opoku, A. et al. (2015) (PDF) organizational leadership role in the delivery of Sustainable Construction Projects in UK, Organizational leadership role in the delivery of sustainable construction projects in UK. Available at: https://www.researchgate.net/publication/275252756_Organizational_leadership_role_in_the_delivery_of_sustainable_construction_projects_in_UK

Perry, F. (2020). *What "Net-Zero Carbon" Really Means for Cities.* [online] www.bbc.com. Available at: https://www.bbc.com/future/article/20200922-how-london-could-achieve-net-zero-carbon-by-2050.

Pitt, M. et al. (2009) Towards sustainable construction: Promotion and best practices, Researchgate. Available at: https://www.researchgate.net/publication/250915619_Towards_sustainable_construction_Promotion_and_best_practices

Rahmawati, Y., Anwar, N. and Utomo, C. (2014). A Concept of Successful Collaborative Design towards Sustainability of Project Development. *International Journal of Economics and Management Engineering*, [online] 7(4), pp. 1042–1048. Available at: https://publications.waset.org/9997067/a-concept-of-successful-collaborative-design-towards-sustainability-of-project-development.

Rademaekers et al. (2012). *Study on Incentives Driving Improvement of Environmental Performance of Companies.* [online]. Available at: https://ec.europa.eu/environment/pubs/pdf/Incentives_Ecorys.pdf.

Rydin, Y., Amjad, U. and Whitaker, M. (2007). Environmentally Sustainable Construction: Knowledge and Learning in London Planning Departments. *Planning Theory & Practice*, 8(3), pp. 363–380.

Sev, A. (2009). How Can the Construction Industry Contribute to Sustainable Development? A Conceptual Framework. *Sustainable Development*, 17(3), pp. 161–173.

Shan, M., Hwang, B.-G. and Zhu, L. (2017) A Global Review of Sustainable Construction Project Financing: Policies, practices, and research efforts, MDPI. Available at: https://www.mdpi.com/2071-1050/9/12/2347

Shankleman, J. and Moralws, A. (2021). *U.K. Government Cuts Funding for Keeping Homes Warm.* Bloomberg.com. [online] 10 Feb. Available at: https://www.bloomberg.com/news/articles/2021-02-10/u-k-government-cuts-funding-for-keeping-homes-warm.

Shepheard, M. (2020). *UK Net Zero Target.* [online] www.instituteforgovernment.org.uk. Available at: https://www.instituteforgovernment.org.uk/explainers/net-zero-target.

Smith, K.T. (2022) *Construction industry responsible for 18% of UK's large particle polution, Chartered Institution of Environmental Health.* Available at: https://www.cieh.org/ehn/environmental-protection/2022/november/construction-industry-responsible-for-18-of-uk-s-large-particle-pollution/

Smith, S. (2014). *Construction 2025 – From Ambition to Reality.* [online] Available at: http://www.infrastructure-intelligence.com/article/oct-2014/construction-2025-%E2%80%93-ambition-reality [Accessed 2 March 2021].

Spence, R. and Mulligan, H. (2005). Sustainable Development and the Construction Industry. *Habitat International*, [online] 19(3), pp. 279–292. Available at: https://www.sciencedirect.com/science/article/pii/0197397594000719.

Sridhar, K. and Jones, G. (2012). The Three Fundamental Criticisms of the Triple Bottom Line Approach An Empirical Study to Link Sustainability Reports in Companies Based in the Asia-Pacific Region and TBL Shortcomings. *Asian Journal of Business Ethics*, [online] 2(1), pp. 91–111. Available at: https://link.springer.com/article/10.1007%2Fs13520-012-0019-3 [Accessed 11 Jan. 2021].

Srivastava, A.K., Dixit, S. and Srivastava, A.A. (2021). Criticism of Triple Bottom Line: TBL (With Special Reference to Sustainability). *Corporate Reputation Review*.

Tan, Y. et al. (2011) Sustainable construction practice and contractors' competitiveness: A preliminary study, Sciencedirect. Available at: https://www.sciencedirect.com/science/article/abs/pii/S0197397510000676

The Construction Index (2017). *Industry Targets Are Not Being Measured*. [online] Theconstructionindex.co.uk. Available at: https://www.theconstructionindex.co.uk/news/view/industry-targets-are-not-being-measured.

Thematic (2019). *Coding Qualitative Data: How To Code Qualitative Research (2020)*. [online] Available at: https://getthematic.com/insights/coding-qualitative-data/#:~:text=Why%20is%20it%20important%20to.

UK Green Building Council (2009). *Making the Case for a Code for Sustainable Buildings* [online]. Available at: https://www.ukgbc.org/sites/default/files/Executive%20Summary%20-%20Making%20the%20case%20for%20a%20Code%20for%20Sustainable%20Buildings%20(1).pdf [Accessed 25 Mar. 2021].

UKGBC (2017). *Climate Change – UKGBC – UK Green Building Council*. [online] UKGBC – UK Green Building Council. Available at: https://www.ukgbc.org/climate-change/.

UKGBC – UK Green Building Council. (2018). *We Need True Collaboration to Create a More Sustainable Built Environment*. [online] Available at: https://www.ukgbc.org/news/we-need-true-collaboration-to-create-to-a-more-sustainable-built-environment/ [Accessed 11 Mar. 2021].

Vaughan, A. (2015). *Government Kills off Flagship Green Deal for Home Insulation*. [online] Available at: https://www.theguardian.com/environment/2015/jul/23/uk-ceases-financing-of-green-deal.

WCED (1987). *Sustainabledevelopment.un.org. 2022*. [online] Available at: https://sustainabledevelopment.un.org/content/documents/5987our-common-future.pdf [Accessed 26 Jul. 2022].

West, B.T. and Blom, A.G. (2016). Explaining Interviewer Effects: A Research Synthesis. *Journal of Survey Statistics and Methodology*, [online] 5(2), p.smw024. doi:https://doi.org/10.1093/jssam/smw024.

Williams, S. (2023). *Sustainable Building Regulations: How Do They Work?* [online] Available at: https://www.mesh-energy.com/insights/sustainable-building-regulations-what-is-the-code-and-how-does-it-work [Accessed 16 Aug. 2022].

World Green Building Council (2009). *The Benefits of Green Buildings*. World Green Building Council. [online] Worldgbc.org. Available at: https://www.worldgbc.org/benefits-green-buildings.

Wright, J.D & Marsden, P.V, (2009). Question and Questionnaire Design. [online] Available at: https://web.stanford.edu/dept/communication/faculty/krosnick/docs/2009/2009_handbook_krosnick.pdf.

15 Pathways to a Sustainable Built Environment

A Knowledge-Based View from Malaysia

John Peter Rogers

15.1 Introduction

This chapter considers how the Malaysian construction sector may transition towards providing a sustainable built environment. The mechanisms used for making predictions about potential future scenarios are clearly explained, making it easier for other researchers to replicate the study.

15.1.1 Construction, Knowledge, and Sustainability

The construction sector's gross domestic product (GDP) contribution to an economy is estimated by Alaloul et al. (2022) as 15.3% for the UK, and 6.9% for China. Malaysia measured a 6.1% contribution to GDP for Q2 2022 (Government of Malaysia, 2022). As a significant contributor to GDP, the construction sector has a correspondingly large effect on resource consumption, including energy. The Brundtland Report (United-Nations, 1987) marked a watershed moment in the recognition of sustainable development as an urgent global priority. Since then, the construction industry has experienced some transformation in its approach to sustainable construction and development. This transformation has been driven by an increasing awareness of the environmental, social, and economic implications of the built environment, as well as the need to mitigate the adverse effects of climate change and resource depletion.

The growing awareness has led to the development of frameworks and guidelines for creating inclusive, healthy, and resilient built environments. The United Nations Sustainable Development Goals (SDGs) have provided a global blueprint for addressing social, economic, and environmental challenges through sustainable urban development. Other advancements in sustainable construction include the development and implementation of green building standards and certifications. These guidelines, such as the Leadership in Energy and Environmental Design (LEED) and the Building Research Establishment Environmental Assessment Method (BREEAM), have provided a comprehensive framework for assessing and promoting environmentally responsible and resource-efficient building practices. Such certifications have not only incentivised sustainable design and construction but have also tentatively facilitated the integration of innovative technologies and strategies into the mainstream building process.

DOI: 10.1201/9781003317890-20

Professional institutes depict construction as a temporal process of knowledge enrichment, beginning with a market-responsive business case and progressing through design, construction, and asset optimisation to reclamation as a resource for building materials in future projects (RIBA, 2019). The necessity for information collation and provision to successor stages is becoming more explicit and contractually mandated, compelling integration among previously segregated subsectors. Pryke (2009, p. 2) characterises the construction value creation process as a supply chain, integrating internal and external capabilities across a network of clients, consultants, contractors, and suppliers connected by knowledge transfer, information exchange, and financial and contractual relationships.

In tandem, the construction industry started to incorporate integrated design concepts, including the collaboration of interdisciplinary teams from the early stages of a project. This holistic approach provides better opportunity for sustainability considerations to be amalgamated into the design and construction process (Dehdasht et al., 2021). Moreover, the increased application of digitalisation tools, such as building information modelling (BIM), has facilitated the integration of sustainable design principles and performance analysis in the planning, design, construction, and operational phases of a project (Ahmad and Thaheem, 2022).

Despite these advancements, the construction sector still trails other industries in knowledge management maturity, largely due to its fragmented and intricate structure. The transient, project-based nature of construction, coupled with the involvement of multiple, often independent, stakeholders, hampers the implementation and execution of consistent knowledge management practices. This inherent complexity poses challenges in establishing unified standards, fostering trust, and facilitating unhindered knowledge flow across organisational boundaries (Bin Tayib et al., 2022).

Environmental sustainability can be considered a catalyst for change within the construction sector, prompting a transformation in the industry's knowledge landscape. As societal and regulatory pressures heighten the demand for greener practices, construction firms will be compelled to innovate and adapt to remain competitive. Accordingly, this would necessitate the development and acquisition of new knowledge related to sustainable materials, energy-efficient design, and innovative construction techniques. Consequently, the construction sector may be witnessing a shift in knowledge dynamics, with firms prioritising the cultivation of advanced sustainability knowledge to enhance their market position, meet evolving client demands, and contribute to a more sustainable future. This research considers whether this shift is occurring, and which firms may capitalise from it.

15.1.2 Malaysia and Sustainable Construction

Malaysia is an active adopter of UN SDGs, aligning its national development plans (11th, 12th, and 13th Malaysia Plans), with the SDG goals (Malaysia-Gov, 2021b) noting that, *"The alignment of SDG and national development is realised through ... the integration of the national development plan's action plans, initiatives and outcomes to the SDGs' goals, targets and indicators."*

The current 12th Malaysia Plan explicitly states, "The national development priorities of the Twelfth Plan will continue to be aligned to the Sustainable Development Goals (SDGs) of the 2030 Agenda for Sustainable Development (2030 Agenda)" (Government of Malaysia, 2021, p. 1–3). Such commitments were reinforced in Malaysia's Nationally Determined Contributions (NDC) submission to the UN in July 2021, by the unconditional pledge to reduce carbon intensity by 45% of 2005 levels by 2030, amongst other commitments (Malaysia-Gov, 2021a). Similarly, a national adaption plan was established to further the adoption and integration of the SDG goals by industry sectors, including infrastructure and water resources (Bernama, 2022). It is evident that UN SDGs are enshrined in Malaysia's national policy development.

15.1.3 Malaysia's Construction Sustainability and Resilience

The public sector of Malaysia has been actively developing and promoting sustainability and resilience in the built environment. Beyond the national development plans, there are specific development plans for the construction industry, Construction Industry Transformation Plan (CIDB, 2022a) and Construction 4.0 Strategic Plan (2021–2025) (CIDB, 2020), the former focusing on sustainability, the latter on digitalisation and modern methods of construction.

Malaysia's Construction Industry Development Board (CIDB) has also established the Sustainable Construction Excellence Centre to drive sustainable construction in the country with innovative initiatives. CIDB provides a number of tools, such as *MyCREST (Malaysian Carbon Reduction and Environmental Sustainability Tool), a comprehensive certification tool aimed at reducing carbon emissions and promoting sustainability in construction. It covers a building's entire life cycle, providing a framework for best practices and facilitating international competitiveness* (CIDB, 2022a). Also from CIDB is Sustainable INFRASTAR, a certification and rating system aimed at promoting sustainable practices in infrastructure projects. It evaluates environmental impact, efficiency, and social responsibility across the project's lifecycle (CIDB, 2022b) along with several reports and guides for resiliency in conjunction with the Construction Research Institute of Malaysia (CREAM and CIDB, 2022).

The professional institutes' bodies also provide the well-regarded Green Building Index. Green Building Index is an initiative started in 2008 by Malaysian Institute of Architects (PAM) and the Association of Consulting Engineers Malaysia (ACEM) to provide "*a localised Green Building rating tool*" (PAM and ACEM, 2009).

Malaysia has a long history of research into sustainable construction (Zainul Abidin N, Khalfan M., and Kashyap, M, 2003) with themes evolving from awareness (Zainul Abidin, 2009) to implementation (Ezani et al., 2018), and on to more specific and granular themes, including firm capabilities (Bamgbade et al., 2022), technologies (Bamgbade et al., 2019), barriers to implementation (Tharim, 2019), and the prediction of the rise of sustainability services firms (Papargyropoulou et al., 2012). Recent publications have reviewed the greening of existing buildings using BIM (Lim et al., 2021), and the commercial and sustainability context (Goh et al., 2020). This selection suggests an active, holistic, and mature Malaysian construction sustainability research capability.

Despite the above efforts to inculcate sustainability capabilities within the construction sector, the media and civil society feel that tangible implementation could be accelerated (BERNAMA, 2021; Chai Yee Hoong, 2021; The Edge Malaysia and Fatin Najihah, 2022).

15.2 Theoretical Framework

Considering "Pathways to a Sustainable Built Environment" requires forecasts about the future, and projections about how firms will need to adapt to a new business knowledge environment. Being inherently predictive, it is beneficial to deploy a framework of ideas to structure thinking about possible future pathways, thereby enabling other researchers to understand and evaluate the context, scope and particulars of the research, and making the research reproducible. A framework of ideas is a theory, or a theoretical framework, that Neuman (2006, p. 50) describes as "a system of interconnected ideas that condenses and organizes knowledge about the social world." As such, a theory or framework of ideas that includes how firms utilise knowledge in a dynamic business environment is needed. Such a theory is the knowledge-based view (KBV) of the Firm.

15.2.1 The Knowledge-Based View of a Firm

The precursor of the KBV is the resource-based view, a framework that considers the strategic resources a firm could optimally exploit to achieve a sustainable competitive advantage (Barney, 1991). Barney (2001) defines firm resources as "all assets, capabilities, organizational processes, firm attributes, information, knowledge, etc. controlled by a firm that enable the firm to conceive of and implement strategies that improve its efficiency and effectiveness." The KBV refines this by making knowledge the principal resource (Curado and Bontis, 2006; Grant and Phene, 2021). The fundamental assumption of the KBV of the firm is that the critical input in production and primary source of value is knowledge (Grant, 2002).

Kogut and Zander (1996) opine, "what firms do better than markets is the sharing and transfer of the knowledge of individuals and groups within an organization." They suggest that the firm is a collective of capabilities for combining individual and social knowledge into economically valuable products. Grant and Baden-Fuller (1995) considers the firm as an institution for integrating knowledge, and that organisational capability is the outcome of this. Nonaka (1994) considered that the knowledge creation process is a firm capability that leads to commercial value and is hard to duplicate. Combining knowledge and resources produces capabilities, and deploying these well gives a firm a competitive advantage (Kaplan et al., 2001).

There have been debates about the KBV and whether it is sufficiently cohesive to be considered a theory (Grant, 2002; Nickerson and Zenger, 2004; Curado and Lupi, 2006; Felin and Hesterly, 2007; Håkanson, 2010). In addressing these issues, Kaplan et al. (2012, p. 16) evolved the theory to become the integrated KBV (iKBV), providing greater structure and categorisation of concepts, thereby enabling improved repeatability and consistency in empirical observation. The iKBV is posited as a mechanism to describe, explain, and predict the utility of knowledge for firms, and it asserts that firms with superior capability to utilise relevant knowledge will attain superior performance and a competitive advantage in their markets.

The essential structures of the iKBV are three pairs of knowledge capabilities. The following definitions are taken from Kaplan et al (2012):

"Pair One

Creation - Capacity to combine knowledge with knowledge or resources to produce output that exceeds the value of the sum of the inputs
Destruction - Capacity to eliminate elements of knowledge or disassemble the interconnectedness of knowledge

Pair Two

Integration - Capacity to secure a supply of knowledge and resources for the firm
Absorption - Enables the use of knowledge and resources, and realizes their potential

Pair Three

Replication - Capacity to transfer knowledge from a transmitting entity to a receiving entity without loss of information
Protection - Capacity to control the replication process including how many times knowledge can be replicated and whether it exits the firm"

The iKBV also proposes an explanation as to a firm's boundaries. By identifying the transition between external and internal knowledge, the scope of what a firm does becomes more evident.

In the context of the construction industry, iKBV provides a valuable lens for understanding whether firms are effectively managing their knowledge resources and capabilities to adapt to a rapidly changing business environment. The construction industry, characterised by its complexity, fragmented, project-based, adversarial nature, is undergoing a transition driven by the growing demand for environmentally responsible and resource-efficient practices. The iKBV predicts that firms with comparatively better knowledge capabilities will be economically rewarded for adapting to changes in regulations and technological advancements. The iKBV, in emphasising the importance of developing mechanisms for acquiring, storing, and sharing knowledge within construction firms to support effective decision-making, problem-solving, and innovation, provides a theoretical framework to assess the relative capabilities of firms to adapt to changing knowledge demands of producing a sustainable built environment.

The KBV theory posits that the strategic management of knowledge assets within an organisation is crucial for gaining a competitive advantage. In the context of transitioning towards sustainable built environments, the KBV theory takes on heightened significance. Organisations involved in architecture, engineering, and construction often possess specialised knowledge in sustainable design, energy-efficient technologies, and eco-friendly materials. Moreover, the KBV theory addresses the complexities of knowledge sharing and intellectual property. In summary, the KBV theory provides a robust framework for understanding how the strategic management of knowledge can catalyse the transition towards more sustainable built environments. It offers insights into optimising internal processes, fostering innovation, and navigating the complexities of intellectual property, all of which are vital for advancing sustainability goals.

As sustainability initiatives evolve, they provide considerable opportunity for change to the construction industry. This should necessitate firms to introduce new human resource capabilities, new processes, and new technological capability, which it can do by internal development or via out-sourcing. Furthermore, sustainability inherently creates new knowledge, such as simulation and environmental analysis, which raises issues over intellectual property ownership. The iKBV considers these issues and whether the alignment and deployment of knowledge capabilities, respective to sustainability, will provide sustainable competitive advantage.

15.3 Research Methodology

Goundar (2012, p. 9) describes research methodology as "the procedures by which researchers go about their work of describing, explaining and predicting phenomena." This section describes this research's methodology and its component parts.

15.3.1 The Research Methodology

Many schools of thought exist and there continues to be much debate about what constitutes a research methodology (Scott, 2014, p. 23). For the purposes of this research, Saunders et al.'s (2019, p. 130) research onion has been selected as a prescriptive framework to guide the research methodology, terminology, constituents, and decision-making. The Saunders Research Onion offers a layered, systematic approach to research design, providing clear guidance through its hierarchical structure. For this research, more so than comparable frameworks, it integrates philosophy, strategy, and methods, ensuring a more cohesive and comprehensive research plan.

Cognisant of the ontological, epistemological, and axiological assumptions, a review of the research area identifies it as broad, dynamic, immature, and in need of exploration. It is unlikely that objective data can be obtained; thus, subjective data, elicited from individuals, will be acceptable.

15.3.2 Research Philosophy

This research is future forecasting, drawing opinions from a diverse array of participants, each with their own circumstances and perspectives on what is happening within their firms, their markets, and their environment. As such, each participant has a unique reality that aggregates to a socially constructed overall reality. This prevents an objective study in which the precise nature of reality can be reasonably defined and empirically measured. As such, it discounts positivism and critical realism as research philosophies, as the knowledge of individual firms cannot be measured in the same way that the height of a steel column or the density of a brick can be measured.

To guide future forecasting, this research adopts the iKBV as a central framework. Philosophically, interpretivism and postmodernism generally do not encourage theory as a starting point, instead they both allow new theories to emerge from the data that is collected.

Pragmatism, however, affords that an existing theory is utilised to guide the data to be collected, and the collected data to be used to develop the theory in an iterative process. Pragmatism considers that an overall reality, the reality external to individuals, is complex, rich, and the "practical consequence of ideas, a flux of process, experiences and practices" (Saunders et al., 2019). It is suited to problem-solving, and informed future practice. Accordingly, pragmatism describes the philosophy for this research.

15.3.3 Approach to Theory Development

As guided by Kaplan et al. (2012), there are a plethora of contributions to the iKBV of a firm, yet without a consistent or common ontology it is difficult to contrast and compare research findings (Grant and Phene, 2021). As such, iKBV has not reached a level of maturity that allows for deductive testing against a broad research problem.

A well-developed, existing theory would allow for the testing of it by collecting data specifically acquired to examine hypotheses derived from the theory. The hypotheses would be deduced from the theory, and the corresponding collected data would be assessed against them. If the data findings agree with the hypotheses, the theory is strengthened; if not, the theory is reviewed. If there was no theory to guide this research, the collected data would be assessed, and if strong patterns emerged, it could be considered that new theory had been induced from the data.

Beyond induction and deduction as approaches, over the past century a third approach, abduction, has emerged from the works of Sanders (1997) that allows iteration between theory and empirical observation, from inductive to deductive approaches. The abductivist approach is aligned to pragmatist philosophy, which was also first developed by Sanders, and is adopted in this study.

15.3.4 Methodological Choice

A qualitative multi-method methodology was selected to suit the availability of topic data. A qualitative multi-method methodology can employ various qualitative techniques, such as interviews, observations, and document analysis, to examine a research question, providing nuanced, context-specific insights unattainable through a single method. Incorporating two qualitative methods can enhance the depth and richness of a study's findings by offering complementary perspectives and enables the triangulation of results from the collected data, to enhance both the credibility of the process and the impact of research (Amandeep, 2014; Reis et al., 2017, 2018).

15.3.5 Research Strategy

Pryke's (2009) depiction of the construction sector supply chain was used to identify the categories of participants from whom opinions should be sought. The framework delineates the construction sector supply chain into key components: clients, consultants, main contractors, subcontractors, suppliers, and regulatory agencies. Because the area of research covers an exceptionally large population – approximately 5% of the workforce – resource constraints prohibited probability or random sampling. While authorities concur that external validity is ceded, for qualitative research non-probabilistic approaches are more suitable for in-depth understanding of complex social phenomena. Semi-structured interviews offer a balance between flexibility and focus, allowing for in-depth exploration of research themes while permitting emergent topics. This method yields rich, context-specific data that can enhance the validity and comprehensiveness of qualitative research. Semi-structured interviews were therefore selected as the method to survey the population sample.

The second selected method was ethnographic case study. Case studies focus on "a particular phenomenon with a view to providing an in-depth account of events, relationships, experiences or processes occurring in that particular instance" (Denscombe, 2007, p. 32). Furthermore, it is ethnographic, written from the researcher's experiences. Early examples of ethnographic case study include Fairhurst and Good (1991) researching a business environment, who observe "*This technique provides an experiential approach to solving company problems and understanding corporate culture.*" Côté-Boileau et al. (2020, p. 9) provide a more current implementation, noting that ethnographic case studies provide in-depth closeness to organisations, which are timely, actionable, and contextualised, resulting in increases of the credibility of the research, and increases in the potential impact of the research.

15.3.6 Time Horizon

This research is cross-sectional as it takes readings from a single point in time, whereas longitudinal studies make comparisons over time. Interviews were conducted between 29 August and 26 September 2022.

15.3.7 Data Collection

15.3.7.1 Semi-Structured Interviews

This research addresses complex issues necessitating the input of individuals knowledgeable about the matters under investigation. Consequently, judgemental sampling was employed to select a sample of professionals appropriate for the research. Judgemental sampling, also known as purposive or expert sampling, is a non-probability sampling technique in which the researcher selects participants based on their expertise, experience, or knowledge in a specific area relevant to the research question.

This method is particularly effective for gathering insights from individuals who possess a deep understanding of the phenomenon being studied, allowing for the collection of rich, contextualised data. In this context, judgemental sampling is appropriate when seeking information from a specific subset of professionals whose expertise directly pertains to the research topic.

While judgemental sampling offers certain advantages, its limitations include potential researcher bias and the restricted generalisability of findings to a broader population. However, when applied judiciously and transparently, judgemental sampling can yield valuable insights from professionals possessing direct experience and knowledge in the area of interest, thus enhancing the understanding of the research topic.

Drawing upon contacts from industry, the researcher identified knowledgeable persons to take part in this research based upon the following criteria:

- Availability for interview
- Minimum 12 years' experience of built environment projects
- Experience of developing procedures, processes, and workflows at project and organisational level
- International experience for comparison and reflection of the construction's sustainability current situation
- Minimum degree educated (professional certification is not always appropriate for some roles, such as entrepreneurs)
- Aware of legislative requirements and changes
- Knowledge of good sustainability practices
- Knowledge of BIM or digitalisation tools or projects
- Previously interviewed by the researcher and known to provide candid, open responses
- Working in industry, not academia

The participants were provided with interview guides, and ethics management information. Interviews were conducted individually, and utilised videoconferencing, each interview was recorded, and transcripts were produced. To maintain their anonymity, only salient data is stated. The participant particulars are provided in Table 15.1.

Table 15.1 Participant particulars

Firm size	*Firm type*	*Role*	*Discipline*	*Code name*
MNC	Design and build contractor	Digital transition lead	Architect	Digital_Transition_Architect
SME	Design, build, operate, transfer	Director	Management	Director_Manager_Operator
SME	Sustainability consultant	Practice head	Sustainability engineer	Sustainability_Expert
Med-large	Client/investor	Business analyst	Civil engineer	Investment_Analyst
MNC	Engineering and consulting	Pipeline expert	MEP engineer	Pipes_Municipal_Fluids_Expert
MNC	MEP consultants	Director	MEP engineer	MEP_Director_BIM_Early_Adopter

15.3.7.2 Case Study

Guided by Fusch et al. (2017) the researcher collected data using the following techniques:

1 Participant observation: The researcher was immersed in a range of organisations and the environment being studied, observing events, relationships, experiences, or processes in real time.
2 Field notes: During the observation period, the researcher took detailed field notes, capturing both descriptive and reflective information about experiences and observations.
3 Semi-structured interviews: To gain deeper insights into participants' perspectives and experiences, the researcher undertook unrecorded unstructured interviews with key informants, such as government ministers, senior civil servants, heads of professional institutes, and other stakeholders.
4 Document analysis: To complement the observations and interviews, the researcher analysed relevant organisational documents, such as internal reports, meeting minutes, or policy documents, to provide additional context and support for the emerging findings.
5 Reflexivity: Throughout the data collection process, the researcher engaged in reflexivity, continually reflecting on his role, biases, and assumptions, and considered how these factors might influence the research findings.
6 Data triangulation: The researcher synthesised data from multiple sources, including observations, interviews, and document analysis, to enhance the credibility and robustness of the findings.

The researcher was a founding member of both the Malaysian chapter of the United Nations Global Compact, and building SMART Malaysia, thus gaining experience in sustainable development and BIM advisory policy development. The researcher has more than three decades of experience across many sub-sectors of the construction industry. The researcher works on built environment projects.

15.3.8 Analysis

Data was tagged in Taguette, a qualitative data analysis software tool (Rampin and Rampin, 2021), and framework analysis, or structured coding, was used. As this research is abductive, the framework emanated from the iKBV theoretical model. Abductive reasoning also allowed additional ad hoc codes to be added. These additional codes can be considered grounded theory as they emerged from the data during analysis.

The semi-structured interview transcripts were analysed and systematically classified according to a predetermined scoring scheme, as presented in Table 15.2. Initially, the presence of concrete evidence was investigated, encompassing documented instances of re-engineered business processes. In cases where such evidence was not apparent, descriptions of ongoing efforts, such as the formation

Table 15.2 Observed evidence scoring

Score (low to high)	*Observed evidence*
1	No activity nor awareness
2	Aspirational only
3	Active but little impetus
4	Active, committed, work in progress
5	Documented, applied business process change

of working groups designated to devise novel processes and systems, were pursued. In the absence of tangible evidence, records of meetings and deliberations that indicated an intention to embark on related initiatives were examined. In instances where such indications were lacking, indications of topic awareness were explored, which encompassed ad hoc references or individual endeavours to generate new knowledge. Ultimately, there were instances in which no discernible evidence of tangible or aspirational activities, nor any indications of awareness, could be found.

The analysis was performed on each respondent's data and is provided in the next section.

15.4 Results and Discussion

Approximately 12 hours of interviews produced 580 coded statements. Quotes, abridged for brevity and edited for readability, are selected as representative of the interviewees' considerations for each primary code category.

The iKBV description of each capability is re-provided for convenience. They are presented in the sequence used for the interviews. Selected quotations from interviewees and case study are then provided for context and narrative sample. A discussion is then provided that considers, firstly, the finding, and then sustainability aspects and theoretical aspects.

15.4.1 Creation

The Digital_Transition_Architect articulated that they were compelled to engage in rigorous enquiry and analysis to enhance the sustainability of their professional activities. This individual further noted that considerable emphasis has been allocated to the architectural division to innovate and implement various methodologies in sustainable design. Concurrently, the Pipes_Municipal_Fluids_Expert indicated their involvement in collaborative industry research aimed at formulating new standards and best practices. This expert also revealed that their organisation has undertaken internal initiatives that are conducive to the cultivation of specialised expertise, with the financial backing for such research being self-generated. Lastly, the MEP_Director_BIM_Early_Adopter stated that their organisation convenes a technical committee, comprising individuals with significant expertise and recognition in their respective domains, on a monthly basis to deliberate on emerging trends and to formulate organisational strategies in response to these developments.

15.4.1.1 Case Study

The researcher works with a materials supplier that has innovated a long-length, flexible HDPE pipe. This has required considerable development with new technological knowledge being created. It has also required creative approaches and solutions to challenging legislation that has not been conducive to new sustainable products entering the market. The researcher reviewed and reflected upon the processes and procedures, standardised and ad hoc, that were followed during the course of the pipe's business venture, and rated them according to the research scales.

Respondent	*Score*	*Comment*
Digital_Transition_Architect	4	Knowledge creation is an evident strategic goal. This is a work in progress while undergoing restructuring.
Investment_Analyst	1	No evidence of tangible knowledge creation was reported, although there appears to be an aspiration for knowledge creation capabilities
Director_Manager_Operator	2	Little tangible evidence but a stated aspiration
Sustainability_Expert	3	There is some concerted effort to create new knowledge, but piecemeal in approach
Pipes_Municipal_Fluids_Expert	5	Commissioning and participation in industry sector research, and evidence of internal production of new processes
MEP_Director_BIM_Early_Adopter	5	A mature internal process for creating and utilising new knowledge
Case_Study	4	Creates and utilises new knowledge, but a consistent approach is not evident
Total	24/30	

There was a unanimous consensus among respondents that creating new knowledge is an indispensable capability. Certain firms exhibit proactive policies geared towards knowledge creation at both the organisational and industry levels, while others restrict new knowledge generation to individual endeavours. However, it was observed that while some firms provided tangible evidence of internal processes designed to achieve knowledge creation, others lacked such processes, rendering this capability merely an aspiration rather than a productive process for them.

For firms possessing a tangible capability, considerable effort is directed towards the development of sustainability knowledge, with BIM perceived as a tool for facilitating this progress.

From a theoretical standpoint, this capability is deemed desirable and regarded as a strategic advantage. Notably, it was predominantly the larger, well-established firms that exhibited superior capability in this domain. All firms exhibited a knowledge creation capability, frequently manifested through the amalgamation of knowledge obtained via their integration capability and its repurposing in the context of local regulations and culture.

It may be induced that small and medium-sized enterprises (SMEs) adopt a more short-term perspective on knowledge creation, accompanied by less structured processes in support of it. Typically, larger firms undertake projects of longer duration, which may allow for a more extended strategic view. It is important to note that the iKBV does not differentiate based on firm size, length of existence, or the maturity of existing processes, suggesting that further research in these areas may be warranted to gain a more comprehensive understanding of the factors influencing knowledge creation capabilities in the industry.

15.4.2 Destruction

The respondents recognised that existing knowledge, as working practices or mindsets, needed to be destroyed before superseding knowledge can be implemented. The Sustainability_Expert identified a necessary change in employees' mindsets, describing a need to let go of past beliefs, and changing the expectation of customers who retained a legacy mindset. Resistance to change was considered an impediment by several respondents.

Respondent	*Score*	*Comment*
Digital_Transition_Architect	5	As the firm restructures for digitalisation, old processes have been replaced with new processes
Investment_Analyst	1	Minimal replacement of legacy processes has occurred
Director_Manager_Operator	3	Ad hoc changes and replacement of old practices are observed
Sustainability_Expert	3	At firm level changes have been made, but the individual is unable to effect change at project level
Pipes_Municipal_Fluids_Expert	4	Changes are evident, but resistance to change is also evident, preventing the application of initiatives
MEP_Director_BIM_Early_Adopter	5	Significant change to organisational and project systems is reported, including the superseding of legacy practices.
Case_Study	4	Firm-level changes have been executed, necessitating the elimination of historical processes. Nonetheless, resistance to change and adherence to traditional methods by external project parties restrict progress and require the concurrent operation of new and historical processes, thereby causing significant duplication of work.
Total	25/30	

The responses regarding knowledge destruction were varied. Although all respondents acknowledged its occurrence, it was often an incidental rather than a planned activity. All firms demonstrated an awareness of the necessity to dismantle existing processes and knowledge to facilitate their replacement. This was regarded as an essential phase in the renewal process. Notably, larger, established firms exhibited a higher capability to execute knowledge destruction in a structured and planned manner, potentially indicating that these firms possess greater resources and more specialised personnel to undertake such transitions.

It is widely acknowledged that current systems are inadequate for generating new sustainability knowledge, and several measures have been taken to supersede them. The replacement of antiquated systems spans from minor incremental improvements to comprehensive societal transformations. The destruction of existing knowledge, including processes and workflows, was generally perceived as a requisite act of renewal; however, significant challenges were encountered due to individuals' resistance to change and human error in implementing novel processes and workflows.

From a theoretical standpoint, when a clear business case for superiority exists, as in the case of advanced scanning technology, it destroys and supplants existing methods without obstruction. However, advanced scanning represents an independent work task, devoid of backward or forward dependencies, rendering it dissimilar to a construction project. In this context, larger and better resourced firms consistently yielded superior outcomes. These findings suggest that firm size, resource allocation, and the nature of the tasks involved may all play a role in determining the extent and effectiveness of knowledge destruction processes within the industry. Further investigation into these factors could provide valuable insights to guide firms in their pursuit of innovation and knowledge renewal.

15.4.3 Integration

The Digital_Transition_Architect stated that his organisation hires consultants to innovate knowledge with them. The Director_Manager_Operator considered that auditing service providers provided a dual functionality of assessment, and the opportunity to learn and share knowledge.

15.4.3.1 Case Study

In the process of designing new buildings, the researcher has developed deep, functioning strategic alliances with a core team of design practices within Malaysia. Recent projects include a chemical treatment facility, a hospital emergency department and wards, and several residential buildings. The researcher reviewed and reflected upon the processes and procedures, standardised and ad hoc, that were followed during the course of building design projects, and rated them according to the research scales.

Respondent	*Score*	*Comment*
Digital_Transition_Architect	5	During its digital transformation, the firm is engaging external consultants to augment the supply of knowledge and resources
Investment_Analyst	1	Little evidence of any intentional new knowledge integration is observed
Director_Manager_Operator	2	Ad hoc efforts are undertaken on a piecemeal basis
Sustainability_Expert	3	Although there is strong aspiration, there is little evidence for organisational or project integration
Pipes_Municipal_Fluids_Expert	5	The firm funds and participates in external research at industry and professional institute levels
MEP_Director_BIM_Early_Adopter	4	The firm collaborates with professional bodies
Case_Study	4	The firm participates in relevant research but has notable resource constraints
Total	24/30	

Larger firms demonstrated a higher degree of interaction with other knowledge-generating entities, such as professional bodies, including commissioning and participating in research projects. These larger firms exhibited more defined and focused supplies of knowledge and resources and were also more likely to engage external consulting specialists to support their endeavours. For example, the Pipes_Municipal_Fluids_Expert states, "[Redacted][we] as a company has taken part in industries studies for many years now and these studies are either sponsored by a standards organization…We do participate in these joint industry studies. There's also some internal work being done where we see an advantage developing more specialized knowledge and funding the research ourselves." In contrast, while all respondents acknowledged the availability of knowledge sources, smaller firms exhibited a less structured and less convincing approach to incorporating this knowledge within their organisation.

Integration was identified as possessing a significant cultural component, with soft skills being considered more valuable than pure design capabilities by some respondents. This highlights the importance of fostering a culture of collaboration and effective communication within firms to facilitate the integration of knowledge and resources.

The capacity to secure a supply of knowledge and resources for a firm, as outlined by Kaplan et al. (2012), appears to be influenced by factors such as firm size, the degree of interaction with knowledge-generating entities, and the presence of a collaborative culture. Further research into these factors could provide valuable insights for firms seeking to enhance their integration capabilities and improve their overall knowledge management strategies.

To varying degrees, firms have identified resources to obtain up-to-date, high-quality information on sustainable products and processes, in accordance with the integration capability component of the iKBV. Smaller firms demonstrated less planning and control over this procedure, frequently adopting an ad hoc, fragmented approach, as articulated by the Investment_Analyst who states, "They may pull people in from the other parts of the group … that have some knowledge in this but … all that has to be managed." Larger firms employed a strategic approach, often engaging external specialists to bolster their efforts. Certain firms commissioned and engaged in industry research. Integration can be regarded as the initial logical stage in a comprehensive iKBV process. Methodological frameworks for creating sustainable buildings, such as LEED, BREEAM, and GBI, were typically perceived as relevant at the project level, rather than as a framework for a firm to re-engineer its internal knowledge processes. This might represent a gap in firms' access to knowledge regarding restructuring to develop capabilities suitable for sustainable projects or insufficient new organisational frameworks intended to provide firms with a pathway to operate within a sustainability paradigm.

15.4.4 Absorption

The Digital_Transition_Architect conveyed that their organisation is endeavouring to cultivate distinct areas of expertise in sustainable design within the internal structure of the company. The Director_Manager_Operator posited that the incremental transition of sustainability knowledge from external out-sourcing to internal teams constitutes an effective strategy. The Sustainability_Expert advocated for the integration of sustainability and resilience into the very fabric of an organisation's operational ethos, asserting that this approach is most efficacious in effecting change, as it subsequently permeates into operational procedures and manifests in financial gains. The Pipes_Municipal_Fluids_Expert contended that harmonising engineering approaches through the consistent updating and company-wide application of pertinent design guidelines is the optimal strategy for an organisation employing 30,000 engineers. Finally, the MEP_Director_BIM_Early_Adopter disclosed that their in-house sustainability team has been reabsorbed into the engineering divisions, and that educational programmes covering various facets of design, including sustainability, are now being offered.

15.4.4.1 Case Study

Although sustainability is core to the researcher's business DNA, it is recognised and accepted that little is codified or embedded into standard operating procedures. This is viewed as a failing but also a reality of small enterprises.

Respondent	*Score*	*Comment*
Digital_Transition_Architect	3	This is a stated aspiration but remains a work in progress during the restructuring programme
Investment_Analyst	1	There is little reported evidence of intentional activities to maximise the use of knowledge and resources
Director_Manager_Operator	3	Despite a clear awareness of the need for leveraging new knowledge, initiatives are ad hoc in nature

(*Continued*)

Respondent	*Score*	*Comment*
Sustainability_Expert	2	Although this is a primary objective of the firm, accomplishing real-world project change is elusive
Pipes_Municipal_Fluids_Expert	4	Strident efforts are made and are embedded in working practices; however, resistance to change and individual human error limits optimised implementation
MEP_Director_BIM_Early_Adopter	5	The firm has robust processes to ensure absorption and has re-integrated previously out-sourced teams back in-house to ensure assimilation of new knowledge
Case_Study	3	It remains a work in progress to optimise knowledge and resources
Total	21/30	

Overall, nearly every firm encountered challenges in absorbing knowledge into re-engineered processes and procedures. Reported issues included resistance to change and human error in implementing new workflows. Smaller firms tended to operate in an ad hoc, piecemeal fashion, whereas larger firms typically adopted a more structured approach. There is mixed evidence of sustainability knowledge being absorbed into firms. Larger firms appear to have a more intended and structured approach. The respondents unanimously expressed the opinion that sustainability is beginning to be demanded by the market, however, the ability to demonstrate internalised sustainability capability is more varied. Being inherently less agile, possibly larger firms, having longer time horizons, need to take more concerted steps at absorbing new knowledge.

This critical transition represents a juncture at which firms can leverage and experience the benefits of the precursor activities of creation, destruction, and absorption. From a theoretical perspective, it is essential to consider these precursor activities, as absorption cannot proceed without them. The successful integration of new knowledge and resources into a firm's processes and systems is contingent upon the effective execution of these preceding activities.

Therefore, it is crucial for firms to not only focus on the absorption of knowledge but also address the challenges and barriers that arise during the creation and destruction phases. By doing so, they can facilitate a smoother transition towards the implementation and realisation of the potential benefits associated with new knowledge and resources. Further research into the factors influencing these precursor activities and the absorption process could provide valuable insights for firms seeking to enhance their knowledge management strategies and overall performance.

In conjunction with destruction, absorption was impeded by resistance to change and human error.

15.4.5 Replication

The Digital_Transition_Architect indicated that their organisation has established knowledge hubs designed to centralise informational resources and augment the organisation's capacity for replicability. This initiative is further bolstered by the implementation of knowledge-sharing sessions and design critique workshops. Meanwhile, the Director_Manager_Operator acknowledged that recent advancements in collaborative practices have served to facilitate more effective replication within the organisation.

15.4.5.1 Case Study

The researcher reviewed and reflected upon the processes and procedures, standardised and ad hoc, that were followed during the course of building design projects, and rated them according to the research scales. The researcher ensures that all BIM models are available in .ifc (industry foundation class) formats. This is enabling better replication with the extended project teams and supply chain who do not possess expensive licences for proprietary software.

Respondent	*Score*	*Comment*
Digital_Transition_Architect	4	This is a primary goal of the firm's restructuring for digitalisation
Investment_Analyst	2	The firm relies upon traditional methods
Director_Manager_Operator	4	The firm has transitioned to utilise cloud-based collaboration platforms and reports significant improvement
Sustainability_Expert	2	Little evidence of developed capability was reported
Pipes_Municipal_Fluids_Expert	4	Although the firm has not fully embraced digital working processes, it has robust, proven existing practices
MEP_Director_BIM_Early_Adopter	5	The firm utilises Level 2 BIM authoring and collaboration tools on a consistent basis
Case_Study	4	Although the firm uses Level 2 BIM authoring and collaboration platforms, the requisite capability is often not present with project team members and the project resorts to legacy systems. It is reported that cost and licencing issues remain sources of difficulty.
Total	25/30	

In order to replicate knowledge, participants articulated a range of responses, from cultural embedding and robust quality assurance methods to employing BIM as the mode for replicating knowledge, with BIM starting to become the central data source for a project, and thereby the mechanism for knowledge replication. Replication was widely regarded as an essential organisational skill. A majority of firms reported an increasing reliance on cloud-based collaboration platforms and expressed great satisfaction with their utility and the resulting improvement in the firm's capability to replicate knowledge. Firms' competence in replicating sustainability knowledge to project stakeholders was deemed satisfactory.

Although all firms expressed a capability to replicate knowledge, the widespread adoption of BIM has yet to occur. Developing a firm's ability to exchange information as data-rich BIM models in a Level 2 environment is not yet common.

While various methods for knowledge replication are employed across firms, the increasing use of cloud-based collaboration platforms and the potential for BIM adoption provide promising avenues for enhancing this critical aspect of knowledge management. Further research into the factors influencing replication success could offer valuable insights for firms seeking to improve their knowledge transfer capabilities.

Emerging digitalisation tools, such as cloud-based collaboration platforms and data-rich BIM models, were beginning to be employed. The complexities of transitioning from unstructured data to semi- and fully structured semantic data were not addressed, and while nascent digital tools were highly appreciated, it remains to be seen how these firms will adapt to future data structuring modifications. Nationally supported initiatives, such as the NBS Toolkit and National BIM Library, exist, with analogous initiatives emerging in Malaysia; however, they were not explicitly mentioned by interviewees, suggesting that increased public support could be provided to the institutions and departments producing these resources, ideally at a supra-national level.

15.4.6 Protection

The Digital_Transition_Architect expressed reservations about the sharing of digital models, noting that the protocols surrounding what and how information is disseminated remain ambiguous and not fully comprehended within the industry. The Investment_Analyst emphasised the necessity of full disclosure, particularly when seeking certification, stating that certifying bodies would require comprehensive understanding of the methodologies and technologies employed. The Director_Manager_Operator highlighted the complexities associated with patenting intellectual property, describing the legal landscape as particularly challenging to navigate. The Sustainability_Expert lamented the tension between the need for open communication of sustainable practices and the desire to protect such knowledge as intellectual property, identifying this as a significant impediment to sustainability efforts. The Pipes_Municipal_Fluids_Expert revealed that certain contracts were not undertaken due to intellectual property concerns, specifically when clients requested soft copies of data for independent calculations, which the organisation was reluctant to provide due to the sensitive nature of the intellectual property involved. The MEP_Director_BIM_Early_Adopter acknowledged the inevitability of their developed technologies and methodologies being disseminated within the industry, given that they are already in the hands of contractors, clients, and, to some extent, suppliers.

15.4.6.1 Case Study

The researcher reviewed and reflected upon his company's processes and procedures and rated them according to the research scales. Beyond commercially and security sensitive information, the researcher willingly shares information. The researcher is of the opinion that it is a struggle to get even paid project participants to properly read information, and even harder to get them to change working practices, so it is extremely rare that non-paid actors will.

Respondent	*Score*	*Comment*
Digital_Transition_Architect	3	The firm acknowledges the need to protect information, and does so with confidential information, but finds it difficult to control digital model replication at present
Investment_Analyst	3	The firm considers this aspiration difficult to operationalise
Director_Manager_Operator	3	The firm has non-disclosure agreements but notes their limitations and challenging enforcement
Sustainability_Expert	3	The firm is reticent to protect non-confidential data and recognises that protection is restrictive and difficult to achieve
Pipes_Municipal_Fluids_Expert	4	The firm has withdrawn bids on projects because of disclosure requirements that breached the firm's intellectual property policies
MEP_Director_BIM_Early_Adopter	3	The firm operates procedures to limit access and loss when employees leave; however, the firm recognises that protection capabilities are of limited utility and that information will leak out
Case_Study	2	Outside of confidential information, the firm considers that protection is futile and actively encourages sharing
Total	18/30	

Protection emerged as the lowest scoring capability, with all firms expressing concerns regarding the adequacy of existing methods for safeguarding their intellectual property. No clear consensus emerged on the best approach to knowledge protection, except for financial and security data. While some parties believed that increased sharing was beneficial, others maintained that knowledge would inevitably emerge, with only a minority taking effective steps to limit knowledge loss.

The "sustainability paradox" emerged as a significant challenge, as highlighted by Brown (2012), Olsson et al. (2019), and Luo et al. (2020). This dilemma arises due to the inherent conflict between the desire to protect the environment, which requires knowledge sharing, and the need to safeguard a firm's commercial interests. Balancing these competing objectives presents a difficult task for firms striving to contribute to environmental sustainability while maintaining their competitive advantage in the marketplace.

The construction industry, being project-based and inherently collaborative due to legally demarcated project roles, often exhibits transient knowledge. Known for its conservative nature (Latham, 1994; Egan, 1998), the construction industry conducts little fundamental research and is disconnected from upstream material extraction and fabrication industries. From a theoretical perspective, this suggests that, excluding financial information, much knowledge in construction possesses low strategic value, rendering its protection less critical. Further research could explore ways of reconciling the sustainability paradox and improving knowledge protection methods within the industry.

No firm demonstrated mastery in intellectual property protection, despite the vast majority considering it of significant importance to the firm. It was commonly regarded as a capability prone to failure, and some even questioned its entire purpose. Given the nature of the industry – fragmented, siloed, project-based, collaborative, dependent, and heavily regulated – it was deemed inevitable that information would be challenging to protect, necessitating disclosure or natural leakage. Construction is notable for its lack of research and development; it may be worthwhile to investigate whether this absence of IP protection deters firms from investing more substantially in R&D, particularly considering the need for enhanced materials, technologies, and practices to achieve a sustainable built environment.

15.4.7 Emergent Matters

This research employs an abductivist approach and pragmatist philosophy to examine emerging matters in the context of sustainability and the construction industry. Firms unanimously agree that sustainability is impacting their business environment, with a strong trend towards embracing it. Being focused solely on knowledge attributes, the iKBV does not consider the wider operating business environment, unlike other business theories, suggesting further research to strengthen the theory by incorporating a contextual component.

All participants concur that superior sustainability knowledge enhances business opportunities and performance. There is a prevailing belief that the market will increasingly demand advanced sustainability knowledge, with anecdotal evidence suggesting a strong causal link between knowledge skills and superior firm performance.

Responses varied regarding the firms' operating scope or boundaries. Some perceive no significant changes to the scope of services that they provide, while others believe sustainability engineering will expand product offerings. Digitalisation is seen as an opportunity to reposition a firm's direction. However, the evolving knowledge landscape and capabilities do not greatly affect the current boundaries of respondent firms, except for the case study.

BIM and digitalisation are perceived as nascent yet potentially transformative. BIM is critical for achieving sustainability goals, and its digital twin product may represent a confluence of iKBV knowledge capabilities. However, it is unclear whether this proxy for superior knowledge capabilities is adequately valued in the marketplace to deliver superior financial performance.

Culture emerges as a vital factor in firms' capabilities with knowledge. Individual soft skills and organisational behaviour are decisive in developing or applying knowledge. Firms are embedding sustainability within their culture, though the iKBV does not consider culture as a factor.

15.4.8 Discussion Summary

This study explores the iKBV's capabilities of knowledge creation, destruction, integration, absorption, replication, and protection within the context of increasing sustainability in the construction industry. There is a consensus that creating new knowledge is indispensable; however, some firms lack internal processes to achieve this. Larger firms exhibit superior capabilities in knowledge creation and destruction, often due to greater resources and specialised personnel. SMEs tend to adopt a more short-term perspective on knowledge creation.

Knowledge destruction, a necessary phase in renewal processes, faces challenges due to individuals' resistance to change and human error. Firm size, resource allocation, and the nature of tasks influence the extent and effectiveness of knowledge destruction. Integration of knowledge, a critical component of the iKBV is influenced by factors such as firm size, interaction with knowledge-generating entities, and a collaborative culture. Absorption is often impeded by resistance to change and human error, while replication is facilitated by digitalisation tools such as cloud-based collaboration platforms and data-rich BIM models.

Protection emerges as the lowest scoring capability, with concerns over the adequacy of existing methods for safeguarding intellectual property. The "sustainability paradox" poses a challenge, as it highlights the conflict between environmental protection, which requires knowledge sharing, and commercial interests. Further research could explore ways of reconciling the sustainability paradox, improving knowledge protection methods, and investigating the impact of intellectual property protection on research and development investment in the industry.

This research, utilising an abductivist approach and pragmatist philosophy, examines the impact of sustainability on the construction industry and explores the iKBV. Following analysis and deductive and inductive reasoning, it emerged that there is a unanimous consensus that sustainability is affecting firms, with a trend towards embracing it. The iKBV, unlike other business theories, does not consider the business environment, and further research could strengthen it by incorporating a contextual component. Superior sustainability knowledge is shown to enhance business opportunities and performance, with anecdotal evidence suggesting a link between knowledge skills and firm performance. Technology emerged as a factor, as BIM and digitalisation are perceived as nascent yet potentially transformative, with BIM seen as crucial for achieving sustainability goals. Culture, both individual soft skills and organisational behaviour, emerges as a vital factor in firms' capabilities with knowledge, although the iKBV does not consider culture as a factor. Legislation is shown to be a key driver for knowledge change. The three factors of legislation, culture, and technology are not previously identified in the iKBV. This research demonstrates their significance to the model, as each factor presents a notable contribution to the context, boundaries, and mechanisms with which a firm operates.

15.5 Conclusion

The dual purpose of this research is to, firstly, add to the empirical understanding of how the construction sector may transition to a more sustainable built environment. Secondly, it is to make a theoretical contribution to the purpose of a firm and how knowledge is among its most fundamental assets.

The research has utilised a knowledge framework, the iKBV view of a firm, to provide a snapshot in time of where the Malaysian construction industry is regarding its knowledge capabilities with regard to sustainability, and how to attain superior performance and competitive advantage by optimising them. The respondents and case study represent a vertical cross-section of the industry, from client to design team, to contractor, to supplier that represents the construction supply chain structure.

In so doing, it may provide a basis of a framework for longitudinal study whereby periodically asking the same questions would provide better indication of direction, and a comparator for identifying which firm's strategy delivered best performance. It has shown that each country, sector, firm, and individual have their own unique perspective on sustainability and knowledge. The reader is thus encouraged to utilise this framework to explore, describe, and maybe predict their own circumstances.

This research has further demonstrated that the iKBV is a reliable, valid yet imperfect theory by which to explore, describe, and possibly predict firm behaviour. The research also describes how knowledge in general, and sustainability knowledge in particular, are critical components of the pathway to a more sustainable built environment. Accordingly, suggestions have been made to further research and strengthen the theory.

References

Ahmad, T. and Thaheem, M.J. (2022) 'LCIA parameters and the role of BIM towards sustainability: regional and temporal trends', *Buildings*, 2022, *12*(5), p. 700. Available at: https://doi.org/10.3390/buildings12050700.

Alaloul, W.S. *et al.* (2022) 'Assessment of economic sustainability in the construction sector: evidence from three developed countries (the USA, China, and the UK)', *Sustainability*, 14(10), p. 6326. Available at: https://doi.org/10.3390/su14106326.

Amandeep, T. (2014) *Market Research Methodologies: Multi-Method and Qualitative Approaches: Multi-Method and Qualitative Approaches*. IGI Global. Available at: ISBN: 978-1-4666-6372-5.

Bamgbade, J.A. *et al.* (2019) 'Building sustainability in the construction industry through firm capabilities, technology and business innovativeness: empirical evidence from Malaysia', *International Journal of Construction Management* [Preprint]. Available at: https://doi.org/10.1080/15623599.2019.1634666.

Bamgbade, J.A. *et al.* (2022) 'Building sustainability in the construction industry through firm capabilities, technology and business innovativeness: empirical evidence from Malaysia', *International Journal of Construction Management*, 22(3), pp. 473–488. Available at: https://doi.org/10.1080/15623599.2019.1634666.

Barney, J. (1991) 'Firm resources and sustained competitive advantage', *Journal of management*, 17(1), pp. 99–120.

Barney, J.B. (2001) 'Is the resource-base "View" a useful perspective for strategic management research? Yes', *Academy of Management Review*, 26, pp. 41–56. Available at: https://doi.org/10.5465/AMR.2001.4011928.

Bernama (2022) *Climate change: Govt to formulate national adaptation plan, The Edge Markets*. Available at: http://www.theedgemarkets.com/article/climate-change-govt-formulate-national-adaptation-plan (Accessed: 30 June 2022).

BERNAMA (2021) *Promote usage of green buildings – The way forward for Malaysia, Bernama*. Available at: https://www.bernama.com/en/thoughts/news.php?id=1952767 (Accessed: 15 June 2022).

Bin Tayib, M.A.A. *et al.* (2022) 'Industry 4.0 in the Malaysian construction industry and its adoption challenges', in Belayutham, S., Che Ibrahim, C.K.I., Alisibramulisi, A., Mansor, H. and Billah, M. (Editors), *Proceedings of the 5th International Conference on Sustainable Civil Engineering Structures and Construction Materials*. Springer, Singapore. Available at: DOI: 10.1007/978-981-16-7924-7_79.

Brown, P. (2012) 'The "sustainability" paradox-interview with Paul Ehrlich', Scientific American Blog Network. Available at: https://blogs.scientificamerican.com/guest-blog/the-sustainability-paradox/.

Chai Yee Hoong (2021) *We need to move beyond green buildings, says Malaysian architect at WCSC 2021, The Edge Markets*. Available at: http://www.theedgemarkets.com/article/we-need-move-beyond-green-buildings-says-malaysian-architect-wcsc-2021 (Accessed: 15 June 2022).

CIDB (2022a) *MyCrest | Construction Industry Development Board*. Available at: https://www.cidb.gov.my/en/construction-info/sustainability/construction-sustainability/mycrest (Accessed: 15 June 2022).

CIDB (2022b) *Sustainable INFRASTAR | Construction Industry Development Board*. Available at: https://www.cidb.gov.my/en/construction-info/sustainability/construction-sustainability/sustainable-infrastar (Accessed: 6 September 2022).

CREAM and CIDB (2022) *Development of Flood Risk Assessment and Flood Vulnerability Index (FVI) Malaysia.pdf*.

Côté-Boileau, E. *et al.* (2020) 'Organizational ethnographic case studies: toward a new generative in-depth qualitative methodology for health care research?', *International Journal of Qualitative Methods*, 19. Available at: https://doi.org/10.1177/1609406920926904.

Curado, C. and Bontis, N. (2006) 'The knowledge-based view of the firm and its theoretical precursor', *International Journal of Learning and Intellectual Capital*, 3(4), pp. 367–381. Available at: https://doi.org/10.1504/IJLIC.2006.011747.

Curado, C. and Lupi, R.M. (2006) 'The knowledge based-view of the firm: from theoretical origins to future implications', Instituto Superior de Economia e Gestão. Departamento de Gestão. 1-06, pp. 18. Available at: http://hdl.handle.net/10400.5/725.

Dehdasht, G. *et al.* (2021) 'Trends of construction industry in Malaysia and its emerging challenges', *Journal of Financial Management of Property and Construction*, 27(2), pp. 161–178. Available at: https://doi.org/10.1108/JFMPC-08-2020-0054.

Denscombe, M. (2007) *The Good Research Guide: For Small-Scale Social Research Projects*. Open University Press. Available at: ISBN: 0-335-22022-3.

Egan, J. (1998) 'Rethinking construction, construction task force report', Department of the Environment, Transport and the Regions, London. Available at: https://doi.org/10.1002/9780470758526.ch13.

Ezani, N. *et al.* (2018) 'Sustainability: Assessment of green procurement implementation in the construction industry of Malaysia', *undefined* [Preprint]. Available at: https://www.semanticscholar.org/paper/Sustainability%3A-Assessment-of-green-procurement-in-Ezani-Alias/43f1b09349027dba32dafe04062d2e429e31c04f (Accessed: 15 June 2022).

Fairhurst, A.E. and Good, L.K. (1991) 'The ethnographic case study: an experiential approach to teaching retail management', *Developments in Business Simulation and Experiential*

Learning: Proceedings of the Annual ABSEL conference, 18. Available at: https://absel-ojs-ttu.tdl.org/absel/article/view/1699.

Felin, T. and Hesterly, W.S. (2007) 'The knowledge-based view, nested heterogeneity, and new value creation: philosophical considerations on the locus of knowledge', *Academy of Management Review*, 32(1), pp. 195–218. Available at: https://doi.org/10.2307/20159288.

Fusch, P. *et al.* (2017) 'How to conduct a mini-ethnographic case study: a guide for novice researchers', *Qualitative Report*, 22, pp. 923–941. Available at: DOI: 10.46743/2160-3715/2017.2580.

Goh, C.S. *et al.* (2020) 'Revisiting triple bottom line within the context of sustainable construction: A systematic review', *Journal of Cleaner Production*, 252, p. 119884. Available at: https://doi.org/10.1016/j.jclepro.2019.119884.

Goundar, S. (2012) 'Research methodology and research method', *Cloud Computing, Research Gate Publications*. Available at: https://www.researchgate.net/publication/333015026_Chapter_3_-_Research_Methodology_and_Research_Method.

Government of Malaysia (2021) 'Twelfth Malaysia Plan, 2021-2025'. Economic Planning Unit, Prime Minister's Department. Available at: https://www.epu.gov.my/sites/default/files/flipping_book/TwelfthPlan/mobile/index.html.

Government of Malaysia (2022) 'Department of Statistics Malaysia Official Portal'. Available at: https://www.dosm.gov.my/v1/index.php?r=column/cthemeByCat&cat=77&bul_id=OFBHZ0FWSUVtaUNJb3NDdzd1K0g2UT09&menu_id=OEY5SWtFSVVFVUpmUXEyaHppMVhEdz09.

Grant, R.M. (2002) 'The knowledge-based view of the firm', in Choo, C.W. and Bontis, N. (Editors), *The Strategic Management of Intellectual Capital and Organizational Knowledge*. Oxford University Press, New York, pp. 133–148. Available at: Available at: https://study.sagepub.com/massingham/student-resources/further-readings/chapter-05-knowledge-strategy

Grant, R. and Baden-Fuller, C. (1995) 'A knowledge-based theory of inter-firm collaboration', *Academy of Management Proceedings*, 1995, pp. 17–21. Available at: 10.5465/AMBPP.1995.17536229.

Grant, R. and Phene, A. (2021) 'The knowledge based view and global strategy: Past impact and future potential', *Global Strategy Journal* [Preprint]. Available at: https://doi.org/10.1002/gsj.1399.

Håkanson, L. (2010) 'The firm as an epistemic community: the knowledge based view revisited', *Industrial and Corporate Change*, 19(6), pp. 1801–1828. Available at: DOI: 10.1093/ICC/DTQ052.

Kaplan, S. *et al.* (2012) 'Knowledge-based theories of the firm in strategic management: a review and extension', *undefined* [Preprint]. Available at: https://www.semanticscholar.org/paper/Knowledge-based-theories-of-the-firm-in-strategic-%3A-Kaplan-Schenkel/1a0762442607f602e2458665c892bba6b171b4c5 (Accessed: 22 July 2022).

Kogut, B. and Zander, U. (1996) 'What firms do? Coordination, identity, and learning', *Organization Science*, 7, pp. 502–518. Available at: ISSN: 1047-7039.

Latham, M. (1994) 'Constructing the team: final report of the government/industry. Review of Procurement and Contractual Arrangements in the UK Construction Industry.' HMSO ISBN011752994x.

Lim, Y.-W. *et al.* (2021) 'Greening existing buildings through Building Information Modelling: A review of the recent development', *Building and Environment*, 200, p. 107924. Available at: https://doi.org/10.1016/j.buildenv.2021.107924.

Luo, B.N. *et al.* (2020) 'Corporate sustainability paradox management: a systematic review and future agenda', *Frontiers in Psychology*, 11. Available at: https://www.frontiersin.org/articles/10.3389/fpsyg.2020.579272.

Malaysia-Gov (2021a) 'Malaysia NDC Updated Submission to UNFCCC July 2021 final'. Available at: https://unfccc.int/sites/default/files/NDC/2022-06/Malaysia%20NDC%20Updated%20Submission%20to%20UNFCCC%20July%202021%20final.pdf.

Malaysia-Gov (2021b) *Sustainable Development Goals | Official Portal of Economic Planning Unit*. Available at: https://www.epu.gov.my/en/sustainable-development-goals (Accessed: 13 June 2022).

Neuman, W.L. (2006) *Social Research Methods - Qualitative and Quantitative Approaches*. 6th edn. Boston: Pearson Education Inc.

Nickerson, J.A. and Zenger, T.R. (2004) 'A knowledge-based theory of the firm—the problem-solving perspective', *Organization Science*, 15(6), pp. 617–632. Available at: http://pubsonline.informs.org/doi/10.1287/orsc.1040.0093.

Nonaka, I. (1994) 'A dynamic theory of organizational knowledge creation', *Organization Science*, 5(1), pp. 14–37. Available at: DOI: 10.1287/orsc.5.1.14.

Olsson, E.G.A. and Gooch, P. (2019) 'The sustainability paradox and the conflicts on the use of natural resources', in *Natural Resource Conflicts and Sustainable Development*. Routledge. Available at: ISBN: 978-1-351-26864-6.

PAM and ACEM (2009) 'GBI ORGANISATION – Green Building Index', Malaysia. Available at: https://www.greenbuildingindex.org/how-gbi-works/gbi-organisation/ (Accessed: 6 September 2022).

Papargyropoulou, E. *et al.* (2012) 'The rise of sustainability services for the built environment in Malaysia', *Sustainable Cities and Society*, 5, pp. 44–51. Available at: https://doi.org/10.1016/j.scs.2012.05.008.

Pryke, S. (2009) *Construction Supply Chain Management: Concepts and Case Studies*. Blackwell Publishing Ltd: London. Available at: https://onlinelibrary.wiley.com/doi/book/10.1002/9781444320916.

Rampin, R. and Rampin, V. (2021) 'Taguette: open-source qualitative data analysis', *Journal of Open Source Software*, 6(68), p. 3522. Available at: https://doi.org/10.21105/joss.03522.

Reis, J. *et al.* (2017) 'Breaking barriers with qualitative multi-method research for engineering studies: pros, cons and issues', *Proelium, Academia Militar*, 7, pp. 275–292. Available at: http://hdl.handle.net/10400.26/17622.

Reis, J. *et al.* (2018) 'Guidelines for qualitative case research in operations management: a banking service industry perspective', *17th European Conference on Research Methodology for Business and Management Studies*, Rome, Italy. Available at: https://www.researchgate.net/publication/326060211_Guidelines_for_Qualitative_Case_Research_in_Operations_Management_A_Banking_Service_Industry_Perspective.

RIBA (2019) 'The sustainable outcomes guide'. Available at: https://www.architecture.com/knowledge-and-resources/resources-landing-page/sustainable-outcomes-guide

Saunders, M. *et al.* (2019) *Research Methods for Business Students*. Pearson, Harlow, England, Eighth Edition. Available at: ISBN: 978-1-292-20878-7.

Scott, L. (2014) 'Proceedings of ARCOM Doctoral Workshop on Research Methodology', Dublin Institute of Technology. Available at: https://www.researchgate.net/publication/282912642_Proceedings_of_ARCOM_Doctoral_Workshop_on_Research_Methodology.

Tharim, A.H.A. (2019) 'Developer's Perception on the Barriers of Green Rating System Implementation in Malaysia', *Jurnal Arsitektur dan Perencanaan (JUARA)*, 2(2), pp. 138–157. Available at: https://doi.org/10.31101/juara.v2i2.991.

The Edge Malaysia and Fatin Najihah (2022) *Green initiatives need to be made compulsory, The Edge Markets*. Available at: http://www.theedgemarkets.com/article/green-initiatives-need-be-made-compulsory (Accessed: 15 June 2022).

United-Nations (1987) 'Our common future (The Brundtland Report)'. Available at: https://sustainabledevelopment.un.org/content/documents/5987our-common-future.pdf

Zainul Abidin, N. (2009) 'Sustainable construction in Malaysia – Developers' awareness', *World Academy of Science, Engineering and Technology*, 53.

Zainul Abidin N, Khalfan M., and Kashyap, M. (2003) 'Moving towards more sustainable construction', in. *The RICS Foundation Construction and Building Research Conference*, University of Wolverhampton, pp. 88–102.

Special Section

The Development of Sustainability Practice in the Built Environment Post the COVID-19 Pandemic

16 Enhancing Sustainable Build Environments through Ecotourism Villages in Mount Bromo Tengger Semeru National Park, Indonesia

Wiratno, Benny Hutahayan, and Adji Achmad Rinaldo Fernandes

16.1 Introduction

Humans live within their environment and interact with the components around them. These interactions occur among biotic, abiotic, and socio-cultural components. At first, the interaction between humans and the environment runs harmoniously, balanced, and in equilibrium. However, recently, the relationship has become increasingly imbalanced. With their scientific and technological abilities, humans are more exploitative to nature, so various environmental problems arise (Waluya, 2008). The emergence of ecological issues is caused by human activities that violate environmental ethics. Therefore, humans should be able to maintain a good and harmonious relationship with nature by supporting and caring for the environment around humans. The environment always exists in an area, both urban and rural areas (Waluya, 2008). One of the environments, namely the natural environment, can be used as a human-built area to support all human activities. Humans realize that the environment interacts with components and the limitations of the environment's capabilities (Waluya, 2008). Human efforts to extend the life of the environment can be referred to as the creation of the built environment. The built environment is a designation/term for the area where a group of people already lives by creating a residence in the form of a building and its complementary infrastructure, even if it is simple (Pangarso, 2019). The artificial environment is formed by human intervention by utilizing both simple and modern technology, which aims to meet the needs of their life (Environmental Services, 2020). The principle of creating a built environment is to reforest the forest area, manage wastewater, and plant trees around the house. With the creation of the built environment, there are several benefits that humans get, namely as a source of electricity generation, as a means of education, as a means of fulfilling needs, as a source of income, as a place for development and research, as a place to store water reserves, and as a place for recreation (Environmental Services, 2020).

Indonesia is a rich country with natural resources consisting of oceans, the sun, beaches, and land, which, if appropriately managed, can significantly benefit the country (Setiawan, 2015). These resources are the main attractions for domestic and foreign tourists. Especially now, Indonesia has many exotic and stunning destinations (Rahma, 2020). Many foreign tourists make Indonesia their tourist

DOI: 10.1201/9781003317890-22

destination because Indonesia has many exciting tourist objects to visit. Mount Bromo is one of the favorite tourist attractions for foreign tourists (CNN Indonesia, 2022). The global pandemic COVID-19 has spread worldwide, affecting almost all countries and regions (Pokhrel & Chhetri, 2021).

This outbreak was first identified in December 2019 in Wuhan, China. Countries around the world are warning people to be careful. Since then, this virus has spread worldwide, and as of May 20, 2020, it has infected 4,806,299 people and caused 318,599 deaths worldwide (Ciotti et al., 2020). Public service strategies include washing hands, wearing masks, maintaining physical distance, and avoiding crowds and mass gatherings. Lockdown and stay-at-home strategies have been implemented to reduce the spread of the disease (Sintema, 2020). It cannot be denied that the COVID-19 pandemic has impacted various sectors of life. Tourism was one of the sectors that experienced the most significant losses (Binus University, 2022). Tourism is one of the sectors affected by the COVID-19 pandemic.

In 2020, the number of international tourists visiting Indonesia was 4.02 million, which is a decline of 75.03% compared to the 16.11 million visits recorded in the same period in 2019 (Central Statistics Agency, 2020). The coronavirus pandemic has had a broad and profound impact on the tourism industry in Indonesia due to reduced demand from domestic and foreign tourists. The drastic drop in demand was due to the introduction of various travel restrictions by the Indonesian government, which was trying to control the spread of the virus (VOA, 2020). The government has determined various health, economic, social, and political considerations. Thus, the tourism sector must be reopened. Opening tourism access means that only some tourism activities can operate. There are several considerations, limitations, and priorities for tourism activities permitted by the government by fulfilling the indicators of cleanliness, health, safety, and environmental sustainability. Of the many types of tourism, domestic tourism and ecotourism are tourism activities that have received approval to return to their activities in the post-COVID era.

The Ministry of Tourism and Creative Economy specifically stated that ecotourism is a type of tourism that has the potential to survive in the post-COVID era. This is due to the character of ecotourism tourist activities (ecotourists) who tend to refrain from gathering and spreading out from a health perspective. Ecotourism fulfills the physical distancing rules that match the recovery phase, the second phase of the three phases of the tourism strategy set by the Ministry of Tourism and Creative Economy in responding to the COVID-19 pandemic. This policy is also motivated by research results, which state that COVID-19 has shifted tourist behavior toward sustainability from what was originally more unsustainable (Brouder, 2020).

Ecotourism opportunities as a solution for tourism certainly should be well-spent because it can improve the economy of the local community and the country. East Java, a province with several ecotourism destinations, will likely become a leading destination. Moreover, Bromo has been designated as one of the priority tourist destinations for the past few years (Rejogja, 2021). Mount Bromo Tengger Semeru is designated as a national park because it has a unique ecosystem in the form of a caldera, within which is an active volcano within the caldera of Mount Tengger with a sea of volcanic sand around it. In addition, the government has developed a tourist village consisting of natural, cultural, and tourism related to understanding the natural environment and managed with sustainable principles to become a sustainable assisted village, hoping to increase the number of tourist visits to Bromo.

The development of tourism in Bromo is very appropriate because with many tourist visits, especially in the village, as a buffer for tourism objects, it will impact the economic activities of the village community, and this can certainly improve the welfare of the people around the tourist object (Hermawati et al., 2019). This research is essential for analyzing the conditions of ecotourism on Mount Bromo, the relationship between the ecotourism village and the sustainable built environment, and the impact of COVID-19 on the Mount Bromo ecotourism village. Mount Bromo was chosen as the research object because it is well-known as a tourist attraction in East Java, which is still an active volcano and is famous for its golden sunrise in Penanjakan, so it has the potential to survive in the post-COVID era if ecotourism is implemented that can comply with physical distancing rules. With Mount Bromo Tengger Semeru National Park's advantages, many domestic and foreign tourists are interested in visiting Mount Bromo Tengger Semeru National Park, which is suitable for analyzing its sustainable built environment through ecotourism villages on Mount Bromo. The analysis results will show that ecotourism can create a sustainable built environment both during and before the pandemic so that the government and residents can maintain the excellent condition of the ecotourism village.

Based on the description above, this book chapter will explore the potential of Mount Bromo as an ecotourism village and analyze the impact of the pandemic on tourist numbers at the park. Additionally, it will briefly discuss the designation of the Bromo Tengger Semeru – Arjuno area in Indonesia as part of the UNESCO World Network of Biosphere Reserves. The analysis will provide an overview of the benefits of establishing an ecotourism village, particularly during the COVID-19 pandemic.

16.2 Literature Study

The literature study method is a series of activities related to collecting library data, reading and taking notes, and managing research materials (Zed, 2008). The literature review is a comprehensive overview of the research that has been done on a specific topic to show readers what is already known about the subject and what is not known, to seek a rationale for research that has been done or for further research ideas (Denney & Tewksbury, 2013). Literature studies can be obtained from various sources, including journals, books, documentation, the Internet, and libraries.

16.2.1 Advantages of Ecotourism Villages

Ecotourism, or ecological tourism, is eco-friendly and closely related to conservation principles. In this form of tourism, through activities related to nature, tourists are invited to see nature up close and enjoy the authenticity of nature and its environment so that they are moved to love nature. Even though this trip is adventurous, tourists can still enjoy it. Ecotourism always maintains the quality, integrity, and preservation of nature and culture by ensuring harmony with the local community. In contrast to mass tourism, implementing ecotourism does not require the availability of modern accommodation facilities equipped with luxurious equipment or excessive artificial buildings.

Ecotourism, which preserves and utilizes nature and people's culture, is much more stringent than sustainability. Environmentally oriented ecotourism development is much safer preserving nature than sustainable development. Ecotourism does not exploit nature; it only uses natural and community services to meet tourists' knowledge and physical needs. In many aspects, ecotourism is a form of tourism that leads to meta-tourism. Ecotourism sells philosophy, not tourist destinations. The ecotourism aspect will not recognize market saturation. If all the principles are implemented, ecotourism guarantees environmentally friendly development from community-based development.

Good ecotourism management will generate several benefits in various aspects. From a macroeconomic perspective, ecotourism has several positive impacts (Yoeti, 2008), namely creating business opportunities, creating job opportunities, and increasing income while accelerating the distribution of people's incomes as a result of the multiplier effect that occurs from relatively large tourist spending, increasing tax revenues, government and regional levies, increasing national income or Gross Domestic Product (GDP), encouraging increased investment from the tourism industry and other economic sectors, and strengthening the balance of payments. If the balance of payments experiences a surplus, Indonesia's balance will be enhanced, and vice versa.

16.2.2 The Benefits of Establishing an Ecotourism Village for Local Communities

The benefits of tourist villages are one form of creative economic activity, especially in creative tourism. The following are some of the benefits of establishing an ecotourism village for local communities:

a Reducing the urbanization rate from villages to cities: Rural areas with distinctive characteristics, such as an unspoiled natural environment, beautiful natural scenery, various kinds of animals and plants, rural residents, and their culture, offer a unique tourist experience. In other words, establishing an ecotourism village can counter urbanization and serve as an alternative to mass tourism. This is because establishing ecotourism villages generally involves relatively small numbers of tourists who seek to experience contact with nature and the villagers they visit. The realization of ecotourism villages creates new job opportunities, allowing villagers to focus on more than just agriculture and animal husbandry livelihoods. They can also focus on making handicrafts as village products or explore other potential businesses. With the existence of an ecotourism village, people will start thinking about how to create new jobs in the village, not just how to find a job. Therefore, to support the success of tourism villages, it is necessary to enhance human resources, which can be achieved by providing regular skills training. In this way, tourism villages can be realized, and the quality of human resources can increase. Improving human resources is also beneficial for owning a tourism village, as good quality human resources can become investment assets for future generations and contribute to the village's progress.

b Promoting local products: Promoting local products is one of the benefits of having a tourist village. In addition to mitigating urbanization, tourist villages can introduce local products, encouraging buying and selling interactions between the community and tourists. Therefore, strong product branding is essential to make local products memorable and attractive to buyers.

c Helping Small and Medium-Sized Enterprises (SMEs): The establishment of a tourist village benefits the growth of SMEs. Micro, Small, and Medium Enterprises (MSMEs) can use local products or natural resources as raw materials for their products, fostering a creative economic climate in the village. Ultimately, the thriving financial business in the village presents the best solution to job availability problems, potentially reducing the urbanization rate as villagers find sufficient income opportunities in their local communities.

d Improving Village Infrastructure Development: Another benefit of having tourism villages is improving village infrastructure development. Adequate infrastructure is crucial in developing tourism villages, ensuring that village economic activities can run smoothly and create a creative economic climate.

e Improving Community Welfare: Establishing a tourist village also increases community welfare. Implementing the tourism village concept stimulates other businesses and increases tourist arrivals, facilitating local product sales and improving the community's economic, social, and quality of life. Ultimately, this can lead to the development of a developed and competitive village.

16.2.3 Community-Based Tourism Development around Bromo

In general, the vision of tourism development is more directed toward developing community-based tourism (local communities) and remaining environmentally sound. Ecotourism is a concept that tries to accommodate both of these aspects. The idea of tourism development, rooted in nature tourism and outdoor tourism activities, is expected to encourage the local community's participation. Most ecotourism destinations are rural-based. Therefore, they deserve particular attention.

The characteristics of ecotourism usually include the main elements, namely conservation, education for participation, and empowerment of local communities. The term ecotourism is also often interpreted as a trip by tourists to remote areas to learn about nature, history, and culture, where the tourism pattern benefits the local community's economy and supports nature conservation (WWF-Indonesia, 2009). Ecotourism patterns should minimize negative environmental and local cultural impacts and increase local communities' economic income and conservation values (WWF-Indonesia, 2009).

Nonetheless, several other factors, such as social media, infrastructure, and tourist area attractions, will later become variables determining tourist visits (Priatmoko, 2017). So, these factors also need to be considered in developing ecotourism areas. If the environment and facilities are maintained, the resources in the ecotourism village will be protected. If tourism continues, it is likely to become ecotourism if people can convince visitors to come to restore the degraded environment. If ecotourism is not culturally acceptable and local people do not benefit from it, they will become disillusioned and seek to undermine it. The ecotourism resource base is destroyed if the environment and its treasures are not maintained. If tourism continues, it is likely to be ecotourism if one can persuade visitors to come to restore a severely degraded environment. If ecotourism is not culturally acceptable and local people do not benefit from its existence, they will be hostile toward it and may work to undermine it. Thus, the economy, environment, and culture are all involved. One is not more important than the other—they are all vital to the successful introduction, operation, and perpetuation of ecotourism.

In general, one of the educations that can be given to the community is to develop ecotourism in tourist villages that do not conflict with the customs or culture of the community. This means that even though many new things will be carried out in the framework of developing a village as a tourist destination, it is hoped that the traditions and customs of the village will still dominate its people's life patterns. Physical development aiming to improve the quality of the village environment for tourism development should remain the same as what already exists there. Instead, it tends to package what is in the village so that it becomes attractive as a tourist attraction. Therefore, in structuring or changing buildings, for example, elements of locality and originality of building architecture must still be considered. The selected landscape pattern and the materials used in the development must highlight the characteristics of the village, reflect the locality and authenticity of the local area, and empower the tourism village community. In addition, the involvement of the village community in every aspect of tourism is also an essential element in developing a tourism village.

For an eco-tourism village to be accepted by the local community, ecotourism providers must pay attention to several critical aspects of implementing ecotourism in a village, including:

a The number of visitors is limited or regulated by the environmental and socio-cultural carrying capacity of the community;
b Eco-friendly tourism pattern;
c Tourism pattern that is friendly to local culture and customs;
d Help the economy of local communities directly;
e The initial capital required for infrastructure is not significant.

16.2.4 The Role of Ecotourism for Environmental Sustainability

Ecotourism is an alternative tourism that is considered a way to improve the economic life of rural communities because it provides job opportunities, business opportunities, and increases business capacity development (Scheyvens, 2000), as well as provides more significant opportunities to control the use of natural resources in certain areas, as one of the assets of ecotourism activities (Ashley & Roe, 1997).

Seeing the many potentials that Indonesia has, it may provide various positive impacts as an effort to empower society in the future for several reasons, as follows:

a To ensure that policy decisions on future ecotourism development collaborate the opinions of tourism actors who are destination visitors or tourists consisting of tourists and travelers who act as consumers.
b To ensure the implementation of good management of ecotourism assets in Indonesia, such as natural resources, because ecotourism is indeed based on the existence of these natural resources.
c To ensure that ecotourism provides economic, social, and cultural benefits to all ecotourism actors (stakeholders).

The application of eco-friendly tourism development and adaptation to local uniqueness and conditions can be measured through sustainable socio-cultural processes involving local communities, sustainable cycles of natural resources, and economic processes providing sustainability. If the ecotourism approach is applied correctly, the tourism industry has the potential to impact the environment positively. The approach is carried out through efforts to protect and conserve the environment. Tourism can be a source of financing for protecting environmental resources and increasing the economic value of the surrounding community's resources.

Ecotourism can be referred to as "Responsible Tourism", which minimizes negative impacts on the natural and socio-cultural environment, such as conflicts that often occur, and maximizes positive effects on the preservation of the wild, socio-cultural, and economic environment, both locally, regionally and nationally to create a tourism life that can last a long time. To make this, Low Choy and Heilbronn (1996) formulated five fundamental limiting factors in determining the main principles of ecotourism, namely:

a Environment: ecotourism relies on the natural environment, the culture which is relatively unpolluted or disturbed;
b Society; ecotourism must provide direct ecological, social, and economic benefits to the community;
c Education and Experience: Ecotourism must be able to increase tourism understanding of the natural and cultural environment with the experience they have;
d Sustainable: Ecotourism can make a positive contribution to the ecological sustainability of the environment, both in the short and long term;
e Management: Ecotourism must be appropriately managed to ensure the sustainability of the natural environment and the culture that aims to improve the welfare of present and future generations.

Based on some of the factors above, ecotourism empowerment efforts will help develop and provide sustainability for ecotourism activities.

Ecotourism villages can be associated with a sustainable built environment because eco-tourism is a responsible tourism activity in natural places that supports efforts to preserve the natural environment and culture and improve the welfare of the local community. In contrast, a sustainable built environment is a sustainable development environment dominated by artificial structures, with a relationship between buildings, people, and the environment. Thus, these two things have the same goal of preserving the environment.

Ecotourism Villages emphasize responsible environmental practices such as waste management, energy saving, and sustainable use of natural resources. Ecotourism Villages also invite tourists to participate in environmental programs that improve the environment. This aligns with the goal of a sustainable built environment, which ensures that physical and environmental development is prepared and managed to minimize negative impacts on the environment and ensure that natural resources can be used sustainably.

Therefore, an Ecotourism Village and a sustainable built environment can complement and strengthen one another. An Ecotourism Village can be an example of how the environment can be managed and developed sustainably. A sustainable built environment can ensure that the environment and natural resources used in an Ecotourism Village are maintained and sustainable.

16.3 Results and Discussion

16.3.1 Tourism Potential on Mount Bromo

Mount Bromo is part of Bromo Tengger Semeru National Park area. A National Park is a conservation area that has a pristine ecosystem, managed by a zoning system that is utilized for research, science, education, supporting agriculture in downstream areas, tourism, and recreation purposes (Law Number 5 of 1990). Mount Bromo tourism is a mountain tourist attraction located in East Java. The unique aspect of Mount Bromo is its tourist attractions, which are different from other mountain tours. Unlike other mountain tourism, Bromo offers many tourist activities, including horse riding, Jeep driving, and dirt biking, and visitors can enjoy the beauty of the white sand. While waiting there, visitors can also enjoy the view of the sunrise through the attractions of Penanjakan 1 and Penanjakan 2 (Seruni Point). Penanjakan 1 is 2,329 meters above sea level, allowing tourists to see the sunrise over Mount Bromo (Azwisata, 2020). Apart from Penanjakan 1, visitors can also see the overall view of Mount Bromo more clearly and broadly from an altitude of 2,400 meters above sea level. From these two points, tourists can enjoy a fantastic view of the caldera and the sea of sand, often referred to as the Sea of Sand, between three mountains. The sunrise on Mount Bromo will be more exotic and attractive when the sky is clear and there are cotton clouds around Mount Bromo. Apart from the two viewpoints mentioned earlier, there are other viewpoints at Bukit Cinta, Bukit Kingkong, and Puncak B29.

Visitors can also enjoy beautiful views of the crater of Mount Bromo with its distinctive volcanic material landscape. This crater is used to hold traditional ceremonies for the Tengger Tribe every 14th of the Kasada month, or what is known as Yadnya Kasada Day. Before reaching the crater of Mount Bromo, visitors can pass through a sea of sand managed by the Whispering Sands tourist attraction. This Whispering Sand location was once popular when it became a shooting location for a movie called "Whispering Sand". Visitors can also enjoy the Sabana tourist attraction known as the Teletubbies Hill, located in the area south of the crater of Mount Bromo, precisely in the Jemplang area, which is a piece of grassland with an area of about 10 square kilometers (km^2). Teletubbies Hill has a calm atmosphere that differs significantly from the general desert atmosphere.

During a tour of Mount Bromo, visitors are spoiled not only by the beautiful natural scenery but also by a tourist attraction in the form of a cultural festival conducted by the Tengger tribe in the form of a religious ceremony held only on certain days. Several religious ceremonies are usually performed or celebrated by the people of the Tengger Tribe, but the most popular among the general public is the Kasada ceremony. The Kasada ceremony is held annually by the Tengger Tribe, who live in 41 villages in four districts in Probolinggo, Lumajang, Malang, and Pasuruan. The ceremony starts at 00.00 Western Indonesia Time (WIB); they move from in front of the Pastor's house and arrive at the sea of sand at Pura Agung Poten around 04.00 Western Indonesia Time (WIB). Then, a ritual is carried out at Poten Temple, which continues to the crater of Mount Bromo, which is 2 kilometers (km) away, to perform a ritual offering consisting of two essential elements, namely the Bungkah Head and the Hanging Head. Bungkah head means fruit from the ground, such as potatoes and cassava, and hanging head implies fruit.

Based on the explanation above, the development of tourism potential in the area around Mount Bromo is quite good. We can see the tourism potential of Mount Bromo, which is unique and cannot be found in other volcanoes. This potential can be seen from the difference in materials released by Bromo and the ceremonies still being held today, namely the Kasada ceremony. Cultural, social, and tourism factors are presented entirely to keep foreign tourists coming. This is evidenced by data on Mount Bromo tourism visitors throughout 2022, reaching 331,215 people (see Table 16.1). This number has increased significantly compared to 2021, when there were only 180,270 visitors. Of these, 180,002 visitors were domestic tourists, and 268 were foreign tourists. From this tourism potential, utilization in other sectors should be intensified to prosper the surrounding communities.

16.3.2 Ecotourism Village in Bromo

Mount Bromo Tourism is one of the leading tours in East Java, attracting many local and foreign tourists daily to enjoy the sunrise at the top of Bromo. The landscape and culture make it an exciting destination for foreign tourists. The Probolinggo, Pasuruan, Malang, and Lumajang routes can be used to reach Mount Bromo.

Over time, Mount Bromo tourism has transformed into a national mainstay tour. The four regencies, the entry points for Bromo, compete to beautify themselves while offering their regional tourism potential to welcome foreign tourists. The development of the tourism potential in the area around Bromo Tengger is quite good. Many foreign tourists come every weekend to enjoy the sand caldera found only around Bromo. This sand caldera is unique to Bromo and is the only one in the world. Usually, the calderas we often encounter are calderas with water potential. However, the tourism potential of Mount Bromo is unique and cannot be found in other volcanoes. This potential is evident from the material differences Mount Bromo emits and the ceremonies still held today, namely the Kasada ceremony. This ceremony attracts a large number of foreign tourists to participate. The cultural, social, and tourism factors that are presented so altogether make foreign tourists continue to visit. From the tourism potential, utilization in other sectors should be intensified to benefit the surrounding community.

In general, there are two concepts: village tourism and rural tourism, both having different meanings. The concept of rural tourism emphasizes the activities of visiting tourist destinations in a village. Rural tourism is functionally rural and provides opportunities for tourists to engage, experience, enjoy, and learn about the unique cultural, natural, and historical attractions and activities provided by local communities in rural areas, with the cooperation of the government and the business world to provide socio-economic benefits without exploiting the environment (Nair et al., 2015). Meanwhile, the tourist village concept emphasizes introducing all aspects of the countryside that reflect its beautiful atmosphere. Tourism villages are relevant to the shifting tourism development model, which focuses on social, ecological, and community-based tourism aspects (Utami et al., 2019). Zebua (2016) stated that tourist villages are rural areas with unique characteristics to become tourist destinations with physical uniqueness and the social and cultural life of the local community as an attraction. It also includes the people's social, economic, and cultural lives.

Mount Bromo is a tourism object based on ecotourism. It is a form of nature-based tourism that seeks to preserve it ecologically, socio-culturally, and economically by providing opportunities for appreciation and learning about the natural environment or other specific elements (Crouch et al., 2003). Ecotourism promotes learning experiences and appreciation of the natural environment, or some of its components, within its cultural context. Ecotourism has advantages in environmental and socio-cultural sustainability, especially in increasing the natural and cultural resource base of destinations and promoting growth (Crouch et al., 2003).

Based on the explanation above, the conceptual difference between village and rural tourism is evident. The most fundamental difference lies in the interaction between tourists and local people in tourist villages, whereas in rural tourism, nothing similar is found. The birth of a tourist village answers the trend of alternative tourism. The concept of tourism emphasizes the relationship between tourists and nature and local people, which makes the tourism village unique compared to other tourist destinations. Tourist villages are alternative tourism destinations and play an essential role in developing local entities, producing commodities, arts and culture, and natural and human resources. To support economic development in the tourism sector, the regional government determines which villages have the potential to become tourist villages, one of which is Ngadas Village, Poncokusumo District, Malang Regency. Ngadas Village has many potential natural, human, and economic resources. Apart from the agricultural sector, tourism is another leading industry that can be developed. Based on its geographical location in the Bromo Tengger Semeru National Park (TNBTS) area, Ngadas Village is unique in culture and natural scenery. It can be appointed as an ecotourism village in Bromo.

16.3.3 Opportunities for Tourist Attractions from Ecotourism Villages in Bromo

Tourism is the most critical sector in all countries because it contributes to foreign exchange. A government can provide significant economic growth and changes in various aspects through tourism. The creation of employment, the income of a country, and changes in people's living standards are part of the economic growth provided by tourism receipts and the changes that occur. Tourism from the ecotourism village in Bromo has potential tourism resources, lying in the uniqueness of its natural phenomena, the uniqueness of flora and fauna, artificial tourism, and its ecosystem.

Mount Bromo was chosen as the research object because it is well-known as a tourist attraction in East Java. It is still an active volcano and is famous for its golden sunrise. It can thrive in the post-COVID-19 era if ecotourism is implemented in a way that complies with physical distancing rules. Besides that, many local and international tourists are interested in visiting Mount Bromo. This interest makes it suitable for analyzing its sustainable built environment through ecotourism villages on Mount Bromo, which is essential for studying the conditions of ecotourism, the relationship between the ecotourism village and the sustainable built environment, and the impact of COVID-19 on the Mount Bromo ecotourism village.

In principle, a tourist village is dedicated to preserving nature, the environment, ecosystems, and the community's traditional culture. The goal is to improve the welfare of residents through tourism businesses. Determining tourist villages needs to be done to maximize the existing natural potential. The potential of other business units will also be developed. In addition, the village has a particular uniqueness not found elsewhere. This village's cultural and arts attractions are unique and different from other ethnic groups, drawing attention from all walks of life and tourists.

16.3.4 The Impact of the COVID-19 Pandemic on Mount Bromo Tourist Attractions

One of the tourist attractions in Indonesia affected by COVID-19 is the natural tourism of Mount Bromo with COVID-19, triggering the cessation of all tourism activities at Mount Bromo. This harms the local community carrying out economic activities on Mount Bromo. Mount Bromo tourism is always crowded with local and foreign tourists every year. However, when the COVID-19 pandemic hit Indonesia, Mount Bromo tourism was closed and did not accept tourists (Liputan 6, 2021). The Bromo Tengger Semeru National Park Center (TNBTS) recorded a drastic decline in tourist visits during 2020 and 2021 due to the effects of the COVID-19 pandemic. In the last nine years, the number of visits in 2021 was the

Table 16.1 Number of Visitors to Mount Bromo Tengger Semeru National Park

Year	*Foreign tourists*	*Domestic tourists*	*Total*
2015	17,016	456,995	474,011
2016	30,409	424,391	454,800
2017	23,568	623,895	647,463
2018	25,076	800,130	825,206
2019	22,061	699,021	721,082
2020	2,589	206,469	209,058
2021	268	180,002	180,270
2022	8,644	322,571	331,215
2023	14,153	397,271	411,424

Source: Bromo Tengger Semeru National Park Center (BBTNBTS) (2024).

worst. This is also supported by data in Table 16.1, namely the number of foreign tourists and domestic tourists at Mount Bromo in 2015–2023.

The data on visitor numbers to Mount Bromo from 2015 to 2023 highlights several trends and impacts, particularly the dramatic effect of the COVID-19 pandemic (see Figure 16.1). From 2015 to 2019, the park saw an overall increase in visitors, peaking in 2018 with 825,206 total visitors. During this period, domestic tourists were consistently the majority, with their numbers notably climbing to over

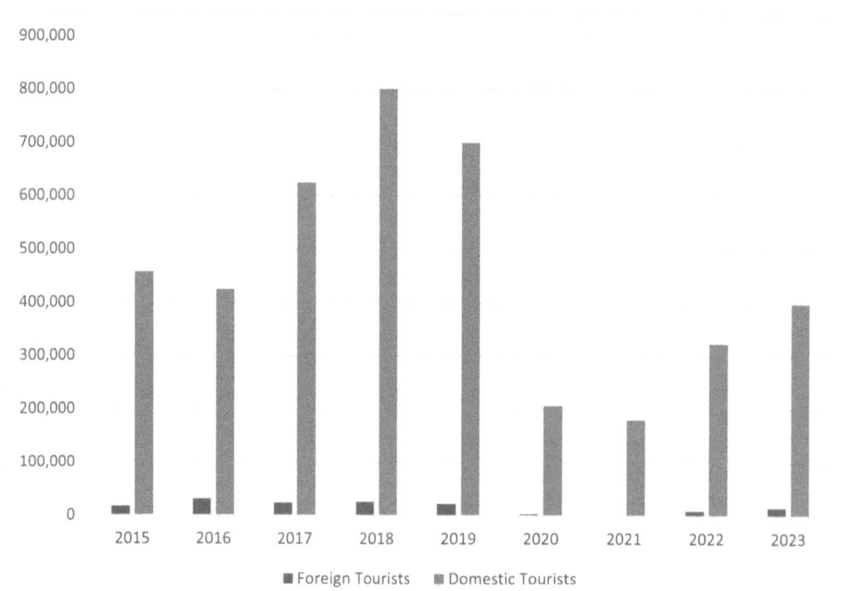

Figure 16.1 Graph of the Number of Visitors to Mount Bromo (2015–2023)

Source: Processed data (2024).

800,000 in 2018. Foreign tourist figures remained relatively stable, with annual counts hovering around 20,000–30,000.

However, the advent of the COVID-19 pandemic in 2020 caused a sharp decline in both domestic and foreign visitors. By 2020, total visitors had plummeted to 209,058, and the downward trend continued into 2021, hitting a low of 180,270 visitors. Foreign tourists were particularly hard hit, with their numbers drastically dropping to just 268 in 2021, indicating severe travel restrictions and high levels of caution among international travelers.

There was a partial recovery observed in the following years. By 2023, the visitor count had rebounded to 411,424, predominantly driven by domestic tourists, who made up 397,271 of these visitors. Although foreign tourist numbers began to recover, reaching 14,153 by 2023, they remained significantly below pre-pandemic levels. This recovery pattern underscores a quicker resurgence of domestic travel and a slower return of international tourists, possibly due to ongoing travel hesitations globally. Overall, the data reflect the significant disruptions caused by the pandemic, followed by a gradual yet incomplete recovery in tourism to the park.

Mount Bromo tourism has an economic impact on the surrounding community. The financial implications arise due to tourist spending during their trip, including expenses for transportation, accommodation, consumption, documentation, buying souvenirs, and others. Spending by tourists can increase the income of the surrounding community. If the local community and the government can maintain and care for the natural attractions of Mount Bromo, the local community's income will be maintained or even increased. In addition, the beauty and condition of Mount Bromo will also be preserved so that its environmental characteristics can last for the next few years.

In anticipation of the spread of COVID-19, the Bromo Tengger Semeru National Park (TNBTS) has temporarily closed the Bromo area and its surroundings. This policy was implemented to prevent the spread of COVID-19 to the public, officers, and tourists and was reopened in August 2020 (Pasuruankab, 2020). However, the managers of TNBTS have set policies for limited tourism activities while still meeting the COVID-19 health protocol standards. In addition, the government has taken advantage of digitization by implementing online ticket purchases through bookingsemeru.bromotenggersemeru.org as an innovation for the TNBTS manager during the COVID-19 pandemic and establishing standard operating procedures to cover nature tourism activities and business actors providing services or natural tourism facilities.

Some businesses affected during the COVID pandemic were hotels, places to eat, travel agents, and souvenir centers. Hotels are the worst affected business because they rely on income from cooperation with travel agents; low income and hotel room occupancy rates impact income received. Some hotels have laid off employees and closed hotel operations to cut further losses. Almost all eating places in TNBTS tourism destinations have closed their operational activities because no tourists are visiting. This is a very realistic choice because the closure of tourist attractions in TNBTS tourism destinations means that there will also be no tourists coming to visit, both local and foreign. Travel agencies are one of the sectors directly affected by COVID-19 through canceling travel plans that were scheduled well in advance. The percentage of cancellations caused by this virus has almost reached 100%, which has resulted in many travel agents also appearing to be in suspended animation and forced to lay off all of their employees because they are bearing much burden due to the cancellation of existing travel plans and many other financial factors caused by COVID-19. Souvenir centers that depend on tourist activities had to be closed because no local or foreign tourists visit TNBTS tourism destinations. Travel restrictions imposed across Indonesia were a factor in the closure of the souvenir center because nearly 70% of its customers were local tourists. Souvenir centers also play a role for MSMEs that supply goods sold by souvenir centers. Therefore, with the closing of the souvenir center, MSMEs also feel the impact and become a factor of successive losses for MSMEs and several tourism service providers working with the souvenir center.

Managing the Mount Bromo area and developing community potential is necessary to create a sustainable built environment. Area management can be interpreted as a process of continuous and systematic participation of human resources in allocating and utilizing natural resources to bring the area to a better condition in the future and resolve current regional problems (Asyiawati, 2014). Area management can be done through community participation, institutions, infrastructure, private involvement, transportation, human resources, regulations and policies, land management, job opportunities, community partnerships, public and private, finance, and promotion management. Meanwhile, developing community potential in the tourism sector is very important as a form of empowerment and introduction of local tourism (Mangion et al., 2005). The impact of COVID-19 has crippled the businesses of people who work in tourism and the lack of visitors engaging in tourism activities, so people are required to have a management strategy and adapt to the new environment because people who want to engage in tourism activities are still very cautious in pandemic conditions (Luo & Lam, 2020). Developing community potential can be done in various ways by assessing the conditions around the place to be empowered and the socio-economic conditions of the community, by strengthening the community by mobilizing and encouraging them to explore their potential and dare to act to improve their quality of life. Empowerment and development are not only individual responsibilities but also collective responsibilities. It is hoped that the management of the Mount Bromo natural tourism area will create a sustainable built environment for nature and society.

16.3.5 Mount Bromo Tengger Semeru National Park as a Biosphere Reserve

The Mount Bromo Tengger Semeru National Park, which has been classified as a UNESCO Biosphere Reserve, represents a significant undertaking in worldwide conservation initiatives. The notion of a biosphere reserve is essential to UNESCO's Man and the Biosphere (MAB) Programme, which aims to harmonize the preservation of biological diversity with its sustainable utilization. Bromo Tengger Semeru, a prominent volcanic region in Indonesia, possesses distinctive ecological traits and cultural legacy. These attributes not only generate considerable scientific curiosity but also present a distinctive framework for sustainable development. The biosphere designation highlights the worldwide importance of the area and establishes a structure for improving ecological research, encouraging international collaboration, and advocating for sustainable practices in the local populations. The integration of various approaches is crucial for preserving the distinct biodiversity of the area and promoting the socio-economic progress of the local communities. This aligns with both the national agenda and global conservation objectives.

The process of designating Mount Bromo Tengger Semeru National Park as a UNESCO Biosphere Reserve was conducted with great care and attention to detail, in accordance with strict international criteria that aim to promote sustainable development by preserving the environment. This designation not only increases the park's worldwide reputation but also obligates it to the two-fold mission of conserving biodiversity while promoting human development. The park is expected to act as a model for sustainable land management by following UNESCO's MAB Programme. This will involve improving both ecological integrity and human livelihoods in a mutually beneficial way. This distinction signifies the beginning of a new period characterized by increased ecological consciousness and global collaboration. It also promotes the use of the most effective methods in conservation research and sustainable tourism.

The Mount Bromo Tengger Semeru National Park is known for its exceptional biodiversity and one-of-a-kind volcanic habitat. It serves as a UNESCO Biosphere Reserve, highlighting its important role in conservation. The park's various habitats have a range of unique species, requiring focused conservation efforts to protect this ecological abundance. Stakeholders, including local communities and international conservation authorities, work together to ensure the adoption of sustainable practices that adhere to global standards for biodiversity protection. These endeavors are essential for preserving the ecological equilibrium and guaranteeing the sustained prosperity of the natural splendor and biological variety of Bromo Tengger Semeru. This area serves as a vital nature reserve and a great asset for scientific investigation and ecological teaching.

The overarching objective for Bromo Tengger Semeru National Park, as a designated UNESCO Biosphere Reserve, is to combine local and global viewpoints in order to strengthen its position as a hub for the preservation of biodiversity and the promotion of sustainable development. The park functions as a dynamic research facility where novel conservation methods can be formulated and evaluated, offering significant knowledge on the sustainable management of natural resources. Therefore, Bromo Tengger Semeru has the potential to make a substantial contribution to worldwide biodiversity conservation initiatives, acting as a guiding light for achieving a balance between human requirements and environmental protection.

16.4 Conclusions

Mount Bromo Tengger Semeru National Park is abundant in natural resources and offers a wide range of tourist activities. The development of sustainable environments around Mount Bromo Tengger Semeru National Park through ecotourism villages holds significant potential for promoting environmentally friendly tourism, preserving biodiversity, and empowering local communities. As a designated UNESCO Biosphere Reserve, the park is dedicated to conserving its unique ecosystems and the traditional culture of local communities, with the goal of improving the well-being of locals through sustainable tourism. However, the COVID-19 pandemic resulted in a cessation of tourism operations, leading to a significant decrease in the number of visitors. In response, the government and park officials have formulated ways to enable tourism to persist securely during the current epidemic, employing health protocols designed to permit tourists while protecting both their well-being and the park's ecological soundness. These activities have a dual purpose: to reduce the immediate effects of COVID-19 and to enhance the park's function as a hub for preserving biodiversity and promoting sustainable development. This will contribute to the recovery and long-term sustainability of the tourism industry.

Abbreviations

BBTNBTS	Bromo Tengger Semeru National Park Center
BPS	Central Bureau of Statistics
KM	Kilometers
MSME	Micro, Small, Medium Enterprise
Rp	Rupiah (Indonesian currency)
SMEs	Small and Medium-Sized Enterprises
TNBTS	Bromo Tengger Semeru National Park
WIB	Western Indonesia Time

References

Ashley, C., & Roe, D., 1997. *Community Involvement in Wildlife Tourism: Strengths, Weaknesses and Challenges*. London: Evaluating Eden Project, International Institute for Environment and Development.

Asyiawati, Y., 2014. Management of spatial utilization in the Mount Bromo area and its surroundings in a sustainable manner in Poncokusumo District. *Journal of Regional and Urban Planning*, *14*(2).

Azwisata, 2020. *Mount Bromo Tourism, The Most Alluring Natural Enchantment in East Java* [online]. Available from: https://www.azwisata.com/2016/01/wisata-gunung-Bromo-pesona-alam.html. [Accessed 25 April 2022].

Binus University, 2022. *The Tourism Industry Affected by the Pandemic, What Should You Do?* [online]. Available from: https://bbs.binus.ac.id/2022/05/industri-pariwisata-terdampak-pandemi-harus-bagaimana/. [Accessed 29 August 2022].

Bromo Tengger Semeru National Park Center, 2024. *Permission to use data on the number of visitors to the Bromo Tengger Semeru National Park* (*Number: S.149/T.8/BIDTEK/KSA.5.1/B/06/2024*). The Ministry of Environment and Forestry.

Brouder, P., 2020. Reset redux: possible evolutionary pathways towards the transformation of tourism in a COVID-19 world. *Tourism Geographies*, *22*(3), 484–490.

Central Statistics Agency, 2020. *The Number of Foreign Tourist Visits to Indonesia in December 2020 Reached 164.09 Thousand Visits* [online]. Available from: https://www.bps.go.id/id/pressrelease/2021/02/01/1796/jumlahkunjungan-wisman-ke-indonesia-bulan-desember-2020-mencapai-164-09-ribu-kunjungan-.html. [Accessed 3 June 2024].

Choy, L.D.C., & Heilbronn, K., 1996. *Ecotourism: An Annotated Bibliography*. Research Report South ROC and Commonwealth Department of Tourism.

Ciotti, M., Ciccozzi, M., Terrinoni, A., Jiang, W.C., Wang, C.B., & Bernardini, S., 2020. The COVID-19 pandemic. *Critical Reviews in Clinical Laboratory Sciences*, *57*(6), 365–388.

CNN Indonesia, 2022. *Frequently Visited by Foreign Tourists, These 5 Indonesian Tourism Destinations Go Global* [online]. Available from: https://www.cnnindonesia.com/gaya-hidup/20220819112255-269-836444/kerap-didatangi-turis-asing-5-destinasi-wisata-indonesia-ini-mendunia. [Accessed 27 August 2022].

Crouch, D., McCabe, S., 2003. *Culture, Consumption and Ecotourism Policies*. In Fennel, D.A., & Dowling, R.K. *Ecotourism Policy and Planning (pp. 77–89)*. London: CABI Publishing.

Denney, A.S., & Tewksbury, R., 2013. How to write a literature review. *Journal of Criminal Justice Education*, *24*(2), 218–234.

Environmental Services., 2020. *The Benefits of the Artificial Environment for Life* [online]. Available from: https://dlh.semarangkota.go.id/manfaat-lingkungan-hidup-buatan-bagi-kehidupan/. [Accessed 25 August 2022].

Hermawati, P., Adisasmita, S., Ramli, M.I., & Hamid, S., 2019. Choices modelsof trip chain and transportation mode for international tourists in tourism destination island. *GEO-MATE Journal*, *16*(55), 195–203.

Liputan 6, 2021. *Bromo Tourism Remains Open, Unaffected by Semeru Eruption* [online]. Available from: https://surabaya.liputan6.com/read/4728869/wisata-bromo-tetap-buka-tak-terdampak-semeru-erupsi. [Accessed 27 August 2022].

Luo, J.M., & Lam, C.F., 2020. Travel anxiety, risk attitude and travel intentions towards "travel bubble" destinations in Hong Kong: effect of the fear of COVID-19. *International Journal of Environmental Research and Public Health*, *17*(21), 7859.

Mangion, M.L., Durbarry, R., & Sinclair, M.T., 2005. Tourism competitiveness: price and quality. *Tourism Economics*, *11*(1), 45–68.

Nair, V., Munikrishnan, U.T., Rajaratnam, S.D., & King, N., 2015. Redefining rural tourism in Malaysia: a conceptual perspective. *Asia Pacific Journal of Tourism Research*, *20*(3), 314–337.

Pangarso, F.B., 2019. *City Architecture*. Jakarta: PT Kanisius.

Pasuruankab, 2020. *Mount Bromo Reopens to Tourists with Strict Health Protocols* [online]. Available from: https://www.pasuruankab.go.id/beritalike/5878/wisata-gunung-bromo-kembali-dibuka-untuk-wisatawan-dengan-protokol-kesehatan-ketat. [Accessed 3 June 2024].

Pokhrel, S., & Chhetri, R., 2021. A literature review on the impact of COVID-19 pandemic on teaching and learning. *Higher Education for the Future*, *8*(1), 133–141.

Priatmoko, S., 2017. The influence of attractions, social media, and infrastructure on tourists' visiting decisions to the Importantsari Tourism Village, Yogyakarta. *Journal of the Treasury of Science*, 8 (1), 72–82.

Rahma, A.A., 2020. Potential of natural resources in developing the tourism sector in Indonesia. *National Journal of Tourism*, *12*(1), 1–8.

Rejogja, 2021. *KSPN Bromo, Tengger and Semeru Program Continues to be Reviewed* [online]. Available from: https://rejogja.republika.co.id/berita/r08qq8327/program-kspn-bromo-tengger-dan-semeru-terus-ditinjau. [Accessed 3 June 2024].

Scheyvens, R., 2000. Promoting women's empowerment through involvement in ecotourism: experiences from the third world. *Journal of Sustainable Tourism*, 8(3), 232–249.

Setiawan, I., 2015. *Potential tourist destinations in Indonesia towards economic independence.*

Sintema, E.J., 2020. Effect of COVID-19 on the performance of grade 12 students: implications for STEM education. *EURASIA Journal of Mathematics, Science and Technology Education*, 16(7).

Utami, M.M., Taufik, H.E.R., & Bhakti, W.N., 2019. Village Tourism: The Implementation of Community-Based Tourism. In *2019 International Conference on Organizational Innovation (ICOI 2019)* (pp. 537–542). Atlantis Press.

VOA, 2020. *The Covid-19 Pandemic Paralyzes Tourism* [online]. Available from: https://www.voaindonesia.com/a/pandemi-covid-19-lumpuhkan-pariwisata-/5501168.html. [Accessed 29 August 2022].

Waluya, B., 2008. *Environmental Management for Kindergarten SMA*. Bandung: UPI.

WWF-Indonesia, 2009. *Community-based ecotourism principles and criteria* [online]. Available from: http://awsassets.wwf.or.id/downloads/wwf_indonesia_prinsip_dan_kriteria_ecotourism_jan_2009.pdf. [Accessed 31 August 2022].

Yoeti, 2008. *Tourism Economics (Introduction, Information, and Implementation)*. Jakarta: Kompas.

Zebua, M., 2016. *Regional Tourism Development Inspiration*. Yogyakarta: Deepublish.

Zed, M., 2008. *Library Research Methods*. Jakarta: Indonesian Torch Library Foundation.

Index

FSC
www.fsc.org
MIX
Papier | Fördert gute Waldnutzung
FSC® C083411

Kolibri 360 GmbH
Ferdinand-Jühlke-Straße 7
99095 Erfurt, Deutschland
produktsicherheit@kolibri360.de